Atlas of fibre fracture and damage to textiles

ATLAS OF FIBRE FRACTURE AND DAMAGE TO TEXTILES

Second edition

J.W.S. HEARLE, B. LOMAS, W.D. COOKE
Department of Textiles
University of Manchester Institute of Science and Technology

The Textile Institute

CRC Press
Boca Raton Boston New York Washington, DC

WOODHEAD PUBLISHING LIMITED
Cambridge England

Published by Woodhead Publishing Limited in association with The Textile Institute
Abington Hall, Abington
Cambridge CB1 6AH
England

Published in North and South America by CRC Press LLC
2000 Corporate Blvd, NW
Boca Raton FL 33431
USA

First published 1989, Ellis Horwood Ltd
Second edition 1998, Woodhead Publishing Ltd and CRC Press LLC

British Library Cataloguing in Publication Data
A catalogue record for this book is available from the British Library.

Library of Congress Cataloging in Publication Data
A catalog record for this book is available from the Library of Congress.

Woodhead Publishing ISBN 1 85573 319 6
CRC Press ISBN 0-8493-3881-6

CRC Press order number: WP3881

Cover design by The ColourStudio
Typeset by Best-Set Typesetters Ltd, Hong Kong
Printed by St Edmundsbury Press, Suffolk, England

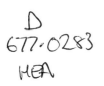

TABLE OF CONTENTS

PREFACE

In the late 1950s, my research students and I were working on the mechanics of twisted continuous-filament yarns, chiefly employed in tyre cords, with twist inserted in order to improve fatigue resistance in use. We therefore decided that we ought to examine the fatigue behaviour of the twisted yarns, in addition to their tensile properties. Dr Tony Booth was the first to work on the subject, but his work showed us that we really needed to know more about the fatigue properties of single fibres. A sequence of research students continued the studies through the 1960s. During this time, we sometimes used optical micros· py to look at broken fibres, but it was difficult to see the form of the break clearly.

In 1967 came the breakthrough. With a grant from the Science Research Council we bought a Cambridge Stereoscan SII scanning electron microscope, and, for the first time, we could observe fibre breaks clearly. This opened up twenty years of exploration, which is still continuing. We have explored the form of single fibre breaks made on laboratory testers. The classification and characterization of these breaks was the first line of research.

Another breakthrough occurred about the same time. The early fibre fatigue studies, using a slow cumulative extension tester, had not produced very illuminating results: usually the fibre either settled down at a certain level of elongation and did not break or it climbed up the load–elongation curve to break at its normal breaking extension. But then Geoffrey Stevens of the RAE, Farnborough, asked us to look at a problem of loss of strength of the cords of brake parachutes deployed behind fighter aircraft on landing. Frequent failures were occurring. One possible cause was fibre fatigue, because the parachute flutters at 50 Hz, each landing lasts 2 minutes, and the cords were used 35 times — which makes 210 000 cycles of tensile loading. Dr Tony Bunsell built a new fatigue tester, which was load controlled and operated at 50 Hz, and uncovered a new fatigue mechanism in nylon and polyester fibres. This started the second line of research: the development and study of fatigue testing methods.

A third important line of research consisted of case studies of fibre failure in use. Many types of product have been examined — shirts, trousers, knitwear, household linen, carpets, ropes, workwear, military webbings, etc. — and characteristic patterns of breakdown have been recognized. In addition to her responsibility for the detailed microscopy, it is in this area that Brenda Lomas has made the major contribution.

As we progressed in our studies, the files of pictures of fibre breaks grew and in 1972 we decided to start publishing 'An Atlas of Fibre Fracture' in the magazine *Textile Manufacturer*, with the thought that the articles might be collected later into a book. However, the magazine ceased publishing and the series ended, but the idea remained. The main problem was how to make the selection, for our files now contain more than 35 000 negatives.

In 1984 Ian Duerden, from the University of Western Ontario, who had been involved in studies of car seat-belt failures, came to spend a sabbatical year at UMIST learning about our work. This was the ideal opportunity for the files to be surveyed and classified and a selection of pictures started. It provided the impetus to produce this book. I finished the selection in the summer of 1986, and Brenda Lomas and I wrote the text, with some more pictures being taken by Brenda Lomas and Bob Litchfield to fill in some gaps. William Cooke contributed Part VIII, arising from his interest in textile conservation. Christine Gisburne gave some advice on the description of scanning electron microscopy in Chapter 1.

The aim of the book is first to report the academic studies of how fibres break in simple laboratory tests, and then to relate this to case studies of failure in use. To a considerable degree, we have tried to let the pictures speak for themselves, supplemented by the necessary information on how the breaks occurred, but we have included comments and explanations, with which the reader may or may not agree.

During the twenty years of these studies, many people at UMIST, staff and students, have contributed to this research. We owe a great deal to all of them. Their names are given in Appendix 1, and, where there have been publications, also in Appendix 2. I apologize for any omissions. The work has been a team effort, which it has been a privilege to lead. I hope that sharing the information with others through this book will make the efforts of everyone involved even more worthwhile.

One of the reasons for the success of the research has been the high standards of the microscopy and the photography. The credit for this rests with the experimental officers who have run the show at the practical level: first, Pat Cross, and then, for most of the time, Brenda Lomas. They have never been content with a picture which is merely adequate, but have always striven for perfection, both in pictorial quality and information content. They have been ably backed up by a succession of scanning electron microscopy technicians — John Sparrow, Alf Williams, Linda Crosby, Creana Green and Bob Litchfield — and encouraged in their high standards by the departmental photographer, Trevor Jones, who has also made most of the prints for this book. The technical staff in the workshop, particularly David Clark, have made major contributions to the development of fatigue testers.

The research has been made possible by generous grants from SRC (now SERC), substantial departmental funding in UMIST, and by contributions from industrial sponsors. We have benefited by discussions with many colleagues and friends inside and outside UMIST, and from organizations that have supplied samples for examination. In a few cases, where we could not draw on our own work, we have used pictures from other sources. All these valuable sources are listed in Appendix 1.

A growing area of fibre usage is in rigid composites. However, we have not studied these materials in our scanning electron microscopy work at UMIST; and a complete account of their fractography would fill another book. Nevertheless, it is right to include an introduction to the subject in Chapter 26. I am appreciative of the opportunity to spend a year as a Distinguished Visiting Professor of Mechanical Engineering in the University of Delaware, associated with the Center for Composite Materials, and am grateful to friends and colleagues there, who taught me more about composites.

Finally, we have been greatly helped in the preparation of the manuscript by secretary, Barbara Mottershead. I also wish to express my personal thanks to the Leverhulme Trust for a research grant as an Emeritus Fellow, which has assisted in the completion of this work.

John Hearle

Mellor, Cheshire
September, 1988.

PREFACE TO SECOND EDITION

The original edition of this book owed much to the encouragement of the publisher, Ellis Horwood, but, coming out as the company which he had built up changed ownership, it soon became unavailable although a demand for copies still existed. We have now been encouraged by Martin Woodhead, another publisher with a personal touch, to produce a new edition. It has been a particular pleasure to work with Patricia Morrison who joined Woodhead Publishing from Ellis Horwood, as Commissioning Editor, and Amanda Thomas in production.

For this new edition we have added more examples from work at UMIST in the 1990s, but we have also drawn more extensively on research elsewhere. Several authors have written their own additional contributions, and other researchers, listed in Appendix 1, have provided pictures and information. For Parts I to VII, this new material from UMIST and elsewhere continues the themes of the existing chapters, and the new information has been added at the ends of the chapters. Part VIII has been revised and augmented by William Cooke. A major addition to this new edition consists of two new parts — on forensic and medical studies. Finally, we have changed the title — always a source of debate between authors and publishers — in order to make it more descriptive of the book.

John Hearle

Mellor, Stockport

ADDITIONAL CONTRIBUTORS
TO THE SECOND EDITION

Dr Franz-Peter Adolf is a forensic scientist at Textilkunde KT 33, Forensic Science Institute, Bundeskriminalamt, Thaerstraße 11, 65193 Wiesbaden, Germany

Dr Ian Duerden is a Professor in the Department of Materials Science, University of Western Ontario, Canada.

Dr Ali Akbar Gharehaghaji, formerly a research student in the Department of Textile Technology, University of New South Wales, Australia, is now a Senior Lecturer in the School of Textile Engineering, Isfahan University of Technology, Iran.

Dr Nigel Johnson, formerly in the Department of Textile Technology, University of New South Wales, Australia, is now Manager of the Physics and Processing Division at the Wool Research Organisation of New Zealand (WRONZ).

Dr Alan McLeod, formerly a UMIST reseach student, has been Research Manager of Surgicraft Ltd and is now Research & Development Manager, Pearsalls Implants, Taunton, UK.

Dr Neil Mendelson is a Professor in the Department of Molecular and Cellular Biology, University of Arizona, USA.

Dr Michael Pailthorpe is a Professor of Textile Technology at the University of New South Wales, Sydney, NSW, Australia.

Dr Leigh Phoenix is a Professor of Theoretical and Applied Mechanics at Cornell University, Ithaca, NY, USA.

Dr William Pelton is a Professor in the Department of Clothing and Textiles, Faculty of Human Ecology, The University of Manitoba, Winnipeg, Manitoba, Canada.

Dr Petru Petrina is a Senior Research Associate in the Department of Theoretical and Applied Mechanics at Cornell University, Ithaca, NY, USA.

Fran Poole is a Detective Senior Constable in the Forensic Services Group, Parramatta Crime Scene Section, NSW, Australia.

Sigrid Ruetsch is a Senior Scientist/Microscopist at TRI, Princeton, New Jersey, USA.

Dr John Thwaites is a Fellow of Gonville and Caius College and was formerly a Lecturer in the Department of Engineering, University of Cambridge, England.

Dr Janet Webster is a Teaching Fellow at the University of Otago, New Zealand.

Dr Hans-Dietrich Weigmann is a former Vice-President of Research at TRI, Princeton, New Jersey, USA.

Part I
Introductory review

1

SINGLE FIBRE FAILURE

FIBRES IN FABRICS

The immediate appeal of a textile fabric lies in its comfort, style and warmth, but ranking equally with these as a measure of quality and value is its durability. How soon will the appearance of the material deteriorate with use? — or, in some perverse instances like blue jeans, become more attractive? How long will it be before the fabric becomes so full of holes or so thin that the garment has to be discarded? The consumer is only concerned with a practical reaction to such questions, but the textile technologist can see that these changes result from the breakdown of the fibres in the fabric.

Textiles are not used only in the traditional clothing and household uses: they have been used for thousands of years in some engineering applications such as ropes, sails, containers and covers. Following the Industrial Revolution, there came a new range of products like conveyor belts, drive belts, filter fabrics and tyres. Today, partly due to a new generation of high-performance fibres such as carbon and Kevlar, an even wider range of advanced engineering applications is opening up: composites, artificial arteries and components of space vehicles are just three examples from a long list. In most of these industrial uses, strength is a major design criterion, and it depends on the resistance of fibres to failure under the imposed combinations of stress. Sometimes thermal resistance is needed. After these initial criteria are satisfied, avoidance of structural fatigue leading to premature failure becomes another design necessity.

There are two possible approaches to design for product performance. The first, which has proved practical and successful, is the craft route of a combination of knowledge and experience applied qualitatively to the selection of raw material and fabric construction, followed by trials and revision if necessary. The second way is engineering design, with mathematical calculation of predicted performance. However, the problems are difficult because textile materials are complicated structures. Nevertheless, it is becoming necessary to move to this approach because of the increasing demands on products and the increasing range of choices. Fig. 1.1 shows what has to be done in basic research and application studies before engineering design can be applied to the strength, and, more important, the life of textile products. This book is concerned with one aspect of the basic research: the study of how fibres fail under stress.

Even with the craft approach, the importance of fibre strength was recognized, and measurement of strength is one of the tests always used to provide entries on a fibre data sheet. But until comparatively recently little was known about the way in which fibres break. The introduction of the scanning electron microscope (SEM), which became commercially available in 1965, opened up the subject.

FIBRE FRACTOGRAPHY

In bulk materials like metals and plastics, the form of fracture can be seen with the naked eye on a large test piece, and a great deal of detail can be observed with light microscopes. Such studies became the science of fractography. For these materials, the SEM was a useful new technique for examining the detail, although the study of shadowed replicas in the transmission electron microscope (TEM) already provided some similar information, though less easily.

With fibres, it was different. It was not until the SEM was in use that it became possible for the fibre scientist to be shown the general form of fracture, let alone the detail. The reason is that fibres are only a few micrometres in diameter, so that to the naked eye a broken fibre is nothing more than a line with an end.

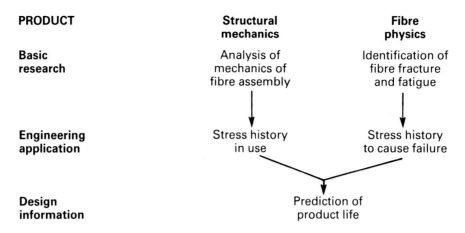

PRODUCT	Structural mechanics	Fibre physics
Basic research	Analysis of mechanics of fibre assembly	Identification of fibre fracture and fatigue
Engineering application	Stress history in use	Stress history to cause failure
Design information	Prediction of product life	

Fig. 1.1 — Past and future design procedures for textile products.

There had been some earlier studies by optical microscopy; but worthwhile results could only be obtained by a dedicated microscopist, such as Gladys Clegg of the Shirley Institute, who devoted long hours of painstaking work to produce beautiful drawings of cotton after swelling, staining and mounting of fibres removed from yarns. There was no way in which a broken fibre could be put directly in the microscope and a picture taken to provide the information needed by an investigator whose principal interest was in the mechanics, and not in the microscopy. The difficulty is not so much the limited resolution of the optical microscope, but is more the lack of depth of focus and the difficulty of obtaining contrast in order to show up clearly the complex shape of a broken fibre end. Furthermore, the problem was not only the great difficulty of establishing reasonable viewing conditions and interpreting what was seen, but that it was easy to be misled.

Transmission electron microscopy was of little help because it was not possible to obtain replicas of complicated fibre breaks with deep re-entrant cavities or multiple splitting. Even the simplest forms of fibre ends would be too three-dimensional to be studied. Surface damage could be examined by replication; and in one study in the 1950s, John Chapman showed frictional wear very clearly by using a conventional electron microscope to observe a fibre directly in the rarely employed reflection mode. However, this is only possible with the electrons directed at a glancing angle, and leads to a strongly foreshortened image.

The available techniques of optical and conventional electron microscopy were thus of very limited value in studying fibre breaks.

The SEM changed the scene, and it became possible for the first time to look at a picture of a broken fibre, in much the same way as one would look at an ordinary photograph of a broken metal bar.

The reasons for this are: (1) the specimens are not transparent to electrons, so that the image formed from the scattered electrons is similar to that seen on looking at a solid opaque object; (2) the great depth of focus means that the whole fibre end is in focus; (3) the usual mode of use of an SEM gives an image which appears to be lit from the side, and this shows up the three-dimensional character very clearly, and only rarely with ambiguity. It is possible to make stereo pairs to give a true three-dimensional image and show depth more clearly within the specimen, but we have not found this to be necessary in our studies of fibre breaks.

Usually, there is no problem in seeing the general form of a break, although sometimes, when there is axial splitting, it may be necessary to take several pictures along the break and join them up to form a montage, in order to give an overall view of the break at a suitable magnification. Finer detail within a break can be seen at high magnification, up to the practical limit of resolution. Instrumentally, the limit is given by the electron beam spot size, and manufacturers now claim 3.5 nm or less; but, in practice, resolution with organic materials is limited by the extent of spreading of the beam as it penetrates into the specimen, and for routine examinations in our SEM is generally about 15 nm. Most fibres are better examined at relatively low beam voltages (between 5 and 10 kV) in order to reduce the penetration and spreading of the electrons, whereas microscopists working with metals, which give a stronger resistance to penetration because of their greater density and atomic number, usually choose a higher voltage (20 kV or more) in order to reduce spot size. The use of a lower voltage also limits loss of surface detail in the image resulting from excessive depth of penetration within the sample.

Most fibres are electrical insulators and therefore charge up in the electron beam. The problems of charging are usually overcome by coating the specimen with metal, though care must be taken that too thick a coating does not mask or distort features of the fracture. Again, the problems are reduced when a lower voltage is used.

TECHNIQUES OF EXAMINATION

The SEM used at UMIST at the time of publication is shown in Fig. 1.2. Electrons are generated by the electron gun at the top of the column, and form a beam, which passes down the column through electromagnetic lenses that control the size of the beam. The final lens focuses the beam on to the specimen surface. The focused spot of electrons is not stationary, since the final lens includes scanning coils, which deflect the beam in a square raster, across a selected area of the surface of the specimen. Electrons have very short wavelengths and are easily deflected and absorbed by other materials and gases, and therefore the gun, column and specimen chamber have to be under vacuum, when the SEM is operating.

The specimens are mounted on special holders, which are fitted on to the stage of the microscope within the specimen chamber. The type of holder depends on the make of the SEM. In our SEM solid aluminium stubs of the type shown later in Fig. 2.3(a,b) are standard; they are 15 mm and 32 mm in diameter. However, the shape and size can vary between makes of SEM, and it should be noted that much larger holders, up to 150 mm in diameter, and maybe larger, can be fitted into the specimen chamber. The choice of holder depends entirely on the type of sample to be examined and on the specimen size limitation of the particular SEM used.

The image formation results from the collection of electrons emitted from the specimen by an electron collector situated to the side of the stage. The collected electrons provide a signal which is amplified and presented on a CRT (cathode ray tube) similar to a TV monitor. The CRT screen is scanned in synchronism with the electron beam scanning over the specimen surface, and the magnification is given by the ratio of screen area to the area of the surface scan. The view of the specimen appears as if the user was looking along the same line as the electron beam, but with the illumination offset in the direction of the electron collector. The particular SEM design and choice of operating conditions influence the image seen.

In order to examine broken fibre ends properly, they must be able to be viewed from all angles, by using the tilt and rotation facilities on the stage of the SEM. This means that single fibres must be held upright, projecting from the stub. A convenient way to do this is to sandwich the fibres between two layers of adhesive copper tape with the broken ends protruding, and then to grip the sandwich in a specially designed split stub. Fig. 1.3 shows a photograph of such an arrangement. If the fibres are fine and straight, about ten fibres (five paired ends) can be mounted on a 15-mm diameter stub, leaving sufficient space between each fibre for clear viewing of the fracture surface; but with crimped, coiled or very thick fibres, the number may be fewer. Coarse monofils can project several millimetres from the edge of the tape, but with fine fibres the distance must be minimized in order to avoid the fibres being bent over or caused to move by the electron beam. However, it is useful to be able to see the fibre surface away from the fracture region, since this can be a source of information on the form of

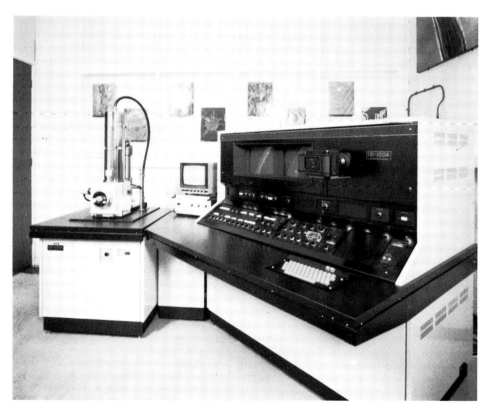

Fig. 1.2 — A scanning electron microscope.

Fig. 1.3 — Fibres mounted on a stub for examination in SEM.

damage. Furthermore, it is fairly common for the effects of fracture to spread some distance along a fibre, either as axial splits, multiple transverse cracks or a generally ragged break. Compromise, and skill on the part of the microscopist, are thus needed.

Most textile fibres are organic polymers and are poor conductors of electricity. Even moisture-absorbing cellulose fibres, such as cotton and rayon, which are moderate conductors in an ordinary atmosphere, become insulators when dried out in the vacuum chamber of the SEM. They are therefore prone to charging by the electron beam. This phenomenon is caused by the fact that electrons do not have a direct path to earth, and, as a result, they remain on the specimen surface, building up electron charge as the electron beam continuously scans the specimen. Charging causes excessive contrast, flaring, banding, streaking and sometimes even image distortion. Charging phenomena are complex and not fully understood, but some materials are worse than others: resin-treated cotton, viscose rayon and wool are particularly bad.

In order to overcome charging problems, the fibres are usually coated with a thin conducting layer after being mounted on the stub. It is necessary to have a coating over all parts of what may be a complicated broken surface and to have a continuous conducting path from the fibre end to the copper tape, and so to earth through the stub. Carbon or various metals can be used for coating, but we find gold to be most suitable for our needs. At first, evaporation of metal in vacuum coating equipment was employed; but now it is more common to use the method of sputter-coating, in which gold atoms are liberated by bombardment of the target by ionized molecules of an inert gas, usually argon. The gold atoms are scattered by the gas molecules, with many eventually settling on the specimen. This scattering of gold atoms ensures a good coverage of the specimen surface. A typical set-up is shown in Fig. 1.4.

The main function of the gold coating is to make the surface of the specimen electrically conducting, in order to prevent charging. The good electron emission property of gold increases the number of electrons emerging from the surface of the specimen, and its density limits to some extent electron penetration and spreading of the spot, thus enhancing the image quality. These advantages must be balanced against the danger of obscuring or falsifying the appearance of the fibre surface details by too thick a coating.

Even in a thinly coated specimen, charging, beam damage and penetration can be prevented or reduced by careful choice of SEM operating conditions. The best conditions must be found by experiment with the particular microscope being used. For our particular requirements, we have found values from 5 kV up to 11 kV to be satisfactory.

Provided care is taken, the above procedures give a good image of textile fibres, without information being seriously lost or changed by artefacts of electron/specimen interaction, the presence of a coating or drying out of the specimen. However, there may be some circumstances, such as the examination of the development of cracks in a fibre being strained in the microscope, where coating is undesirable, or drying changes the situation. Charging may be reduced in these circumstances by treating the fibres with antistatic agents, or by employing special environmental mounts which release water vapour in the immediate vicinity of the specimen. Artefacts due to drying can be reduced by the special environmental mounts or by cryogenic stages, which enable the specimen to be examined in a frozen state.

There are other means of reducing charging effects and of examining uncoated specimens. The normal practice in collecting electrons from the surface of the specimen is to use a secondary electron detector, which collects electrons of various energy levels, but the bulk of the signal is composed of low-energy secondary electrons. It is these low-energy electrons

Fig. 1.4 — Sputter-coating equipment.

which are affected by charging phenomena. High-energy electrons, normally termed back-scattered electrons, are far less influenced by charging problems, and there are now commercially available backscattered detectors of two types, namely the scintillator and the solid-state backscattered detectors. These detectors are very useful in examining poorly conducting or uncoated specimens, which may otherwise have charging problems. They often give a better topographical image, and also show atomic number contrast. The latter is useful in the examination of metals and other situations in which atomic number difference is seen.

If a backscattered detector is not available and metallic coating of the specimen is undesirable, then the specimen should be examined at low accelerating potentials, say 5 kV and lower, in order to reduce specimen charging effects and beam damage. Due to the potentially poor signal-to-noise ratio of the image at low voltages, slow scan rates are usually used, and this in turn may exacerbate problems of charging and beam damage.

The latest development in overcoming these problems is the use of digital image/frame store systems. This technique yields clear images from normal output signals from the SEM. Images can be collected rapidly at scan rates from below 5 MHz up to 10 MHz (television rates), and then stored and processed as required. Consecutive frame scans can be continuously integrated, and this gives a dramatic reduction in noise level in the final displayed image. One or more processed images can be stored and recorded. The processed image can be further enhanced by the sharpening of vertical edges, or by expansion of selected parts of the image grey-scales. It can be displayed in monochrome or pseudo-colour.

Digital image processing has two main advantages over the conventional system when examining uncoated, insulator-type samples: (a) the specimen can be scanned at fast rates, including TV rates; (b) whilst the image is being processed, the electron beam can be switched off so that the specimen is no longer exposed to electrons. Both of these features reduce the possibility of specimen charging and beam damage. Digital image processors and frame store systems are offered as add-on accessories for retro-fitting to SEMs. They are also beginning to be supplied as standard items with new SEMs, and can be used with video-printers that produce high-resolution, hard-copy prints of the processed image.

It is not possible, or appropriate, to provide a complete course of instruction in the scanning electron microscopy of fibres in this book. Further information can be found in the references listed in the bibliography at the end of the book, or, better still, through training in a good laboratory experienced in dealing with fibres, or from specialist training courses.

CLASSIFICATION OF BREAKS

Lord Rutherford once said that science was either 'stamp-collecting or mathematics'. The early years of fibre fracture studies have been 'stamp-collecting': observing the different forms of failure and classifying them into categories. We have now identified 18 distinctly different categories of break and other fibre ends, illustrated in Fig. 1.5 and discussed in greater detail in later chapters of this book. This classification is based on a pragmatic combination of visual form of the break, macroscopic cause and structural mechanism.

Types 1–6 were found in laboratory tensile tests on different sorts of fibre. But our concurrent studies on worn textile materials rarely showed similar breaks. This is not surprising because textiles in use do not usually fail through the application of a single excessive load: they break down after repeated small or moderate loading over a long period of time.

We were therefore led to the study of fatigue testing in the laboratory, with the development of new instruments and test methods. Types 8–12 were distinct forms found with different ways in which repeated stresses can be applied to fibres. The principal methods used have been: (1) tensile fatigue, namely application of cyclic axial stresses on a fibre; (2) flex fatigue by pulling a fibre backwards and forwards over a pin; (3) biaxial rotation fatigue by rotating a bent fibre over a pin so that the material alternates between tension and compression; and (4) surface abrasion. The differences in the way in which the breaks occur illustrate clearly the need to be specific in characterizing fibre fatigue, and to make comparisons between fibres only on the basis of a well-defined test method.

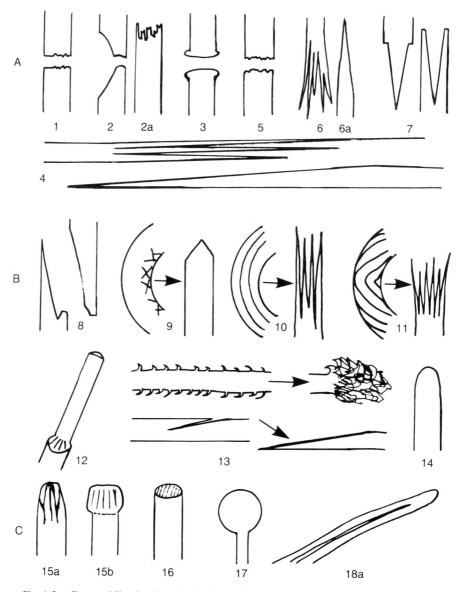

Fig. 1.5 — Forms of fibre break and other fibre ends. A — Tensile failures: (1) brittle fracture; (2) ductile fracture, (2a) variant, light-degraded nylon; (3) high-speed, melt-spun fibre; (4) axial splits; (5) granular; (6) independent fibrillar, (6a) collapsed; (7) stake and socket. B — Fatigue: (8) tensile; (9) flex kinkband; (10) flex split; (11) biaxial rotation, bend and twist; (12) surface wear; (13) peeling and splitting, alternative forms; (14) rounding. C — Other forms: (15) transverse pressure, (15a) mangled, (15b) localized; (16) sharp cut; (17) melted; (18) natural fibre ends, e.g. (18a) tip of cotton fibre.

Type 13 includes several forms of break associated with splitting and peeling due to cyclic shear stresses: this category may need to be subdivided, but the forms are not yet clearly differentiated, and, in some instances, have been studied more in relation to failures in use than to laboratory tests. Type 14 is a rounding of the end of a fibre, which developed after further wear of a fibre which had broken in use.

Type 7 is a tensile failure which has only been found after chemical attack on a fibre. Type 15 results from severe lateral pressure by crushing or blunt cutting, whereas type 16 is from a sharp cut. Type 17 is a form of melting. Finally there are natural fibre ends, type 18.

FRACTURE MECHANICS AND POLYMER PHYSICS

Fracture mechanics is the 'mathematics' of Lord Rutherford's statement. The foundations were laid in 1921 with the classic work of A. A. Griffith on brittle fracture, when he investigated the association of fracture with flaws, either on the surface or internal, which led to stress concentrations. The mechanics can be analysed in two different ways.

In what may seem to be the most direct approach, stress analysis is used to find the stress concentration which is then compared with a material property, namely its inherent strength. The difficulties with this approach are that the stress analysis is complicated and the inherent strength is difficult to measure or calculate. Griffith proposed an energy criterion, with the condition for crack propagation leading to fracture being $dE_m > dS_c$, where dE_m is the elastic energy released in the material when the crack advances, and dS_c is the surface energy of the newly formed crack surfaces. For a crack of unit width advancing a distance dx, the criterion can be given in terms of the elastic energy per unit width E and the surface energy per unit area S:

$$\frac{dE}{dx} > S \tag{1.1}$$

The simple Griffith theory applies only to purely elastic brittle materials, like glass, and is not valid when there is plastic deformation which also absorbs energy and blunts the crack. However, in metals and other materials in which the plastic deformation is limited to a small zone close to the crack tip, equation (1.1) can be modified, by redefining S to include the energy of plastic flow at the crack tip, as well as the surface energy of crack formation.

The zone of plastic deformation will be so small that it has a negligible effect on the elastic energy which is associated with the main bulk of the test specimen. On this basis, fracture mechanics has developed considerable mathematical complexity and power as a means of predicting the conditions for failure of metal structures. These treatments also usually assume small strains and isotropic and ideal elastic–plastic mechanical behaviour. A recent account of the subject is the treatise by Atkins and Mai (1985). The application of fracture mechanics to polymers has been covered by Williams (1983).

For most fibres the situation is more complicated because their structure is highly anisotropic, the stress–strain curves have a more complex non-linearity, deformation is viscoelastic and viscoplastic, and strains are large. Furthermore, the zones of plastic deformation are not restricted to small regions near the crack tip, but often extend over distances greater than the crack depth and may include the whole specimen. These are conditions which have not yet been properly analysed in studies of fracture mechanics.

Most of the explanations of fibre fracture which have been given up to now and are included in this book have been purely qualitative accounts of the sequence of events, with some indications of what is happening to the material structure. However, this is a necessary prerequisite for more advanced work. The observation and classification of the forms of failure provide a challenge for research by theoreticians into fibre fracture mechanics, which combines interesting problems in the applied mechanics of stress and strain with the polymer physics of the material response.

PRACTICAL APPLICATIONS

The development of engineering design procedures, incorporating fibre failure, is a long-term development, perhaps aimed more at the twenty-first century than the twentieth. But there are immediate practical applications of the study of fibre failure.

Even a qualitative understanding of the way in which fibres are breaking down in particular applications can be a great help to the thinking of the engineer, whether concerned with improving fibre properties or with assembling them in ways which will minimize the stresses which cause damage.

More directly, SEM studies of fibre breaks are a tool in the pathology of product failure. For example, the pictures shown in **34G** provided an important clue to the discovery of a particular source of environmental damage to work clothing. Needs also arise in connection with consumer complaints and, even more important, with product liability litigation. For instance, after an accident, it may be necessary to establish whether a component has broken due to a design fault by the manufacturer, or due to misuse by the consumer, or indeed whether it has been cut after the accident. In a very specific area, such as the work on automobile seat belts reported in Chapter 37, comparative studies can be made on the particular product; but, for more general application, there is a need to know how individual fibres fail under different conditions. This can only be based on laboratory testing and the examination of single fibre breaks, which constitute Parts II, III and IV of this book. These individual failures can then be related to the study of wear and damage in products by the techniques which are given in the next chapter and provide the information in Parts V, VI and VII.

ADVANCES IN TECHNIQUES

The SEM techniques for examining the forms of damage and failures in fibres and textiles, as described in the first two chapters of this book, remain essentially unchanged. The most notable advance in SEM technology is a by-product of the digital information revolution. Instead of recording images on photographic film, they can now be digitally stored in a PC with adequate memory. The size of each picture file occupies most of a floppy disk, but easily fits on to a hard disk. Off-line storage of pictures and ancillary information is on CD-ROM. Prints that are adequate for most purposes can be made on an ink-jet or laser printer, as used for printing text. Special laser printers and paper are used for higher quality prints. These changes greatly improve the speed and convenience of obtaining and keeping information.

Other advances are related to more specialised investigations and not just the simple and clear observation of the essential morphology of failure. Environmental scanning electron microscopes allow specimens to be observed in conditions other than a room temperature vacuum. This means that a humid environment is possible, which avoids any changes due to drying of fibres and eliminates the need for coating by providing electrical conductivity in the sample. One application is to observe specimens of fibre or textile as they are being deformed in the SEM. This is illustrated in Chapter 23 by the work of Johnson and Gharehaghaji on the development of damage in wool fibres as they are pulled against wire or pins of the type used in opening machinery.

Coating can also be avoided by the use of newer scanning electron microscopes with field emission electron guns, such as the Hitachi model S-4100 used for **1B(2),(3),(5),(6)**, which are discussed later in comparison with AFM observations. The high power of a modern FESEM gives a high current with small spot size at low beam voltage, so that uncoated samples can be studied at 1 kV with a resolution up to 8 nm, Phillips *et al* (1995). Other examples of the use of an advanced SEM are shown in **8G**. Since these were studies of carbon fibres, electrostatic charging was not a problem and the ultra high resolution available at a higher beam voltage could be used. The many ways in which such a powerful instrument, with a spot size down to less than 0.5 nm at 30 kV, can be used for imaging and chemical microanalysis are described by Boyes (1994).

Transmission electron microscopy (TEM) can be used to study the finer details of crack development as a prelude to failure or other damaging features. Studies reported by Davis (1989), who discusses the problems of difficulty of sectioning fibres without damage and of their low electron density contrast, are shown in **1A(1)–(3)**. An internal crack in a polyester fibre, due to some unknown cause is seen in **1A(1)**. This picture was taken using a negative staining technique, which supports the fibre material during sectioning and increases contrast; the method was developed 20 years earlier, Billica *et al* (1970). Extensive delamination in a stretched polyester film is shown in **1A(2)**. The cracks are crossed by fibrils. Macro- and micro-fibrils are also seen in the replica of the internal splitting which occurs when a strip is peeled off a polyester fibre, **1A(3)**. Such observations are useful in increasing understanding of fracture mechanics and its relation to fibre structure. These three examples all suggest that cracks develop along internal structural boundaries.

Hagege, as reported by Oudet *et al* (1984), has obtained additional information by the use of TEM and electron diffraction in the examination of polyester fibres subject to tensile fatigue, which gives failure of the type shown in **11C**. In this type of fatigue, an initial transverse crack turns and runs down the fibre almost parallel to the fibre axis. A TEM picture of an oblique longitudinal section through such a crack is shown in **1A(4)**. Valuable additional information on the way the crack grows is obtained from a higher magnification view of the crack tip, which is found to be preceded by micropores, **1A(5)**. An oblique transverse section containing a crack tip is shown in **1A(6)**. Away from the crack, the electron diffraction images show patterns which are characteristic of a semi-crystalline material, but in the images near to the crack tip the crystalline arcs are missing, which suggests that the material is amorphous. The reduction in crystallinity was confirmed by infra-red spectroscopy and X-ray diffraction.

Atomic force microscopy (AFM) is a technique which has become available since the first edition of this book was published. A fine probe is scanned over the surface and topographical, mechanical or other information is recorded to produce an image of the material surface with an optimum resolution of 1 nm or less. Phillips *et al* (1995) describe the use of AFM to examine the surfaces of wool fibres and AFM pictures of scales are shown in **1B(1a)** and in a 3D view in **1B(2)**. A picture, **1B(3)**, of the same area as the AFM view, **1B(1a)**, shows what can be achieved in an uncoated sample by using an advanced SEM with a field emission gun.

However, as shown in **1B(3)**, better resolution is obtained with coated fibres. The similarity of features, shown in AFM and FESEM, can be seen by comparing **1B(1a)** with **1B(1b),(3)** and, at higher magnification, **1B(2)** with the FESEM image of the same scale edge, **1B(4)**. Another FESEM picture by Phillips *et al*, **1B(5)**, shows a crack opening between the scales of a wool fibre.

An AFM picture of the internal cross-section of a wool fibre, which records the differences in mechanical resistance to probe penetration in different parts of the material, is shown in **1B(6)**. So far, atomic force microscopy has hardly been used to study aspects of fibre fractures, but it clearly has great potential for such studies. Cracks and other deformations of the internal structure of fibres could be examined. Jones (1995) has used AFM to show one form of damage to fibre surfaces, the effect of exposure to ultra-violet light. The difference in surface texture is shown in **1C(1),(2)**.

In some circumstances, at lower magnification, there are benefits in using optical microscopy rather than SEM. Examples of features of damage in ropes are shown in **39M–39P**, both for the larger-scale forms of whole fibres bent at kinks and, in polarised light, for the observation of internal kink-bands, which are not visible, except as surface projections, in SEM views. Another use of optical microscopy, preferably combined with computer-assisted image analysis, is in quantitative studies intended to determine the frequency of various types of damage. Once the SEM has been used to identify different forms clearly, they can be picked out in optical microscopy examination, which is easier and quicker. Studies of patterns of wear in carpet fibres have utilised this procedure, as described in Chapter 33.

The new technique of confocal light microscopy enables detailed views of the internal parts of fibres to be examined. In a confocal laser scanning microscope, a small spot of light is scanned through the specimen and the resultant reflected or fluorescent light picked up by a detector. An image can then be seen on a monitor, or digitally recorded, in the same way as for a scanning electron microscope. Burling-Claridge (1997) at WRONZ has used the technique to follow changes within a wool fibre, as it is deformed by bending. Hamad (1995) has used fluorescence confocal microscopy to study microstructural degradation and fatigue failure mechanisms in wood pulp fibres, as used in paper-making. The cumulative development of cracks is seen in the series in **1C(3)**. The cross-sections in **1C(4)** are from a series taken 30 μm apart, obtained without the disturbance of cutting sections. As the techniques are improved with experience, confocal light microscopy should make it possible to follow the development of cracks and other damage within fibres.

The developments in image analysis in recent years make quantitative analysis of features in images, which may be obtained from many forms of interaction with a specimen, much easier and more powerful. An example comes from UMIST studies on archaeological textiles. By image analysis, the scales on a wool fibre, **1C(5)**, can be reduced to a pattern of sharp lines, **1C(6)**, on which measurements can be made rapidly and accurately. In this example, the object of the investigation was to identify different species by the scale patterns of their hairs. Another example of image analysis on yarn structure is shown in **42D(1)–(3)**.

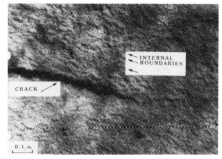

1

2

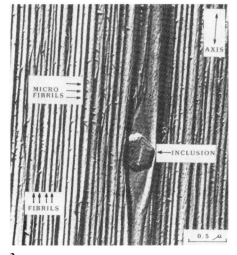

3

4

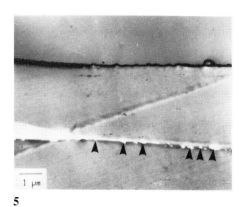

5

6

Plate 1A — Transmission electron microscope observations, Davis (1989).
(1) An internal crack of unknown origin in a polyester fibre. (2) Delamination of uniaxially oriented
polyester film. (3) Platinum replica of a peeled polyester fibre surface.
TEM of tensile fatigue of polyester fibres, Oudet et al (1984).
(4) Oblique longitudinal section with crack. (5) Tip of crack, preceded by micropores. (6) Oblique
transverse section with electron diffraction images.

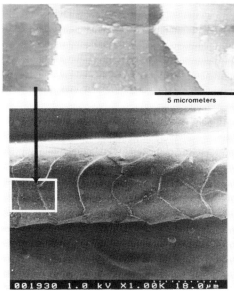

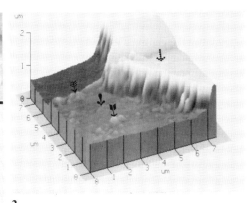

1

2

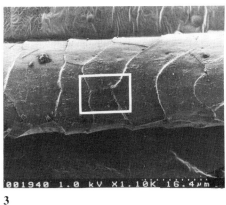

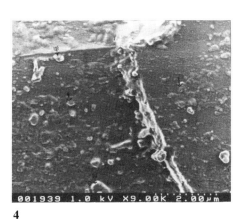

3

4

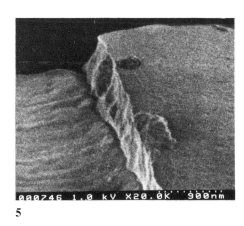

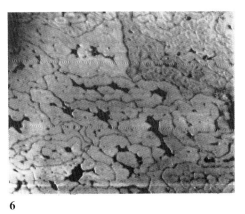

5

6

Plate 1B — AFM and FESEM pictures of wool fibre surfaces, Phillips et al (1995).
(1) (a) Upper: AFM mosaic of the surface of a merino wool fibre. (b) Lower: uncoated FESEM image, including the same area of the fibre surface and showing the same shape of the scale edges and other features. (2) Three-dimensional AFM image of one of the scale edges in (1) at high magnification. (3) FESEM image of the same area of the chromium coated fibre with higher resolution. (4) FESEM image of the same scale edge, chromium coated, at high magnification, showing same features. (5) FESEM high magnification micrograph showing a gap between the two scales.
AFM picture of internal structure of wool, courtesy of M. Huson, CSIRO
(6) Showing separate cells in cross-section.

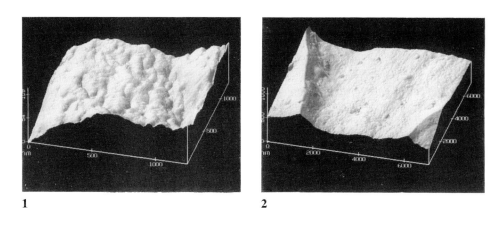

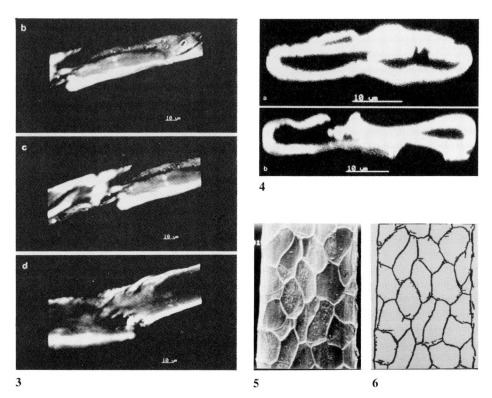

Plate 1C — AFM images of wool fibre surfaces, Jones (1995).
(1) Unexposed wool. (2) Wool exposed to UV. Scales are in nm.
Confocal images of wood-pulp fibres, jack pine RMP, refined at 6.5 GJ/t and cycled in shear, Hamad (1995).
(3) Series along a fibre. (4) Two cross-sectional views separated by 10 μm.
Image analysis of surface of wool fibre
(5) Image as seen. (6) Pattern after image analysis.

2

EXAMINATION OF WEAR IN TEXTILES

INTRODUCTION

When a single fibre has been broken in a laboratory test, the task of examination is a straightforward one of mounting the fibre and viewing it in the SEM. The examination of wear in a textile product is a much more formidable undertaking. The investigator may be presented, for example, with a pair of trousers, a square yard of carpet, or a six-foot length of rope. Each specimen will contain many millions of fibres. Some parts of the material may be almost undamaged; some will show a few broken fibres, which affect appearance but have a negligible weakening effect; some will show severe damage; and finally, in some places, there may be complete failure, in the form of a hole, of pile worn away leaving only the carpet backing, or of a break or tear.

Where should the examination start? And how should it proceed? A careful plan of work is necessary if meaningful results are to be obtained and if the magnitude of the work is not to become vastly time-consuming. In the end, pictures of two or three fibres may be crucial in providing understanding of what is happening, but only if they can be properly placed in the context of the total sample.

What follows in this chapter is an account of the procedures which we have found useful in a large number of studies over a period of over fifteen years at UMIST, with some earlier experience at the Shirley Institute under the guidance of the late S. C. Simmens.

THE INVESTIGATIVE PROCEDURE

Examination of deterioration in any textile product requires some form of magnifying aid to see fibre damage, and the success of the investigation will depend partly on the equipment available within the investigatory laboratory and partly on the expertise of the laboratory staff.

It is not possible to lay down set rules for examination of worn materials because of the diversity of background of laboratories, but some guidelines can be given based on our experiences in examining a wide range of 'textile materials'. Many 'lay' people equate materials with fabrics, but the term textile materials encompasses many types of products, including, to name but a few: garments of many sorts; some footwear; household materials such as towels, curtains, table linen and bed linen; upholstery and carpeting; workwear and environmental protection garments; ropes; conveyor belts, hosepipes, webbings and many other industrial applications.

In the microscopy laboratory in the Department of Textiles at UMIST, there is a range of equipment from general light microscopes to the more sophisticated and expensive scanning electron microscope (SEM) plus other equipment of use in the work.

The SEM is the ISI model 100A, which is near the top of the range as an imaging instrument, but it does not include analytical facilities, which are available elsewhere in UMIST. For more routine studies, one of the simplest SEM models would be suitable. A list of equipment needed in a laboratory which aims to be well set up for the study of worn or damaged textiles is given in Table 2.1.

The range of textile materials that 'wear out' or 'break down' is considerable and various examples of wear have been investigated at UMIST. These have ranged from worn clothing to worn carpeting, and from ship's hawsers to gas meter gaskets. Not every item can be treated in the same way, but in all investigations the same basic rules for examination apply.

The first step in any investigation is to find out as much about the history of the sample as possible. Always endeavour not to work blind. If the required information is not available or,

Table 2.1 — Laboratory equipment for examining worn textiles

1. *Well-lit table for viewing samples.
2. Macrophotography set-up.
3. *Stereomicroscope preferably with zoom magnification control.
4. *Polarizing microscope for general light microscopy, measurement of some optical properties of fibres and for fibre identification. Plus provision for taking photomicrographs.
5. *Sectioning equipment, fibre and yarn cross-sectioning by either the plate method, hand microtome or precision microtome. Cross-sectioning of fibre, yarn and fabrics by low-speed saw or grinding techniques.
6. Hot-stage for microscope — for investigation of thermal behaviour of fibres and for fibre identification.
7. Interference microscopy for measurement of refractive indices of fibres.
8. *Scanning electron microscope plus sputter-coating equipment.
9. *Provision for developing and printing photographic films.

*Necessary Equipment.

for reasons of security, not released, then this should be noted in the final report, and as much information as possible about the sample gleaned during the course of investigation.

Direct, visual assessment must be made of the sample or samples before any specimens are extracted for detailed examination. The sample should be viewed under good lighting conditions and all aspects noted such as appearance of damage, location of damage, details about wear, discoloration, soiling, colour fading or loss, contamination and any other visible features. At this stage it is advisable to make a record of all observed information together with diagrams or sketches showing the location of damage or other features, as indicated in Fig. 2.1. Photographs of the sample are especially helpful: examples are shown in some of the plates in the case studies, such as **33A**, **35A** and **39A**, **39B**. One photograph of good quality is worth many words of verbal description, but the accent must be on high quality. Any good camera can be used to take a general picture of the garment, often most conveniently while being worn by a model; but, in order to show up damage, a camera with either a macro lens, close-up lenses or extension rings is required. Lighting conditions must be good and carefully noted, to avoid ambiguity between samples and to avoid confusion if more photographs are required at a later date. Remember that a microscopical examination requires removal of specimens which can destroy the history of the sample.

It is advisable at this early stage to check fibre content, yarn and weave structure, seam construction, any other detail that may have some bearing on the examination. Fibre identification is most easily carried out by examining the fibres in a polarizing microscope, but solubility tests and melting-point determinations can be used as back-up tests. Staining tests are often rendered useless when the sample is already dyed or is contaminated. Full details of methods of fibre identification can be found in the book edited by Farnfield and Perry (1975).

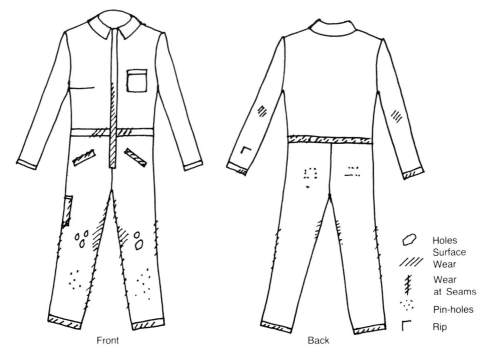

Front   Back

Fig. 2.1 — Sketches illustrating location and type of damage in a worn coverall.

No matter how good one's eyesight, the size and fineness of textile fibres requires magnification for fibres to become clearly visible. An extremely good, first-stage microscope is a stereo microscope, particularly one with a zoom magnification facility. This microscope works in the lower magnification range and gives a three-dimensional image of the sample surface for those people with normal binocular vision. The long working distance between sample and objective lens and the fact that the image is not inverted and reversed as in normal bench microscopes makes examination of damage and checks on fabric structure easy to perform. It also allows easy access, so that the whole product can often be examined, without having to cut out small pieces. If necessary, yarns or fibres can be extracted for detailed study. Again it is advisable to make notes of such features as: damage appearance; location of areas of damage, contamination, discoloration or fading; whether yarn crowns have been flattened; whether the surface is hairy or rubbed-up; and any other interesting detail observed through the stereo microscope. Photography of interesting areas either through the microscope or on a separate macrophotography set-up is an added advantage.

The stereo zoom microscope gives a clear view of the external features of the material, as seen in reflected light, but its use is limited to low magnification. Fibres can be distinguished and the location of broken ends observed, but no detail of damage within the fibre can be resolved.

In order to observe the detailed form of fibre damage it is necessary to go to the higher magnification and resolution of a more powerful light microscope or an SEM. If both are available, it is necessary to decide at this stage which will be of more use, or whether both are needed in the investigation. Each instrument has its advantages and disadvantages and these are summarized in relation to studies of wear of textiles in Tables 2.2 and 2.3. Our experience is that the SEM is usually the most useful tool, and is employed in a simple imaging mode to view the specimen from various directions at appropriate magnifications. In some studies this is usefully assisted by optical microscopy or by X-ray energy analysis in the SEM to make chemical identifications.

Table 2.2 — Advantages and disadvantages of light microscopy

Plus	Minus
1. Internal detail	1. Small depth of focus at high magnifications
2. Surface detail	2. Short working distance at high magnifications
3. Polarised light microscopy (fibre structure)	3. Lower resolving power limit
4. Interference microscopy (fibre structure)	4. Lower magnification limit
5. Colour perception	5. Small samples, mounted in liquid between a cover glass and a microscope slide, and viewed from one direction.
6. Phase contrast and other techniques to introduce image contrast	6. May be confusion between internal and surface features.

Table 2.3 — Advantages and disadvantages of scanning electron microscopy

Plus	Minus
1. Longer working distance	1. Surface detail only
2. Larger depth of focus	2. No colour perception (electron image)
3. Larger specimens	3. Specimens charge up (sputter coating)
4. Higher magnification	4. Possibility of beam damage (sputter coating)
5. Greater resolution	5. Specimen in vacuum chamber
6. Very clear view of surface features	6. Sometimes inability to distinguish between components in a flat (polished) surface
7. Better viewing and handling facilities, e.g. tilt rotation	7. Sometimes inability to differentiate between different types of worn fibres, e.g. between worn wool and nylon fibres.
8. Image processing	
9. Chemical analysis possible by X-ray analysis	
10. Greater surface topography with backscattered detector.	
11. Atomic number contrast with backscattered detector	

SEM STUDY

The range of textile materials is vast, and microscopical techniques of examination vary, depending on the characteristics of the sample. Therefore, for simplicity, let us consider the examination of a worn shirt using the SEM as the main diagnostic tool. The mode of examination can be applied to other garments, and with modifications to thicker larger structures such as ropes and carpets.

The operation of the SEM was described in Chapter 1. It is most useful for examination of the physical effects of damage to a surface. The higher magnifications obtainable, greater depth of field and focus, and large specimen size allow damage to be viewed *in situ*, without disturbing fibre or yarn position. When taking specimens for examination always choose a range of different pieces to cover the full spectrum of damage; whenever it is possible, select from undamaged fabric, then through slight and moderate damage, to severe damage. There are two reasons for this procedure. Firstly, it is necessary to be sure which features are a result of the damage and which were in the original material. Secondly, and most importantly, the region of actual break is usually highly confused, with vast numbers of broken fibres most of which will have failed after the break has started, and may well have been disturbed after the break is completed. The less damaged regions are much more instructive in showing up the sequence of damage and giving clues to its cause. It may be difficult to find undamaged material because, even after only a few wear/wash cycles, the shirt fabric suffers some surface damage which gets progressively worse as the shirt is worn. However, undamaged or relatively undamaged fabric can be found under the collar, in pockets (if any) and in front facings. Slight damage caused mainly by the physical effects of laundering is usually found in the centre back region, and more severe wear is found down the fronts and in elbow regions. The most severely worn parts of the shirt occur along the collar fold, at collar points and along the edges of the cuffs. Rips and tears can occur in various places and are usually accidental or stem from failure of weak places in the shirt.

Care must be taken to avoid bias when taking specimens particularly if a comparison has to be made between several worn garments or items. The way garments wear out depends on working/wearing conditions and also on the individual wearer. Experience has shown that there are variations in wear patterns between users, so that taking specimens at the same places in each garment may solve the problem of personal bias in specimen selection but not cover all the main areas of damage. The positions from which the specimens were taken should be noted either by written notes, or by marking on a sketch of the garment, Fig. 2.2, or on a photograph. The original diagram or photograph of sample damage can be used to mark specimen positions, provided the diagram or photograph does not become too confusing.

The size of the specimen taken will depend on the SEM facilities. We use 15-mm and 32-mm diameter stubs, Fig. 2.3(c,b), in our ISI 100A SEM. The larger stubs will take quite large pieces of material. Alternatively, flat sample holders can be used, Fig. 2.3(a). Specimens are stuck to copper adhesive tape secured to the stub by double-sided adhesive tape. Copper

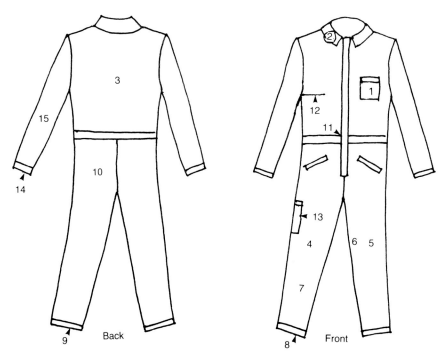

Fig. 2.2 — Sketches showing location of specimens taken for examination. The samples with least wear are (1) inside pocket and (2) under collar.

Fig. 2.3 — Stubs for use in SEM. Clockwise from top: (a) large flat specimen holder; (b) large stub, 32 mm diameter; (c) small stub, 15 mm diameter; (d) split stub for mounting fibres, shown in more detail in Fig. 1.3; (e) hollow stub for mounting yarns.

tape is a good conductor to earth for the sample and the sticky layer does not seep up through the sample, as can occur with wet adhesives through a wicking action by the fibres. This is important if worn or frayed edges such as collar folds or cuffs are being examined. Further precautions can be taken to ensure good conduction to earth, particularly for bulky specimens, by painting round the edges of the specimen with silver conducting paint, but, of course, excluding the areas for examination.

Yarns and fibres can be secured to stubs in a similar manner, but to get a clearer view, without the adhesive layer as a background, yarns are often secured across a hollow stub, Fig. 2.3(e), with a carbon base at least 20 mm from the yarns. This means that no signal is received from the base of the stub; and yarns and filaments are viewed against a black background uncluttered with superfluous details. Single broken fibres are examined in split stubs, Fig. 2.3(d) and Fig. 1.3, as described in the previous chapter for fibres broken in laboratory tests.

Personal preference when starting SEM examination is to view the original material to get some idea of what the fabric surface is like before any damage occurs and record this appearance. The direction of warp and weft yarns should be marked on the stub in such a way that it can be clearly seen and understood when the specimen is in the SEM. All stubs must be carefully tabulated to prevent confusion of specimens at a later date. It is important to examine the whole of the specimen surface before choosing areas for photographic record, and always to be wary of making biassed judgements. Choose areas for photographic record which illustrate the type of fibre damage seen, and record some estimate of the amount, severity and extent of damage in notes made at the time of examination: do not rely on memory recall.

Fabric weave controls the height of warp and weft yarns within the fabric; usually one yarn direction receives most damage, as in twill weaves and poplin weaves. Shirts are usually made from poplin fabrics where the warp yarns are damaged first, and it is only after removal of the warp face that the deeper-seated weft yarns are damaged. It is therefore important to know the direction of warp when examining a worn shirt, and the way the various garment pieces are cut from the cloth. Care has to be taken if blended yarns are used in a fabric, because wear of fibres can make their identification suspect in the SEM. Shirts are often made of polyester/cotton blended yarn fabrics, and the main effect of wear here is the loss of cotton fibres, leaving mainly polyester fibres in the worn areas. However, in lightly worn areas difficulty may be experienced in distinguishing cotton from polyester fibres if the cotton component has been mercerized and is of similar diameter to the polyester fibres. Similar difficulty is encountered in wool/nylon blends used in carpets. Undamaged wool fibres can be recognized by their scales, but attrition of the surface removes the scales, and gives the wool fibres a smooth round appearance similar to that of the nylon fibres.

Once SEM examination has been completed, then evaluation is made of the photographic record together with all notes from both the SEM examination and all previous examinations. After evaluation, a report of the results of the investigation can be written. It may be found that further work is needed to fill gaps in information already obtained. This may mean more SEM work, or that some other technique such as light microscopy is required before the investigation of the worn shirt is complete.

LIGHT MICROSCOPY

Light microscopy has several modes of operation not possible on the SEM. One main one is the observation of colour such as effects of dyeing or discoloration of textile materials. It is possible to see colour variations between fibres and depth of dye penetration within a yarn from examination of yarn cross-sections. Fibre identification is possible with the light microscope from a knowledge of fibre appearance both longitudinally and in cross-section, and from the behaviour of fibres when examined in a polarizing microscope. Two or more different fibre components can be identified in a blended yarn and their relative positions within the yarn checked by examining yarn cross-sections.

Important internal features of a fibre can be viewed in light microscopy, such as voids, cracks or slips in molecular alignment (kinkbands). The content and dispersion of titanium dioxide (a fibre delustrant) can be checked for differences between fibres. Variations in birefringence and other defects are seen in a polarizing microscope, and the refractive indices of fibres measured in an interference microscope. Variations in optical properties between fibres of the same type may suggest that the fibres have different histories or have been through different processing conditions. None of these changes in properties can be detected in the SEM, unless the change has physically altered the external appearance of the fibres.

When looking for internal detail it is advisable to match the refractive index of the mountant to the refractive index of the fibre perpendicular to its axis, known as $n_\perp$. Then if the fibre is aligned with its axis (lengthways) perpendicular to the vibration direction of the polarizer, the surface of the fibre will be invisible and only internal detail observed, such as delustrant particles or voids. Liquids used as fibre mountants must be chemically inert to the fibres. It is most convenient to make a series of mountants, each of known refractive index, by mixing together two chemically inert but miscible liquids in set proportions to give the range of refractive indices required. Two suitable liquids are liquid paraffin and α-bromonaphthalene.

External damage is seen most clearly when yarns or fibres are mounted in a liquid which is of a much higher refractive index than that of the fibres, e.g. diiodomethane. The contrast in refractive indices clearly reveals damage detail on the surface of fibres.

Staining techniques may be used to reveal damage on fibres. The Congo Red test reveals physical damage, localized chemical and heat tendering of cotton fibres. Chemical damage to wool is revealed by tests such as Pauly reagent for alkali damage and Kiton Red G test for chlorinated wool. Other chemical tests are used on wool to reveal alkali damage (Allworden reaction) and acid damage (Kris–Viertel reaction). The results of all these tests are examined in the light microscope.

The selective dyeing of damaged regions in a fibre is a particularly useful way of estimating the extent of damage in a specimen. The piece of fabric is dyed, and then, when examined in the microscope, the damaged zones on the fibre show up and can be counted. A technique of this type was used by W. D. Cooke to elucidate how pills (small tangled balls of fibre) develop on knitted fabrics by alternating damage at the anchor point and roll-up into the pill.

Cellulose fibres such as cotton and viscose rayon are attacked by micro-organisms if the conditions for growth are present. The growth may be visible in the SEM, but its nature is confirmed by optical examination of stained specimens. Mildew can be detected by the Safrannin–Piero Aniline Blue test, and bacteria are stained by Loeffler's Methylene Blue method. The damage caused by these organisms is distinguished in cotton fibres by using the Congo Red test.

As with SEM work, it is always advisable to make a written record of what has been seen in the light microscope, augmented with diagrams, sketches or better still by photomicrography. Once the examination is complete a report of the finding is made. This may be based solely on light microscopy or may be a combination of both light and SEM work, depending on laboratory facilities. Much of our work is a combination of both.

EXAMINATION OF OTHER PRODUCTS

The handling of samples will depend on the nature of the product. Small pieces of carpet can be mounted in the SEM, but the following points should be noted. The pile of a carpet and its backing trap air, which leads to long pump-down times when sputter-coating. When examining a carpet pile in the SEM, only the top of the pile is seen, and therefore either pile yarns have to be examined separately in the SEM or examined in the light microscope to see how far damage has progressed down into the pile. In carpets with blended piles, e.g. wool/nylon or acrylic/nylon piles, wear removes surface features from wool and acrylic fibres, making differentiation from nylon fibres very difficult; however, in the polarizing microscope the fibres are easily distinguished.

Ropes pose problems with regard to size. The smaller ropes can be mounted whole, and samples larger than a 32-mm diameter stub can still be accommodated in the SEM chamber. With our ISI 100A SEM, specimens up to about 100-mm across can be examined. If the sample has to be divided into smaller pieces, careful examination and recording of detail, preferably by photography, is essential. It should also be remembered that it is not just the outside of a rope which gets damaged; but that damage occurs between strands and also between yarns in the strands, and even between filaments within a yarn. Even small-diameter ropes have to be separated down into smaller units. Careful observations and notes of sample appearance before examination are essential.

Samples which have been plastic coated or have received some surface covering finish or are contaminated cause problems. If damage can be seen through gaps or breaks in the surface covering, examine the damage *in situ* before attempting to remove the coating or contaminant by chemical means, since this causes disturbance of damaged yarns and removal of loose material such as broken fibres in the damage sites.

CONCLUSION

The brief account given in this chapter should help to lead investigators to productive studies of damaged products, and this will be further helped by examining the case studies in Parts VI and VII. However, the only real training consists of experience on the job. The guidelines which have been given can be applied to any textile product or sample, provided its size is manageable — and, indeed, even the largest items such as a massive hawser laid out in a rope-maker's yard can be sampled.

The amount of time and effort needed to complete an investigation depends on the nature of the material and the purpose of the investigation. Sometimes it may only be necessary to identify the presence of features which have been extensively studied in earlier investigations, and thus serve to establish the cause of damage. This type of investigation will be quite short. At the other extreme, the first basic studies of particular forms of damage can be extended almost limitlessly in time as more and more detail is shown up, and so it is necessary, implicitly or explicitly, to balance the cost of the investigation and other working constraints against the value of the information obtained.

Sometimes it is not possible to reach a firm and valid conclusion. In these cases one can only report what has been observed, and perhaps speculate on the possible causes of damage.

Part II
Tensile failures

3

INTRODUCTION

The easiest way of studying the mechanical properties of fibres is to take a single fibre, grip the two ends to give a defined test length, and then extend it to break on a tensile tester. The measured load and elongation can be converted to provide the axial stress–strain curve of the fibre. The technique is schematically indicated in Fig. 3.1; and Fig. 3.2 illustrates some of the variety of shape of stress–strain curves observed.

For fuller accounts of the mechanical properties, see Morton and Hearle (1993) and Booth (1961). The end–points of these tests are the tensile breaks illustrated in the following chapters. There are many forms of tensile testers, and a number of test parameters which must be specified and controlled in order to obtain valid quantitative data, although these will not usually affect the qualitative forms of break which are described in this book. Generally, the tensile breaks shown here will have been made on a constant-rate-of-elongation Instron tester, with a test length between 1 cm and 10 cm, a rate of extension selected to give a time-to-break between 10 s and 100 s, in an atmospheric environment controlled at 20°C, 65% r.h. Where there are major departures from these conditions, for example in high–speed or wet testing, this will be specially noted.

A NOTE ON UNITS

Elongation is normalized by division by the initial test length to give *strain*, which is then usually multiplied by 100 to give *extension* per cent.

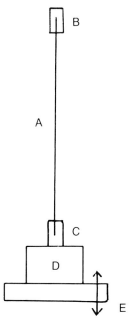

Fig. 3.1 — Typical arrangement of tensile tester: (A) fibre specimen; (B) upper jaws; (C) lower jaws; (D) load–cell; (E) cross–head.

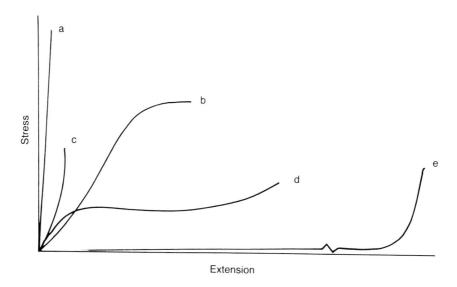

Fig. 3.2 — Typical stress–strain curves of fibres: (a) strong inextensible fibre; (b) tough synthetic fibre; (c) cotton and other plant fibres; (d) weak man–made fibre; (e) elastomeric fibre.

Load (force, tension) may be divided by area of cross-section to give *stress*, but in fibre technology is more commonly and usefully divided by linear density, namely mass per unit length, to give *specific stress*.

Unfortunately, there is a great diversity of units for both stress and specific stress, and an equivalence of quantities, such as energy/mass or stress/density for specific stress. A full conversion chart is given by Hearle (1982). A few examples are:

linear density: 1 tex = 1 g/km
                   1 denier = 1 g per 9000 m
specific stress: 1 N/tex = 1 kJ/g = 1 GPa/(gcm³)
                   1 gf/tex = 9.8 mN/tex
                   1 gf/den = 0.0885 N/tex

4

BRITTLE TENSILE FRACTURE
Glass, ceramic, carbon, elastomeric fibres

Inorganic fibres such as glass have a simple Hookean stress–strain curve, line (a) in Fig. 3.2, and the sharp break is reflected in the clean failure, which may appear as a single flat cleavage plane perpendicular to the fibre axis, **4A(1)**. More commonly, the very smooth region is limited to an approximately semicircular zone centred on the start of crack propagation, and the remainder of the break, while still perpendicular to the fibre axis, has a rougher, hackled appearance, **4A(2)–(4)**.

This form of break is the classical brittle fracture, which obeys the linear elastic fracture mechanics first introduced by Griffith. The surface of the fibre inevitably contains a number of cracks or flaws, as indicated in Fig. 4.1(a). When tension is applied, there is a stress concentration at the tip of each crack, which increases in magnitude with crack depth. As the load on the fibre increases, the largest stress concentration at the deepest flaw eventually exceeds the local tensile strength of the material, which ruptures and so causes the crack to start to grow, Fig. 4.1(b). Since the stress concentration then increases, the growth continues catastrophically to give the smooth 'mirror' region extending radially outwards from the initial flaw, Fig. 4.1(c). This continues until the stress on the part of the fibre ahead of the crack reaches a level at which further crack growth starts from internal flaws to give the rougher region, Fig. 4.1(d). The broken fibre shows the mirror region A and the hackled region B, Fig. 4.1(e).

The fibre strength, which ranges from about 0.75 N/tex (1.9 GPa) up to 1.8 N/tex (4.5 GPa) in the strongest modern glass fibres, depends partly on the inherent structure but is also

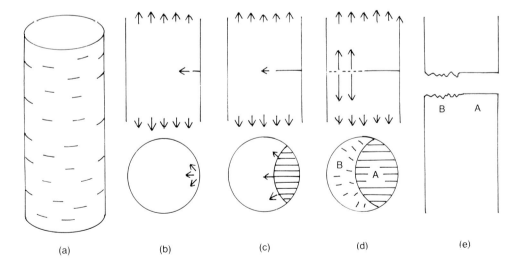

Fig. 4.1 — (a) Glass fibre, indicating surface flaws. (b) When tension reaches a given level, the deepest crack starts to propagate. (c) The crack growth continues catastrophically. (d) Eventually the high stress on the unbroken part starts multiple crack growth. (e) The final failure shows a mirror region A and a hackled region B.

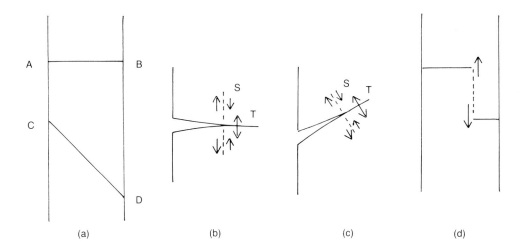

Fig. 4.2 — (a) Plane of maximum tensile stress AB, and maximum shear stress CD. (b) Tensile (T) and shear (S) stresses near a crack tip. (c) Tensile (T) and shear (S) stresses when crack is angled. (d) Two separate cracks linked by a plane of high shear stress.

critically dependent on the state of the fibre surface, with the strength decreasing as damage leads to deeper surface cracks. Glass fibres show greater strength than bulk glass, which commonly contains surface cracks much larger than the few micrometres of a fibre diameter.

An essential feature of the pure form of brittle fracture is that the material strain is everywhere elastic, with no plastic yield. As an alternative to the formulation in terms of stress concentration and material strength, the fracture mechanics may be expressed by the criterion that crack growth will occur when the reduction in elastic strain energy, due to unloading of material near the growing crack, exceeds the surface energy of the newly exposed material (see Chapter 1).

Although breaks perpendicular to the fibre axis are the simplest form and are frequently observed, it is not uncommon to find cracks at other angles, **4A(5)**,**(6)**. There are a number of possible reasons for the changes of direction. The line of maximum tensile stress in the whole fibre is across a plane perpendicular to the fibre axis, and locally it is along the line of crack opening: these two directions usually coincide to give the fractures perpendicularly across the fibre. However, there is also a line of maximum shear stress in the whole fibre at 45° to the fibre axis, and locally there is shear stress perpendicular to the crack. These stresses are indicated in Fig. 4.2(a)–(c) and it is obvious that they could turn the crack into other directions of propagation. Another possibility is that two separate cracks develop, and that these then become linked by shear failure between them, Fig. 4.2(d).

Brittle fracture is also shown by other inorganic fibres, **4B(1)**–**(6)**: these are ceramic fibres with high-modulus linear stress–strain curves, type (a) in Fig. 3.2. The fibre strengths may be comparatively low in fibres intended for uses such as thermal insulation, but will be high in reinforcing fibres. The breaking extension is always very low. Most of these ceramic fibre breaks are very similar to those of glass fibres, but the ICI Saffil (alumina) fracture surface does show appreciable granulation. In other alumina fibres (see Chapter 8), the granulation is much more pronounced, so that the breaks are regarded as falling in a different category.

Another high–modulus fibre, with a stress–strain curve of type (a) in Fig. 3.2, and a smooth fibre brittle failure, is carbon fibre, **4C(1)**–**(3)**, although this form is shown only by certain types of carbon fibre, notably those with a viscose rayon precursor. Other carbon fibres show granular breaks (see Chapter 8).

More surprisingly, brittle failure is also found in fibres at the opposite end of the spectrum: the low–modulus, highly extensible elastomeric fibres, **4C(4)**–**(6)**, with stress–strain curves of type (e) in Fig. 3.2. However, the important point is that the deformation is elastic, and, indeed just before breakage it is also linear and of relatively high modulus. The material is not influenced by the way in which it reached the state near to failure.

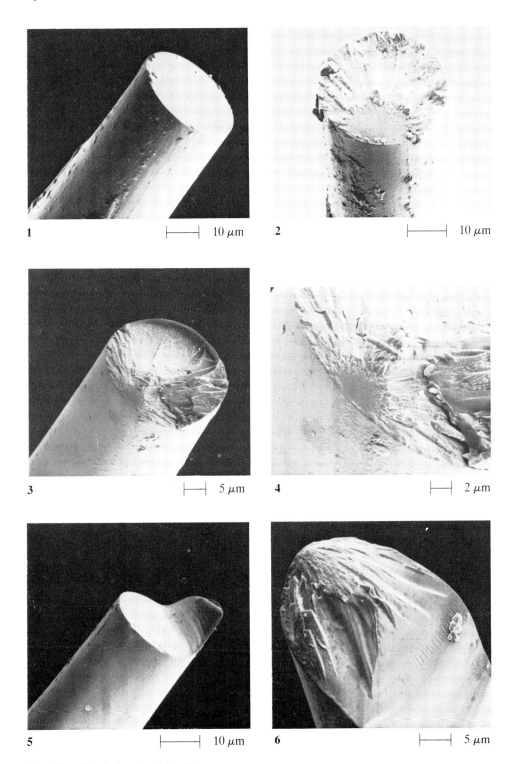

Plate 4A — Tensile breaks of glass fibres.
(1) Single cleavage plane. (2) Fracture starting from front, showing mirror and hackled zones. (3) Fracture with smaller mirror zone. (4) Detail of crack initiation, mirror zone and change to hackled zone. (5) Angular displacement at edge of fibre. (6) Crack propagation with failure in a plane not perpendicular to fibre axis.

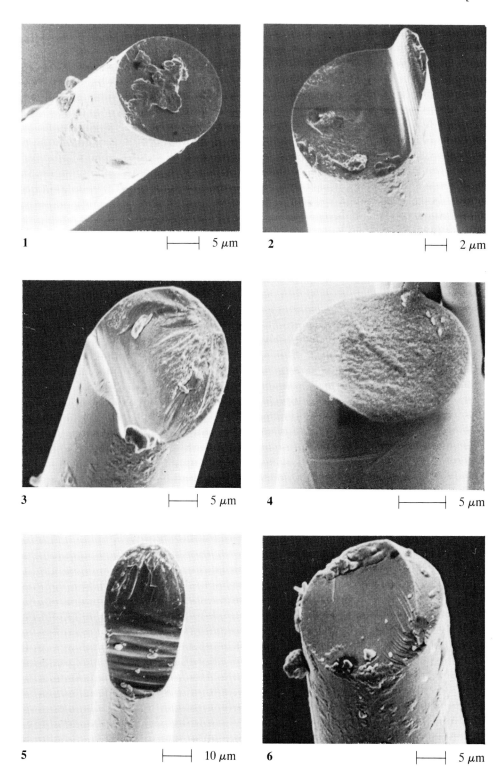

Plate 4B — Tensile breaks of ceramic fibres.
(1) 3M fibre Nextel 312: single cleavage plane with some complication. (2) 3M fibre Nextel 312: crack turning and running along fibre. (3) Sumitomo alumina fibre: crack initiation at top right, leading to angled crack. (4) ICI Saffil alumina fibre: perpendicular crack with slight granulation. (5) Nicolon SiC fibre, NLM 202: angled crack. (6) Nicolon SiC fibre, NLM 202: mixed crack.

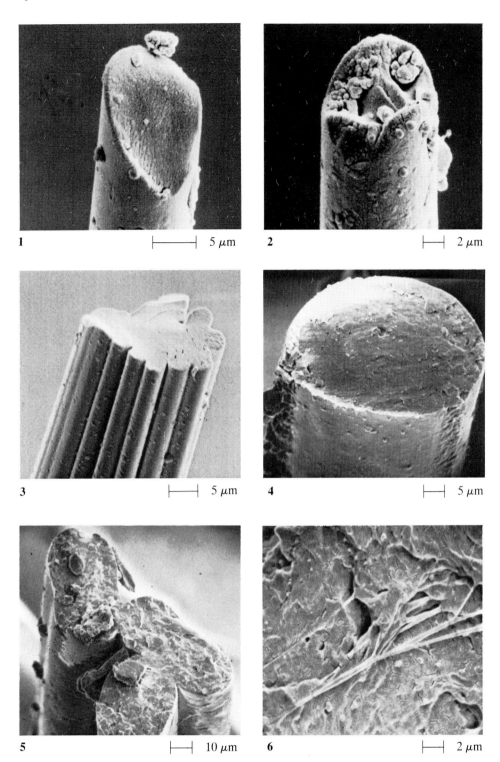

Plate 4C — Tensile breaks of carbon fibres.
(1) Hercules fibre AS46K: angled crack. (2) Hercules fibre AS46K: more complicated break. (3) SCL
(viscose rayon precursor) fibre: perpendicular fracture.

Tensile breaks of elastomeric fibres.
(4) Du Pont Lycra (segmented polyurethane): flat fracture plane. (5) Lycra: break propagated across
three fibres which are fused together. (6) Lycra: detail of fracture surface.

5

DUCTILE TENSILE FRACTURE
Nylon, polyester, polypropylene, etc.

Rupture of the melt-spun synthetic fibres like nylon is dominated by yield. The tensile strength is essentially the yield stress, as shown by the end of line (b) in Fig. 3.2. This final flat portion of the stress–strain curve is really the end of the long draw which can be applied to an unoriented fibre formed on cooling a filament from the melt.

Study of a thick undrawn nylon monofilament shows up clearly the mechanism of ductile crack propagation leading to break. The load–elongation curve is shown in Fig. 5.1, although because this is a thick, short specimen (10 mm long, 1 mm diameter) the strain values may be falsely exaggerated: break usually occurs in undrawn nylon fibres at extensions of around 500–600%. In the nylon bristle, there is an initial elastic extension until yield occurs; the stress then drops after a small overshoot, and the fibre draws at a neck under a constant stress. With further elongation the neck propagates out of the specimen, and then uniform plastic elongation occurs under a steadily increasing tension. Long before the fibre breaks, a skilled operative can detect where break will occur, and subsequently a large crack is easily visible.

The form of break is shown in **5A(1)**, and the way it develops in **5A(2)**. Three main regions can be identified in the break, as illustrated in Fig. 5.2: initiation at A, stable crack propagation at B, and final catastrophic failure at C. These zones are also present in the breaks of finer fibres, but there are some special features to be noted in the thick monofilament. The initiation, shown in **5A(3),(4)**, is due to a development of voids below the fibre surface: these grow and finally coalesce into the crack. The surface of the crack is concave and has a texture of fine voids. It thus appears that cavitation, rather like crazing in glassy plastics, is the detailed way in which a crack forms. The transition from the crack B to the final failure zone C shows a structure of ridges, seen in **5A(5)**, presumably due to an alternation of break and stick. Finally, the far edge of the catastrophic region shows some irregular tearing. An end-on view of the break, **5A(6)**, shows how yielding of the unbroken material leads to a great thinning of the cross–section.

Essentially the same type of break is shown by typical nylon textile fibres, **5B(1)–(4)**, except for some differences in the initiation and transition regions; and studies on films, by Buckley

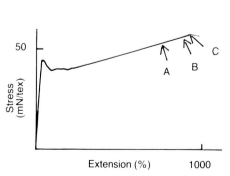

Fig. 5.1 — Load–elongation curve of an undrawn monofilament: crack observed at A; half–way across fibre at B; break at C.

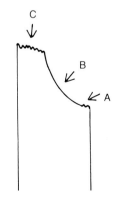

Fig. 5.2 — Ductile break, showing separate regions.

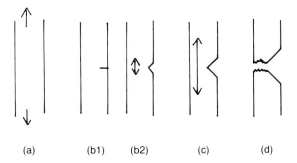

Fig. 5.3 — Sequence of stages in the occurrence of a ductile break. (a) Fibre under load. (b) Crack starts (b1), but immediately opens up (b2). (c) Further crack growth and opening, due to high-yielding extension of the unbroken part. (d) After catastrophic failure.

(1979), **5B(5),(6)**, demonstrate clearly the form of deformation associated with the crack propagation. The mechanisms involved are illustrated schematically in Fig. 5.3. The specimen initially extends uniformly under load, Fig. 5.3(a); but when the tensile stress reaches a certain level, Fig.5.3 (b1), a crack starts to propagate into the specimen, from a surface flaw on a fibre, from a cut deliberately put in the film, or from self-induced voids in the bristle. Plastic yield (drawing) of material causes the crack to open into a V-notch, Fig. 5.3(b2), which propagates steadily into the specimen, Fig. 5.3(c). The discontinuous separation at the open end of the V is linked to the continuous elongation on the other side by the long zone of plastic shear shown in **5B(5)**. Finally catastrophic failure occurs under the high stress on the unbroken part of the cross–section, Fig. 5.3(d). Similar forms of failure have been reported in plastic films by Walker, Haward and Hay (1979). The fracture mechanics is very complicated in stress and strain distribution and has not been analysed. Distortion of the break occurs when the film specimen is cut at an angle to the orientation direction, **5B(6)**: this picture also shows thinning of the film due to the high plastic extension on the opposite side to the crack.

Although the ductile V-notch break is the common form, there are several variants in the form of nylon breaks. Increasing the rate of strain, without going to ballistic impact and the change of mechanism described in Chapter 6, causes the size of the crack region to decrease relative to the final failure region, with changes from over 50% in **5B(3)**, to 40% and less than 20% in **5B(1)** and **(4)**, respectively.

Initiation may be at a crack or flaw perpendicular to the fibre axis, as in **5B(1)**, or at a point, **5C(1)**, or points, a wide line, **5C(2)**, or an angled line, **5C(3)**. The latter distorts the form of the V-notch, and may in extreme cases, such as **5C(4)**, give a multiple final failure zone. Occasionally, cracks develop in two places on the fibre, either opposite one another, **5C(5)**, or axially displaced, **5C(6)**.

In rare circumstances, the break starts internally and not on the surface, as shown in **5D(1)**. This would occur when there is a substantial flaw inside the fibre. The three-dimensional geometry of ductile crack propagation then causes the formation of a double cone within the fibre, leading to the catastrophic failure region. Intermediate forms are shown in **5D(2),(3)**, which are nylon fibres partially oxidized by hydrogen peroxide.

Fibres with a special shape, such as the trilobal nylon in **5D(4)**, show a modified geometry of crack formation.

Although this chapter is concerned with tensile failure, it is worth noting that when a nylon fibre is twisted to break it can show a fracture morphology, similar to tensile breaks, except for a skewing round of the crack, as shown in **5D(5)**. Of more direct relevance is the fact that the tensile failure of a fibre which has been heat set in a twisted state also has a distorted shape. The crack appears to propagate perpendicular to the helical line of the twist, which corresponds to the molecular orientation, as seen in **5D(6)**. There is also some splitting between the lines of orientation, due to shear forces.

Most of the pictures in **5A–5D** are of nylon 66, but other melt–spun synthetic fibres show similar forms of break: polyester (polyethylene terephthalate) in **5E(1)–(3)** and nylon 6 in **5E(4)**. Polypropylene also shows a V-notch leading to a catastrophic region, **5E(5)**, but due to features of chemical and physical structure, which affect its melting and thermomechanical behaviour, always shows a more disturbed final failure region, with pieces of material sticking out from the break. Sometimes the V-notch is almost lost in the confusion **5E(6)**.

In most tensile tests of normal nylon and polyester fibres, with a typical test length, say 5 cm, the failure will usually develop through a crack from one point on the fibre surface. Occasionally, there may be two cracks, and sometimes they are internal. However, in some fibres, where the surface must have been affected in an unusual way, a line of separate cracks appears, with one happening to propagate first to rupture: examples are shown in **5F(1),(2)**.

Studies of crack development on a polyester monofilament confirm that the cracks start to develop and grow shortly before the final rupture occurs. Cracks as they exist at 38% and 40% extension are shown in **5F(3),(4)**, for a fibre which breaks at a little more than 40% extension with the form shown in **5E(1)**.

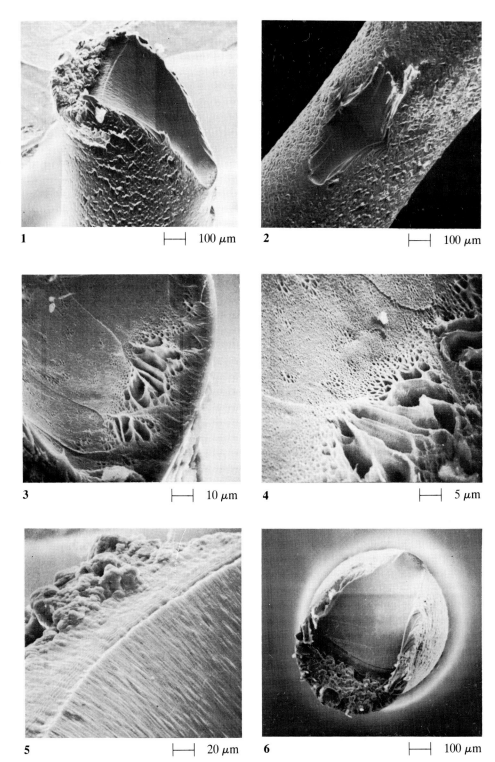

Plate 5A — Rupture of a coarse undrawn nylon 66 bristle, 1 mm diameter, extended at a strain rate of 8 x 10^{-4} s^{-1} (see Fig. 5.1).
(1) Broken end, showing initiation at bottom right (A in Fig. 5.2), ductile crack growth across fibre (B), and final failure at top left (C). (2) Propagating crack, from fibre removed from test before break. (3) Initiation region, showing start of crack at large voids. (4) Detail showing transition from large voids in initiation to small voids in growth region. (5) Transition from region of stable crack growth to final catastrophic failure region. Note also irregular tearing at far edge of crack. (6) End-on view of broken fibre. Note thinning of cross–section due to local drawing of material near break.

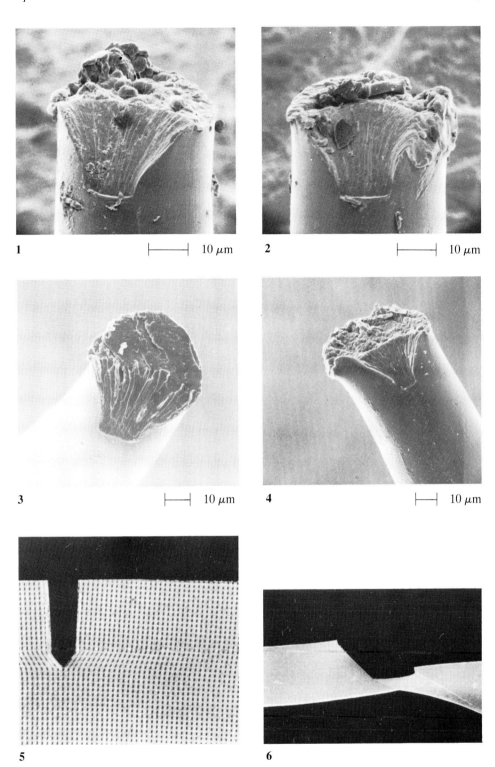

1       $\vdash\!\!\dashv$ 10 μm   2 $\vdash\!\!\dashv$ 10 μm

3       $\vdash\!\!\dashv$ 10 μm   4 $\vdash\!\!\dashv$ 10 μm

5   6

Plate 5B — Tensile break of nylon fibres.
(1) 17 dtex nylon 66 fibre, broken by extension at strain rate of 1.67×10^{-2} s^{-1}. Failure starts from a line perpendicular to the fibre axis. Ductile crack crosses about 40% of fibre thickness before catastrophic failure. (2) Opposite end of same nylon 66 break. (3) Nylon fibre, as (1), broken at low rate of strain, 1.67×10^{-4} s^{-1}, with ductile crack covering 50% of fibre. (4) Nylon fibre, as (1), broken at higher rate of strain, 3.33×10^{-1} s^{-1}, with crack penetrating less than 20% of thickness.
Tensile break of polyester film.
(5) Polyester (PET) film, with crack growth from an initial cut. Grid shows strain distribution. (6) Polyester film extended at an angle to orientation direction, showing distortion of crack growth and yield.

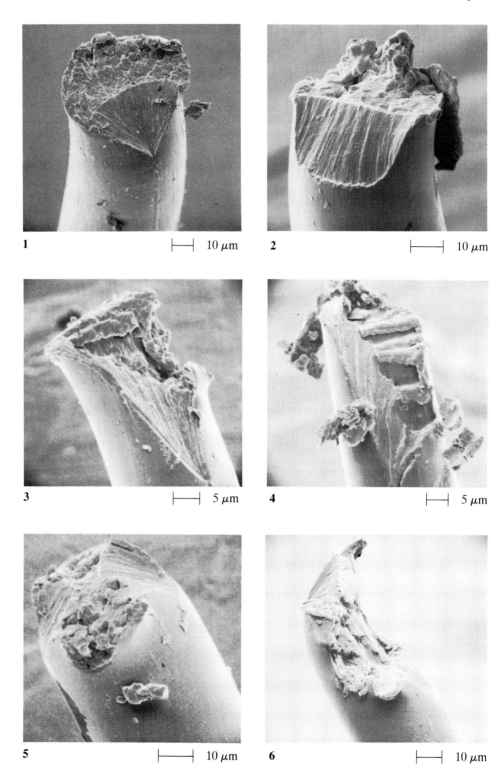

Plate 5C — Tensile breaks of nylon fibres.
(1) Experimental nylon 66 filament, with initiation of break at a point. (2) Break of nylon 66 fibre, previously immersed in 30 vol. H_2O_2 for 2 hours, with wide initiation zone. (3) 4.5 dtex nylon 66 fibre, with break initiated at an angled linear flaw. (4) 4.5 dtex nylon 66 fibre, with inclined initiation leading to multiple catastrophic failure zones. (5) Ductile crack growth, from initiation at two zones at same axial position in 17 dtex nylon 66. (6) Double crack in 17 dtex nylon 66, starting at two different positions along the fibre, with catastrophic failure propagating axially between the two cracks.

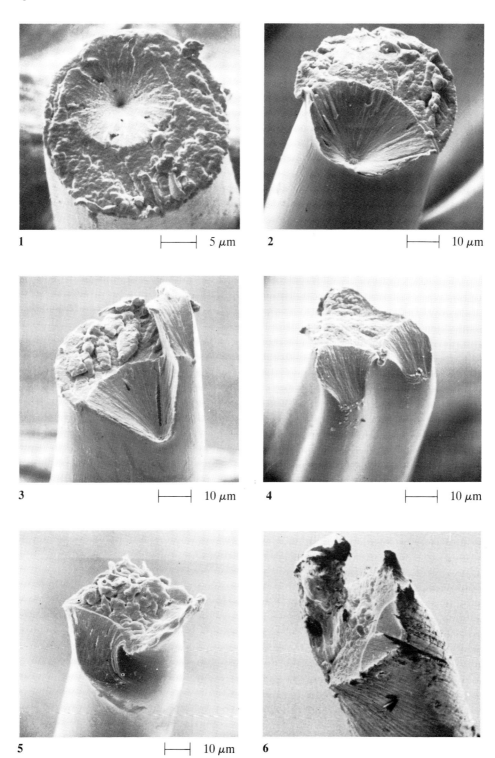

Plate 5D — Tensile breaks of nylon fibres.
(1) Break starting at an internal fault in 4 dtex nylon 66 fibre, to give a cone-shaped ductile crack zone, surrounded by the catastrophic failure region. (2) Break starting near the edge of a nylon 66 fibre, immersed in 100 vol. H_2O_2 for 2 hours, before testing, with a cone developing into a V-notch. (3) A combination of complications in break of nylon 66 fibre, immersed in 30 vol. H_2O_2 for 2 hours, including double initiation, internal and external. (4) Rupture of trilobal nylon 66 fibre, with double initiation on separate lobes. (5) 17 dtex nylon 66 fibre twisted to break at 78 turns/cm under constant tension (see also Chapter 24). (6) Tensile fracture of nylon 6 fibre heat set with a surface helix angle of 34°.

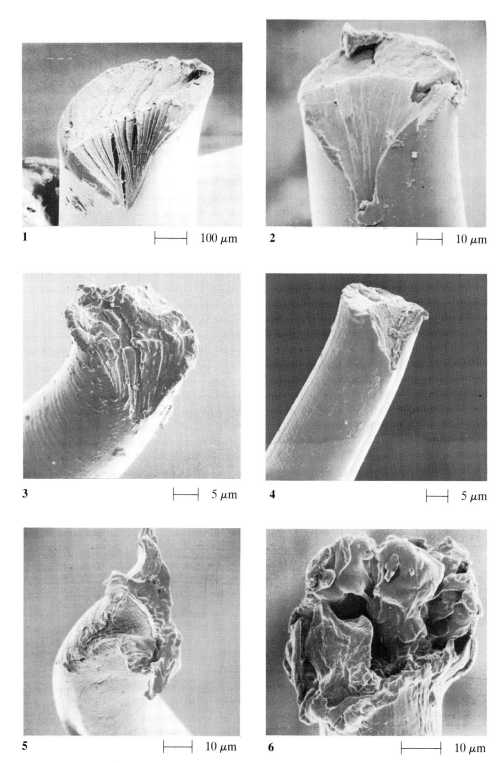

Plate 5E — Tensile breaks of polyester fibres.
(1) Break of a thick (0.55 mm diameter) polyester monofilament. (2) Break of a fine polyester fibre, produced on laboratory melt–spinning equipment and drawn 1.5×. (3) Break of a commercial polyester fibre.

Tensile break of nylon fibre.
(4) Break of a nylon 6 fibre.

Tensile break of polypropylene fibre.
(5) Break of Phillips polypropylene, 8.3 dtex, tenacity of 523 mN/tex, breaking extension of 53%.
(6) Break of 15.7 dtex AKZO polypropylene fibre.

1   2   ⊢——⊣ 10 μm

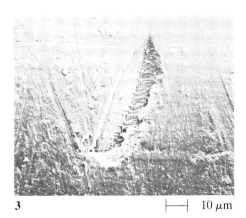

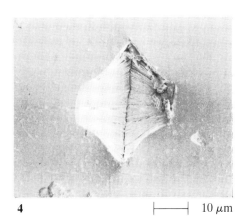

3   ⊢——⊣ 10 μm   4   ⊢——⊣ 10 μm

Plate 5F — Tensile breaks of nylon: multiple cracks.
(1) Nylon 6 fibre from yarn, heat set at 195°C, showing multiple initiation of cracks. (2) Multiple initiation in a commercial nylon 6 fibre with break and partial cracks.
Tensile test of polyester monofilament (0.5 mm diameter) with final break at just over 40% extension crack development.
(3) At 38% strain. (4) At 40% strain.

6

HIGH-SPEED TENSILE BREAK
Nylon, polyester and other melt-spun fibres

It was noted in Chapter 5 that when the rate of extension was increased the size of the V-notch (ductile crack region) reduces. This was demonstrated with ordinary tensile testers, which can give times-to-break down to about 1 s. But a more dramatic change occurs at higher speeds, such as can be imposed by a falling pendulum, with breaks in a fraction of a second, or even faster with ballistic impact. The typical form is a mushroom end, shown for nylon in **6A(1)**. The detail of the broken end in **6A(2)** shows a smooth, partly globular appearance which suggests melting of the material.

The reason for the difference from slow breaks is a change from isothermal to adiabatic conditions. At low speeds heat can be lost to the surroundings, but at high speeds the heat of drawing warms up the fibre in the region of the break. The softening, near melting, presumably then allows break to occur by localized flow of viscous material, so that any initial crack geometry is lost, and the snap-back after break causes the material to collapse into the mushroom cap. The exact shape of the end will depend on details of the thermomechanical forces.

Sometimes, as in **6A(3)**, there is evidence of initiation at the surface and a vestigial V-notch region. In other circumstances there is a more distinct small V-notch, and this is usually so when the break is a little slower, as in **6A(4)**. It is interesting to note that there are two stages in the crack propagation, and the final rupture shows more globular indications of melting or softening. At higher speeds, as in **6A(5)**, the mushroom head may be smaller.

Similar mushroom cap breaks are found in polyester fibres broken at high speed **6A(6)**. This picture also shows transverse striations along the length of the fibre which may be due to snap-back effects.

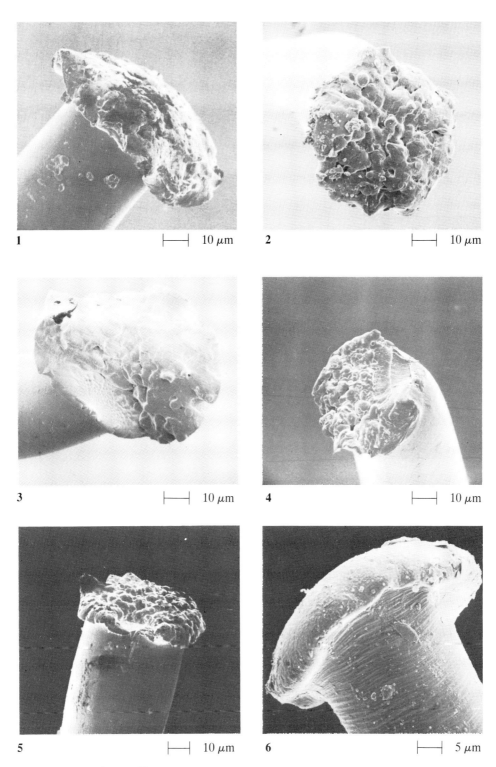

Plate 6A Nylon 66 fibres, 17dtex.
(1) Broken by a pendulum falling from a 90° angle. (2) Broken, as in (1), showing detail of end. (3) Broken, as in (1), showing a small V-notch. (4) Broken by a pendulum falling from a 45° angle. (5) Broken by a relaxation catapult.
Polyester.
(6) Pendulum break showing mushroom head and transverse striations.

7

AXIAL SPLITS

Para-aramid (Kevlar), high-modulus polyethylene (HMPE), nylon at low temperatures

Highly oriented, highly crystalline, linear polymer fibres, such as the para-aramid Kevlar, show very high strength (more than 2 N/tex or 3 GPa) and an almost linear stress–strain curve like line (a) in Fig. 3.2. They break through the development of long axial splits in the filaments, **7A(1),(2)**.

If the fibre was perfectly uniform and subject to pure tension, there would be no stress across planes parallel to the fibre axis, and so no reason for axial splitting. But if there is any discontinuity or defect, either on the surface of the fibre or internally, this will give rise to a shear stress as shown in Fig. 7.1.(a). As the load on the fibre is increased, the shear stress rises, eventually overcoming the transverse cohesive forces and causing an axial crack to form, Fig. 7.1(b). If the crack is slightly off axis, it will eventually cross the fibre and lead to rupture, Fig. 7.1(c). Failure occurs in this way because even a small shear stress will overcome the weak intermolecular bonds between the polymer molecules before the large tensile stress breaks the covalent bonds within the chain molecules, as indicated schematically in Fig. 7.2. The difference may be accentuated by structural discontinuities.

In a majority of tensile breaks, it has been found that one end shows multiple splits, **7A(1)**, whereas the other end shows a single split, **7A(2)**. This is not due to any asymmetry in the material or the test method, but is a consequence of geometry. In Fig. 7.3(a), a split develops from a surface flaw, and then divides by bifurcation to give multiple cracks. However, if all the cracks develop at the same rate, the one on the outside must reach the other edge of the fibre first, giving the one end with a single split, and the other with multiple splits, as in Fig. 7.3(b). Even if one of the inner cracks develops faster, it would not give a simple pattern of multiple splits on each end, but would be a more complicated form, with re-entrant cracks on one end, as shown in Fig. 7.3(c).

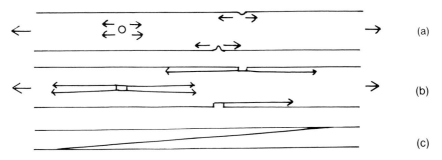

Fig. 7.1 — (a) Shear stresses at discontinuities in a fibre under tensile stress. (b) Shear stresses causing single or multiple cracks. (c) Crack at an angle to fibre axis, running across fibre and forming a break.

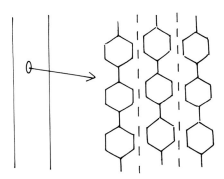

Fig. 7.2 — Schematic illustration of strong covalent bonds parallel to fibre axis contrasted to weak intermolecular bonds.

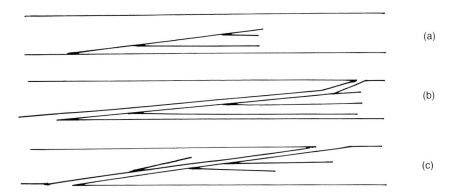

(a)

(b)

(c)

Fig. 7.3 —(a) Split propagated from one side of fibre with bifurcation, with all cracks propagating at the same rate. (b) When split reaches other side, one end of the break is a single split and the other has multiple splits. (c) If one of the inner cracks were to grow faster, a different pattern of splits would be observed in the broken fibres.

Fracture of Kevlar in liquid nitrogen, **7A(3)**, shows a similar long axial split, but interesting transverse striations can be seen along the split surfaces. Another example of the pronounced axial splitting is shown in **7B(1),(2)**.

The release of elastic energy, following rupture, can cause other features to appear. Snapback along unbroken fibre leads to intense kinkband formation, **7B(3)**, due to an internal buckling of the oriented structure. Fibrillated portions may buckle as a whole into helices, **7B(4)**, and more complicated local splitting and buckling may occur, **7B(5),(6)**.

High-modulus polyethylene is another fibre which breaks with axial splitting, **7C(1)–(4)**. This is to be expected, because it is also a very highly crystalline and highly oriented fibre with high strength.

Ordinary nylon fibres broken at low temperatures, by means of immersion in liquid nitrogen while being extended on an Instron tester, also often show breakage by long axial splitting, **7C(5),(6)**. However, long splits were not obtained when special precautions were taken to ensure that the jaws and the test fibre were strictly aligned and there was no twist in the fibre. The splitting thus appears to be caused by the presence of small shear stresses, in addition to the larger tensile stress. At such low temperatures, internal chain mobility through rotation round bonds will be blocked and the material will be glassy. Ductile behaviour changes to brittle, with the load–elongation curve ending sharply without yield, so that a different form of break would be expected. However, it is surprising that it is so easy to cause transverse cohesion to give way before the axial cohesion, even at a moderate degree of orientation.

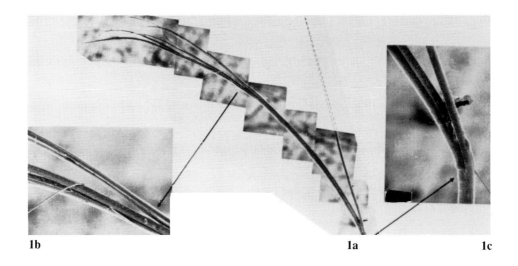

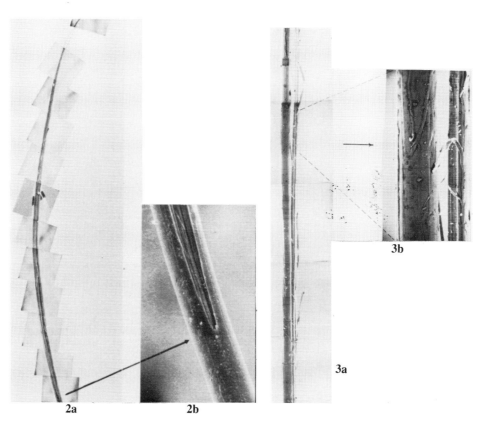

PLATE 7A — Tensile breaks of aramid (Kevlar 29) fibres.
(1a,b,c),(2a,b) Opposite ends of break, one showing multiple splitting and one showing a single split.
(3a,b) Fracture in liquid nitrogen.

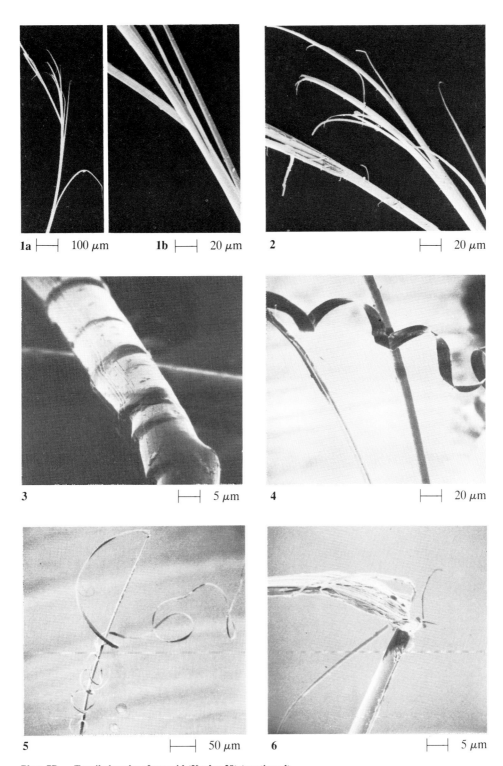

1a |—— 100 μm   1b |—— 20 μm   2 |—— 20 μm

3 |—— 5 μm   4 |—— 20 μm

5 |—— 50 μm   6 |—— 5 μm

Plate 7B — Tensile breaks of aramid (Kevlar 29) (continued).
(1a,b),(2) Break with two details at higher magnification.
Snap-back effects in Kevlar 49.
(3) Kink-bands along fibre. (4) Helical deformation of ribbon-like fibril. (5),(6) Complicated coiling, with detail of splitting.

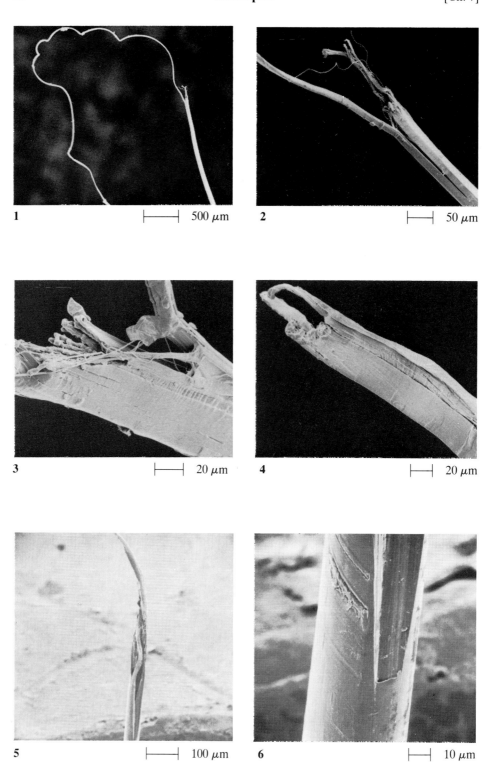

Plate 7C — Tensile breaks of high-modulus polyethylene (Allied Spectra 900) fibres.
    (1)–(3) Breaks of one end at increasing magnification. (4) Break of other end.
Tensile break of nylon 66 in liquid nitrogen.
            (5),(6) Break, shown at two magnifications.

8

GRANULAR FRACTURE
Solution-spun fibres

A group of poorly crystalline fibres, mostly with rather strongly interactive molecules, and all spun from solution, have stress-strain curves like line (d) in Fig. 3.2, showing an initial elastic region which yields at about 2% extension to an easily extensible region of poor recovery. Such fibres include viscose rayon (cellulose), acetate (cellulose diacetate and triacetate) and acrylic (polyacrylonitrile). They break to give a granular surface running more or less perpendicularly across the fibre, **8A(1)**. Similar breaks occur in cotton and wool fibres in some circumstances (see Chapters 18 and 19), and in alumina and carbon fibres, as mentioned at the end of this chapter.

In rayon, **8A(1)–(3)**, and even more in acetate fibres, **8A(4)–(6)**, there are many cracks visible all along the fibre. This means that the whole fibre is everywhere on the verge of failing when one section breaks. Indeed, it has been known for a triacetate fibre to be lost on rupture because it shattered into a cloud of tiny pieces. An incipient form of this sort of break-up is shown in **8A(5)**, where at a point remote from the actual break an acetate fibre has almost broken into two.

Triacetate fibres, **8A(6)**, break in a similar way, but the fracture surface is smoother. The sharpness of the granulation decreases from cellulose (viscose rayon) to secondary cellulose acetate to cellulose triacetate. This change is associated with ease of melting, which determines the effect of the heat generated during plastic deformation on the surface appearance. Cellulose, which is strongly hydrogen-bonded, will not melt, but at a high temperature it chars; secondary acetate melts with some difficulty; but triacetate is truly thermoplastic so that softening due to heat leads to the smoother, rounded surface of the break.

Granular breaks look very like a low-magnification view of the break of an oriented fibre composite, but, for such a material, the magnification can be increased, and the grains are resolved into individual fibre breaks separated by matrix. At a much finer level of structure, similar discontinuities must exist within fibres showing granular breaks. It has, in fact, been shown by Knudsen (1963) that during their spinning acrylic fibres coagulate from solution to give a spongy structure with excess solvent filling voids in the fibre. In the subsequent stretching and drying, the voids elongate, collapse and apparently disappear. But it is likely that the original void surfaces remain as weak boundaries separating the material into separate fibrillar elements.

Fig. 8.1(a) shows an idealized view of such a structure. When the tension reaches a certain level, elements will begin to break, Fig. 8.1(b), but the discontinuity prevents the occurrence of a large enough stress concentration to cause the crack to continue propagating across the fibre. However, there is some cohesion between elements, and excess stress is transferred to neighbouring elements which are thus more likely to break at a nearby position. Eventually the failure becomes cumulative over a cross-section, Fig. 8.1(c), and the granular break results, Fig. 8.1(d). The granular surface is clearly shown in the Courtelle acrylic (PAN) break, **8B(1),(2)**.

There can be deviations from the simple granular form. Sometimes, although the detail on the surface is granular, the overall effect shows evidence of ductile crack propagation, which was shown in its simplest form in Chapter 5, with a V-notch leading to catastrophic failure. This is just detectable in the Courtelle break, **8B(1),(2)**, and is very clear in the Acrilan carpet fibre, **8B(3)**. An incipient development of such a crack is shown in **8B(4)**.

In other fibres the break occurs at two places, linked by an axial split, to give a stepped break. There is a slight indication of this in the viscose rayon fibre, **8A(2)**, but the steps are very distinct in the acrylic fibres, **8B(5),(6)**. The separation between the steps can be many fibre

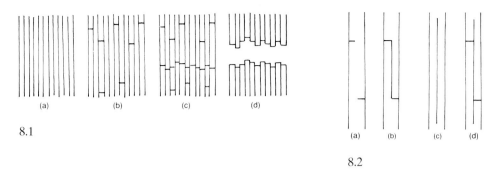

8.1

8.2

Fig. 8.1 — (a) Structure of separate elements. (b) Under tension, elements start to break. (c) Stress transfer causes cumulative break over a cross-section. (d) Granular breaks.

Fig. 8.2 — Mechanism of formation of stepped break. Alternative 1: (a) two breaks form; (b) breaks join up by axial split. Alternative 2: (c) fibre containing axial split; (d) two parts break independently.

diameters, **8B(6a)**. It is not clear whether two breaks form and then join up by the axial split, as in Fig. 8.2(a,b), or whether the fibre is already split axially into two parts, which then break independently, as in Fig.8.2(c,d).

In one special case, **8A(3)**, the break occurred preferentially at a weak place in the fibre caused by the presence of large voids, which reduced the area of material available to carry load.

Bicomponent acrylic fibres may break without showing any very special features, **8C(1)**, but they may also show splitting between the two components, **8C(2)–(6)**, with granular breaks and continuing splits at several different places in the fibre.

Other solution-spun organic fibres which show similar forms of break are polyvinyl alcohol (PVA), **8D(1),(2)**, where the 'granules' on the surface become elongated projections, and the advanced engineering thermally resistant fibre, polybenzimidazole (PBI), **8D(3)**.

Granular breaks are also found in some high-modulus fibres, such as the alumina fibre, DU PONT FP3, **8D(4)**. This fibre is made by extruding an aqueous slurry of alumina and spinning additives, drying, and then heating the fibre to a high temperature. Evidently, the method of manufacture leads to a granular structure which shows up in the break. The influence of the mode of formation is clearly shown in breaks of carbon fibres, **8D(5),(6)**, which have features similar to those of the precursor PAN or PVA fibres. Other forms of alumina and carbon fibre, illustrated in Chapter 4, show different break appearances.

As stated in the first paragraph of this chapter, fibres spun from solution typically show granular breaks, such as those in **8A(1)–(3)** for cellulose fibres produced by the viscose process. Tensile breaks of Tencel, show similar granular fractures; this is a *lyotropic* fibre from Courtaulds, made by a new process for regenerating fibres directly from a solution of cellulose in an organic solvent. The break can be seen to fan out from an incipient initial crack in **8E(1),(2)** and more prominently in **8E(3)**. In **8E(4),(5)**, the break divides into steps, and in **8E(6)** splits to a greater extent.

8F(1),(2) shows granular fractures similar to **8A(2)** and **8D(5)**, in acrylic (PAN) fibres after initial thermal stabilisation. **8F(3)–(6)** shows granular fractures in melt-spun polyester fibres loaded with barium sulphate.

Granular breaks of carbon fibres were shown in **8D(5),(6)**. A recent investigation by Boyes and Lavin of DU PONT has demonstrated how the use of a high resolution SEM, referred to in Chapter 1, can be used to show up differences in the fracture of carbon fibres from different precursors. **8G(1)–(3)** are for high modulus PAN-based carbon fibres. **8G(1)** shows a clean break with a nubby surface texture. **8G(2)** shows a skin-core effect and a braided texture, which may result from aggregations of polymer chains. At the highest magnification, the cross-section in **8G(3)** demonstrates the very small scale of the texture. In contrast to this, the cross-section of a high modulus pitch-based carbon fibre in **8G(4)** shows many long crystal planes parallel to the yarn axis, following zig-zag paths across the fibre. This influences the composite character of the granular break of the pitch-based fibre, as shown in **8G(5)**. At a higher magnification, **8G(6)**, the length and perfection of the carbon crystal planes becomes apparent.

The SEM pictures of ceramic fibres in **8H** were obtained by Bunsell and colleagues at Ecole des Mines de Paris.

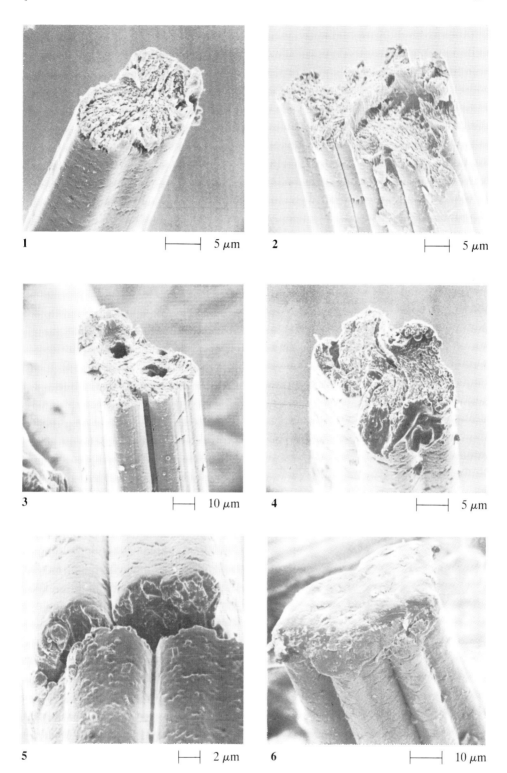

Plate 8A — Tensile breaks of cellulosic fibres.
(1) 3.3 dtex polynosic viscose fibre (Vincel). (2) 14.7 dtex viscose rayon fibre. (3) 20 dtex viscose rayon fibre with holes at point of break. (4) 4 dtex secondary acetate fibre (Dicel). (5) Cracks on surface of broken acetate fibre, near to breakage. (6) Triacetate fibre (Tricel).

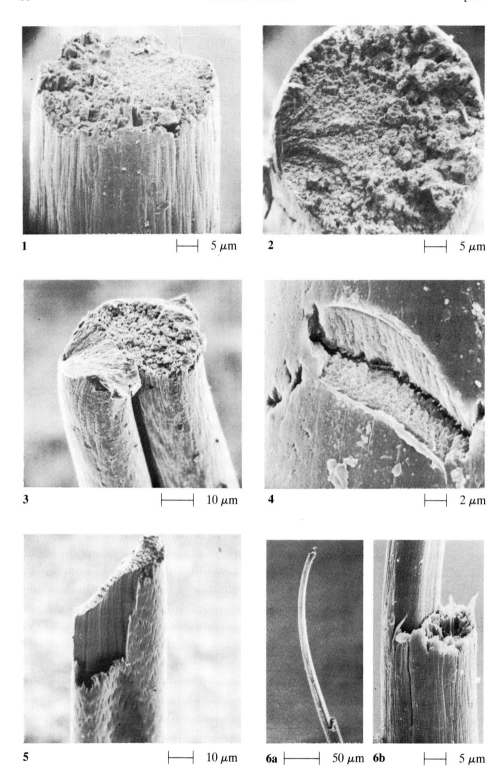

1   ├─┤   5 μm   2   ├─┤   5 μm

3   ├────┤ 10 μm   4   ├───┤ 2 μm

5   ├────┤ 10 μm   6a ├─────┤ 50 μm   6b ├───┤ 5 μm

Plate — 8B — Tensile breaks of acrylic fibres.
(1),(2) 16.7 dtex Courtelle; two views of same break. (3) 16.7 dtex Acrilan. (4) Crack development in
Acrilan, at a position away from the break. (5) 7 dtex Orlon 42. (6a,6b) 3.3 dtex Courtelle.

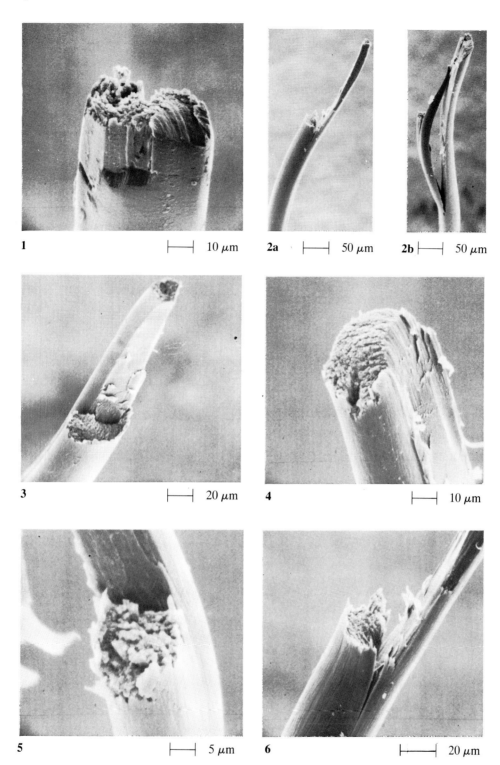

Plate 8C — Tensile breaks of 18 dtex bicomponent Acrilan acrylic fibre.
(1) Relatively simple break, with some evidence of initial crack formation, and a small step at the edge.
(2a,b) Opposite ends of another break. (3) Enlarged view of end (2a). (4) Tip of end (2b), which fits into
step on (2a), shown in (3). (5) Detail of step break on middle of lower part of (2b). (6) Continuing split at
step on end (2a).

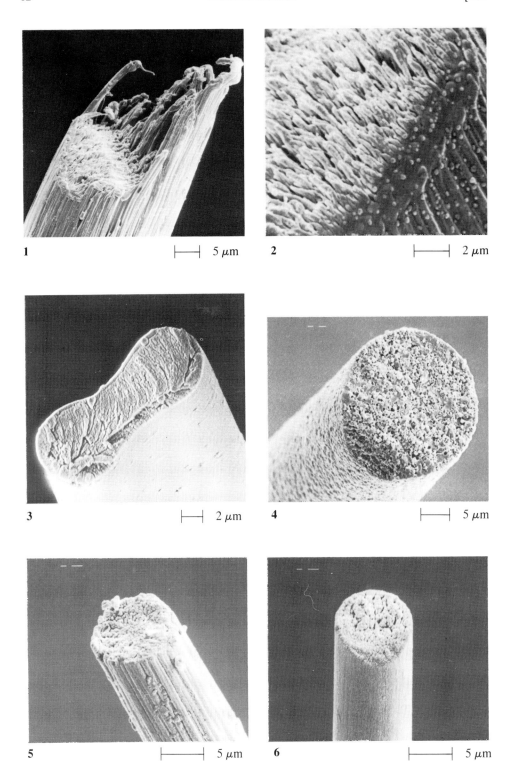

Plate 8D — Tensile breaks of other fibres.
(1),(2) Polyvinyl alcohol fibre (PVA). (3) Polybenzimidazole fibre (Celanese PBI). (4) Alumina fibre (DU PONT FP3). (5) Carbon fibre, PAN precursor. (6) Carbon fibre, PVA precursor.

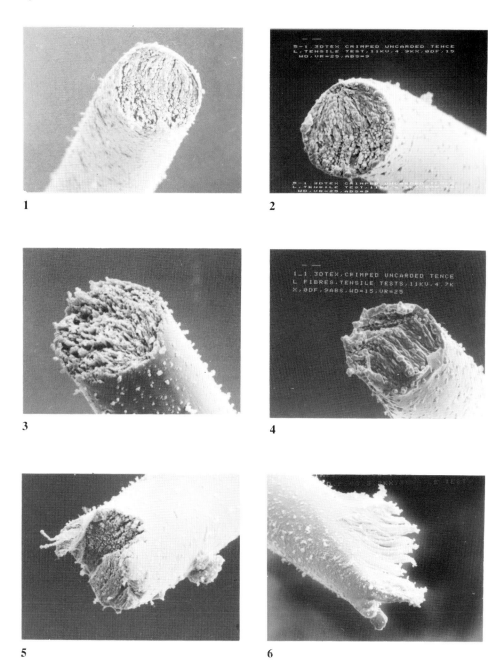

Plate 8E — Tensile breaks of *Tencel* fibres.
(1)–(6) Granular breaks of varying complexity.

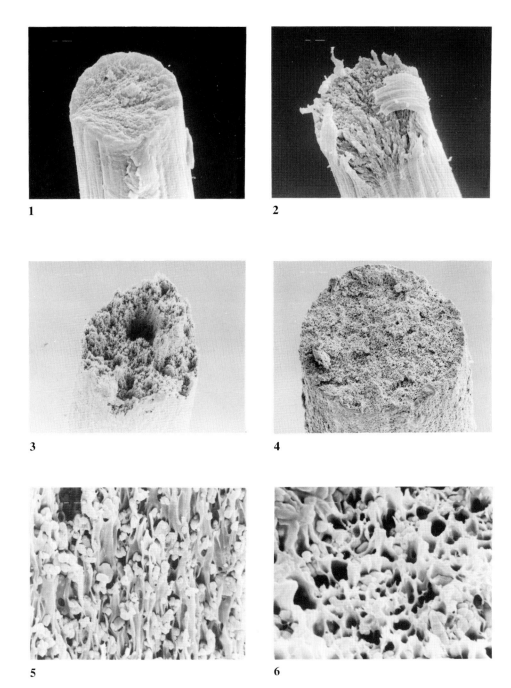

Plate 8F — Tensile breaks of oxidized PAN fibres.
                    (1) Fibre stabilized for 90 minutes at 250°C. (2) Stabilized for 15 minutes.
Tensile breaks of polyester fibres loaded with barium sulphate.
(3) Fibre with 60% barium sulphate. (4)–(6) Fibres with 70% barium sulphate at different magnifications.

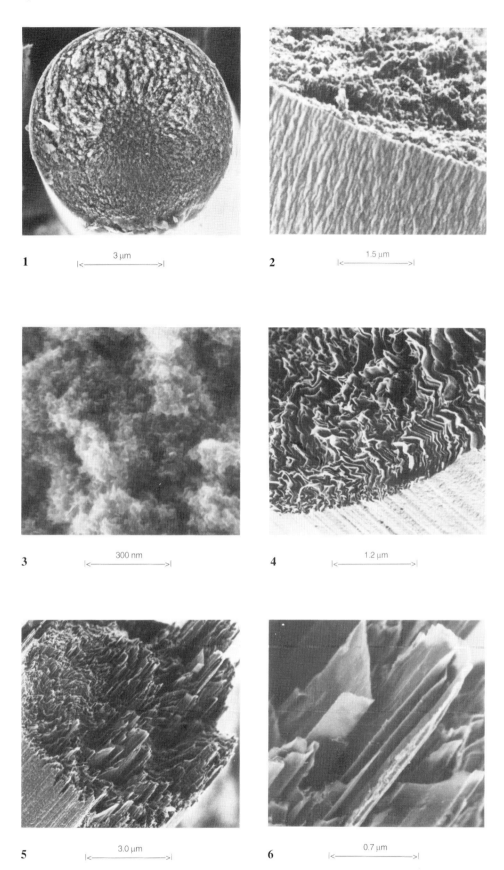

Plate 8G — Carbon fibres viewed in high-resolution SEM, by courtesy of E.D. Boyes and J.G. Lavin, Central R&D, DuPont Co.
(1),(2) Tensile breaks of high modulus PAN-based carbon fibres. (3) Cross-section of PAN-based fibre at high magnification. (4) Cross-section of high modulus pitch-based carbon fibre. (5),(6) Tensile breaks of pitch-based fibres.

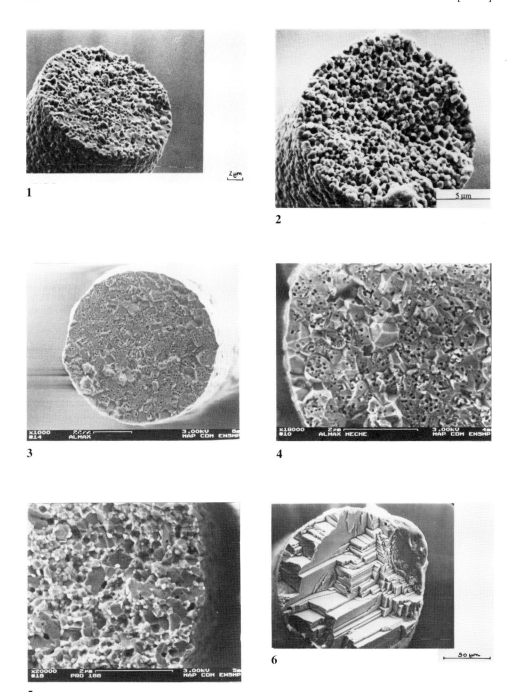

Plate 8H — Breaks of ceramic fibres, by courtesy of A.R. Bunsell, Centre des Materiaux de l'Ecole des Mines de Paris.
(1)Tensile break of FP alumina fibre at room temperature. (2) Creep failure of FP fibre at 1300°C. (3),(4) Break of Almax alumina fibre. (5) Break of PRD 166, alumina with zirconia, fibre. (6) Break of single crystal Saphikon α-alumina fibre.

9

FIBRILLAR FAILURE
Wet cotton

Cotton breaks in tension in different ways, depending on the humidity and the chemical treatments applied to the fibre. This diversity is explored in Chapter 18. However, one form of break is introduced here, because it is a distinct identifiable mode of separate fibrillar failure, **9A(1)–(4)**.

Cotton is known to be an assembly of crystalline microfibrils, and when wet the fibrils will be separated by layers of absorbed water molecules. The interaction between the fibrils will be very weak, and they break independently, as indicated in Fig. 9.1(b). When all the fibrils have broken, Fig. 9.1(c), the two ends separate, Fig. 9.1(d): the fibre has broken. The ends of the break will be a collection of fibrils, or groups of fibrils, which may collapse into a tapered end under the surface tension of the water as the fibre dries, Fig. 9.1(e). These features can be seen in **9A(1)–(4)**.

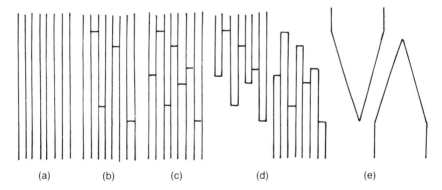

(a)   (b)   (c)   (d)   (e)

Fig. 9.1 — Schematic representation of independent fibrillar break. (a) Structure of separate fibrils, only weakly linked. (b) Under sufficient tension, fibrils begin to break. (c) Finally all fibrils have broken and the ends can separate. (d) Two broken ends. (e) Possible collapse to tapering ends.

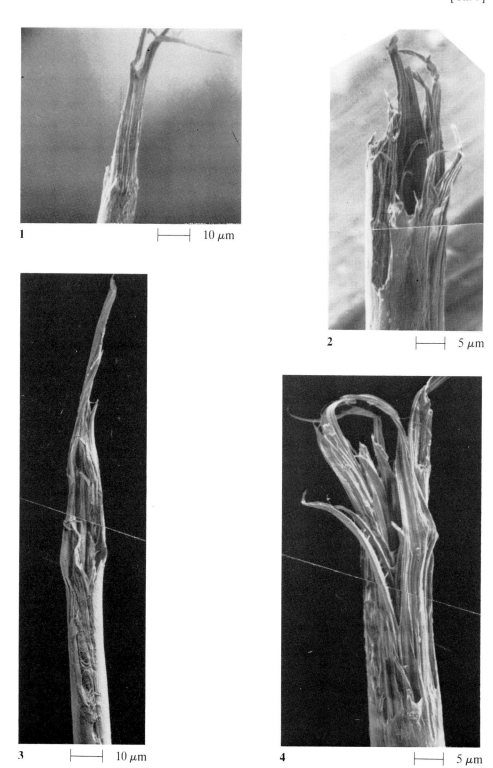

1 ├─────┤ 10 μm

2 ├─────┤ 5 μm

3 ├─────┤ 10 μm

4 ├─────┤ 5 μm

Plate 9A — Tensile breaks of wet cotton.
(1) Raw cotton. (2)–(4) Mercerized cotton.

Part III
Fatigue

10

INTRODUCTION

FAILURE DUE TO REPEATED LOADING

Tensile breaks of single fibres with the forms shown in Part II are not representative of the damage which commonly occurs in textile materials in real-life wear. Failure then is most often due to repeated loading over a long period of time, and the forms of break found, as illustrated by the case studies in Parts VI and VII, are usually different. This leads to the need to study fibre 'fatigue' in the laboratory.

Fatigue may be defined as the failure of material after repeated stressing at a level less than that needed to cause failure in a single application of stress.* This was a problem which caused troubles for the railway engineers of the last century and the aircraft engineers of this century. They designed with what seemed to be adequate margins of strength, but then got unexpected catastrophic failures after months or years of use.

Owing to these problems, metal fatigue has been widely investigated. Fatigue cracks occur during repeated deformation within the elastic region of a brittle or, more commonly, an elastic-plastic material as shown in Fig. 10.1. The fatigue may be either tension–tension (T–T), compression–compression (C–C), or tension–compression (T–C), but is always a result of recoverable elastic deformation, except very close to the tip of a fatigue crack.

FIBRE FATIGUE

In fibres, the conditions for fatigue testing are less easily arranged. Firstly, the material cannot be put into axial compression because the fibre buckles. Secondly, in the general-purpose textile fibres, which are semi-crystalline polymers, there is no well-defined elastic region followed by plastic yielding: there is a viscoelastic response, which gives a stress-strain curve

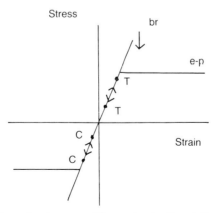

Fig. 10.1 — Cyclic deformation between different levels, T and C, will cause fatigue in elastic–plastic (e–p) or brittle (br) materials.

---

*The term *static fatigue* is sometimes used to denote failure after the single application of constant load for a long period of time, but our preference is to term this *creep failure*, and reserve the word *fatigue* for cyclic application of stress.

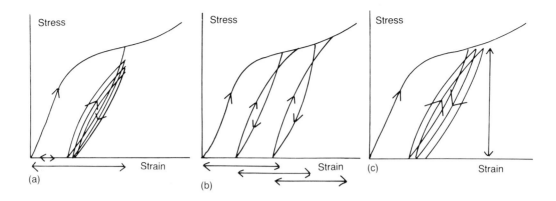

Fig. 10.2 — Typical fibre stress–strain responses. (a) Simple extension cycling. (b) Cumulative extension cycling. (c) Load cycling.

typically of the form shown in Fig. 10.2. Thirdly, there is the anisotropy of fibres which means that the directions of the cyclic stress will have a major effect on the fatigue.

The easiest experimental arrangement gives simple extension cycling, Fig. 10.2(a), in which the fibre is repeatedly stretched between its initial length and a fixed elongation. However, the imperfect recovery means that the specimen is slack for much of the time, and furthermore stress relaxation means that the maximum load steadily decreases. Experience has shown that simple extension cycling does not lead to failure unless the imposed extension is close to the normal breaking extension.

In order to overcome this problem, cumulative extension cycling was adopted. In this method, the slack is removed after each cycle, by unclamping and reclamping a rod attached to the lower jaw holding the fibre, and the selected elongation is imposed on the new straight length, as in Fig. 10.2(b). There are two main responses in this type of test: (i) the fibre may climb steadily up the load–elongation curve until failure occurs at the normal breaking extension, so that the test is no more than a rather complicated tensile test sequence; (ii) the fibre may settle down to an oscillation between two levels, with the recovered extension equalling the imposed extension, and does not break, at least in any practical time. A large amount of work carried out with this method did not yield very interesting results, although it did appear that there was a very narrow band of imposed extensions in which break was occurring at extensions less than the breaking extension and in a different form, which suggested the influence of fatigue. In rayon the band was from about 2% to 2.5% extension and in a typical nylon from about 10% to 10.5%.

Interesting tensile fatigue results came only when the experimental problems of cycling between given load levels were overcome in a new tensile fatigue tester developed by Bunsell, Hearle and Hunter (1971). The cycling has the form shown in Fig. 10.2(c). This tester also operated at a higher frequency, usually 50 Hz, than the old cumulative extension testers. The results of studies with this tester are described in Chapter 11.

FLEXING AND TWISTING

Two modes of deformation which fibres commonly suffer in use are bending and twisting, and fatigue due to these causes has to be investigated.

The common method of testing flex fatigue is to pull a fibre, held under a small tension, backwards and forwards over a pin, as shown in Fig. 10.3(a), so that fibre elements alternate between straight and bent forms. If the tension is imposed by a hanging weight, there can be complications due to inertial effects and the fibre can swing round so that different sides go into tension or compression somewhat irregularly. Better control is exerted when the tension is imposed by an elastic string, and one side of the fibre then always goes into tension and the other into compression.

Although this type of test is easy to set up, it does not expose the fibre solely to pure bending moments. There are normal forces and frictional forces at the contact between fibre and pin, which can cause surface wear. Attempts have been made to minimize this wear by mounting the pin so that it is free to rotate. In addition, the abrupt change of curvature from bent to straight means that there are shear stresses, which have their maximum value at the centre of the fibre.

Another method, which we have used to a very small extent, is to buckle the fibre repeatedly, as illustrated in Fig. 10.3(b). In the initial deformation this gives a smoothly varying curvature, which could be calculated from elastic theory. But, owing to the non-linearity of recovery, a sharp kink usually develops, giving an unknown level of high curvature.

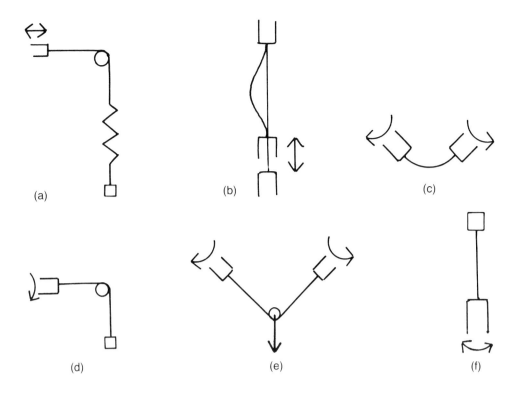

Fig. 10.3 — Bending and twisting fatigue tests. (a) Flex fatigue over a pin. (b) Buckling flex fatigue. (c) Free biaxial rotation. (d) Rotation over a pin. (e) Biaxial rotation over a pin. (f) Torsional fatigue.

Yet another mode of cyclic flexing is biaxial rotation testing. In this technique a bent fibre is rotated so that material on the outside of the fibre goes through a cycle of tension and compression. This is much the same as bending backwards and forwards, except that the extremes are reached by swinging round in a circle instead of by moving in a plane: there is the difference that all positions around the fibre suffer the same cycle of deformation.

Biaxial rotation testing was first used on thick monofilaments, where the specimen can be bent freely between inclined rotating jaws, Fig. 10.3(c). However, for fine fibres, it is not possible to get tight enough curvature unless the bend is concentrated round a pin. The first procedure adopted, Fig. 10.3(d), was to drive the rotation from one end with a weight which was free to rotate hanging from the other end. However, this is not ideal because of the inertial and drag torque from the weight. Various configurations were then adopted in order to drive both ends while maintaining the fibre under a controlled low tension. The best form is shown in Fig. 10.3(e), with the tension imposed by a load on the pin support and appropriate gears driving both ends together from a single motor.

The deformation in this test appears to be solely in bending, as indicated in Fig. 10.4(a), with no torque present except from any frictional drag on the pin. This is true for perfectly elastic materials. But even with the freely bent monofilaments, where there is no friction, twist develops in opposite directions from either end as shown in Fig. 10.4(b). The reason is the hysteresis in the tension-compression relation, shown in Fig. 10.4(d), in contrast to the elastic response of Fig. 10.4(c). Work must be done in each cycle to overcome the energy loss, and this can only come from the torsional drives, which must impose a torque on each end. In terms of force and moments, the explanation is that stress and strain are in phase in an elastic material, but out of phase in a material with hysteresis. Consequently, as seen in Fig. 10.4(e,f), the bending moment shifts in direction and balances a torsional moment on the end of the fibre.

The biaxial rotation test has been very much used in our work, because the forms of failure are similar to those found most commonly in the real wear of textile materials. Both this test, and the flex to and fro on a pin, are easily adapted to be performed in different environments.

Pure twist cycling, Fig. 10.3(f), is easy to formulate in principle, but less easy to put into practice because of the very large number of turns which have to be inserted in each cycle. We have only studied it in one investigation, using a highly geared system, which suffered nearly as much damage as the fibre when driven at high speed.

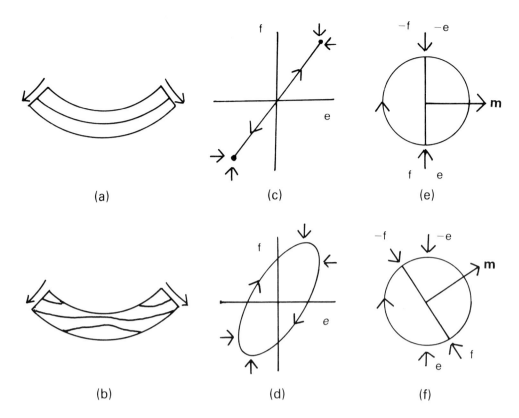

(a)   (c)   (e)

(b)   (d)   (f)

Fig. 10.4 — (a) Simple bending of elastic material, even in rotation. (b) Bending plus false twist in a material with hysteresis. (c) Stress–strain relation of elastic material. (d) Stress–strain relation with hysteresis, showing locations of peak stress and strain. (e) Cross-section of rotating fibre showing peak stress, f, and strain, e, in phase. (f) Peak stress and strain out of phase which shifts the direction of the bending moment vector, **m**.

SURFACE ABRASION

Fibres in use can also be subject to surface shear. One way of simulating this is to hang a fibre under tension over a rotating rod or pin as illustrated in Fig. 10.5. We have done a limited number of tests by this method, but more research on it is needed.

Another method is to twist two yarns together, or two parts of the same yarn, and then pull them backwards and forwards, as illustrated in Fig. 10.6. This test has yielded very interesting results, which are dominated by surface shear, although complicated by the presence of many filaments in the yarn. There will be some fibre bending, but this will be of low curvature if the yarn diameter is large. The results from this test are discussed in Part V, Chapter 24, since it is a yarn and not a fibre test. An analogous procedure with single filaments twisted together would give rise to large bending and twisting deformations.

CONCLUSION

The various test methods described in this chapter do provide useful ways of investigating the forms of fatigue failure of fibres, and illuminate what is found in case studies of wear. However, from the viewpoint of mechanics and basic theory, they have the defect that they all include complicated combined stresses. The subject is still wide open for further research, both experimental and theoretical.

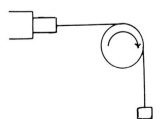

Fig. 10.5 — Surface wear by a rotating pin.

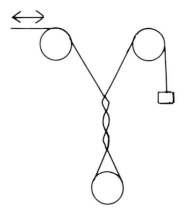

Fig. 10.6 — Yarn-on-yarn abrasion test.

Quantitative studies of fatigue life involve carrying out many tests since there is always considerable statistical variability. The best that can be achieved is usually about a tenfold range between the shortest and the longest life in any set of tests under nominally identical conditions. This does, however, suffice to show up marked differences between different fibres and test conditions.

The general experience is that increasingly rapid breakdown is observed as one goes from tension–tension cycling, to tension–compression in flex testing, to tension–compression plus torque in biaxial rotation testing. Any cyclic shear stresses lead to breakdown by axial splitting.

DEVELOPMENT IN FATIGUE TESTING

There has been a significant advance in fatigue testing since 1989. Figure 10.3(b) shows a buckling test, which had been used for a few tests on single fibres at UMIST. This has now been adapted by TTI (Tension Technology International Ltd) as a yarn buckling test for the evaluation of axial compression fatigue in yarns used in ropes, as described in the report on FIBRE TETHERS 2000 (1994,1995), and has also been used on wool yarns intended for use in carpets. In the unrestrained test method, a length of yarn is clamped between jaws a short distance apart, and then subject to cycling between the zero tension position, as mounted, and a reduced spacing. The yarn buckles and, after a number of cycles, the bend concentrates into a sharp kink. Yarns are removed after given numbers of cycles and their residual strength is measured.

A closer simulation of the yarn kinking which occurs in ropes, with lateral restraint from neighbouring yarns, is obtained by testing a number of yarns within shrink-tubing. After a suitable number of yarns have been inserted into the plastic tube, shrinkage is activated by heat. The tube is then mounted in the tester and cyclic buckling applied. The test distinguishes well between different yarns in their sensitivity to axial compression fatigue. In one set of restrained tests, strength loss was detectable in aramid yarns at 1000 cycles but not until 50 000 cycles in polyester yarns, and became severe at 20 000 and 1 000 000 cycles, respectively.

In the new parts of this edition, examples of breaks of buckled yarns are included in Chapter 12 and similar effects in ropes in Chapter 39 and in carpets in Chapter 33.

11

TENSILE FATIGUE

Nylon, polyester, Nomex, polypropylene, acrylic, Kevlar

Several different loading patterns which can be obtained on the fibre tensile fatigue tester, referred to in the previous chapter, are illustrated schematically in Fig. 11.1(a). In reality the data would have appreciable variability, which might lead to some overlapping of the failure conditions. In a simple tensile test in which a nylon fibre is extended at a constant rate, the tension will increase along the line OA, and break will occur at a tension T_0, with the assumption that the rate of extension has been adjusted so that the time-to-break is 20 seconds. If a somewhat lower tension, T_1, is held constant for a long time, creep failure will eventually occur at B: we assume that T_1 has been selected so that the time-under-load to cause break is 1 hour. Neither of these are fatigue situations, and the breaks will be typical ductile tensile failures, with a V-notch leading to a catastrophic break as shown in Fig. 11.1(b).

Now suppose that we impose an oscillating load from some intermediate tension up to the same maximum value T_1, as indicated by C. Experiment shows that break will occur in about

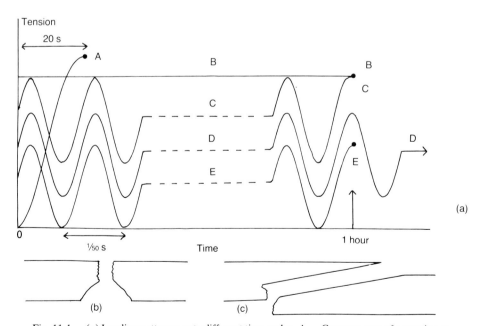

Fig. 11.1 — (a) Loading patterns: note different time scales. A — Constant rate of extension to break in 20 s at tension T_0. B — Constant tension $T_1 < T_0$, to break in 1 hour. C — Tension oscillating at 50 Hz between $0.4\,T_1$ and T_1. D — Tension oscillating at 50 Hz between $0.2\,T_1$ and $0.8\,T_1$. E — Tension oscillating between 0 and $0.6\,T_1$. (b) Ductile tensile failure with loading patterns A, B, C. (c) Tensile fatigue break in loading pattern E.

the same time as with a constant load. Owing to the inevitable scatter of results it is not possible to say definitely whether the time to failure is nearer to the same value as in creep failure, namely 1 hour in this example, or to the longer time which would give the same area under the load–time curve; but these two are not very different, and certainly there is no hastening of break caused by the oscillation. The form of failure is again the ductile V-notch of Fig. 11.1(b). So this test is just a creep failure under a different loading history — and the behaviour can be predicted with sufficient accuracy from a simple constant-load test.

If the maximum load is lowered slightly below T_1, the time to failure will increase. If it is appreciably lowered as in D, break will not occur, at least not within a measured time.

But if the load is lowered still more, as in E, break once again occurs in about the same time, with the precise number of cycles to failure depending on the particular cyclic load. So reducing the severity of loading has unexpectedly led back to a break situation. The criterion which must be satisfied is that the minimum load must be reduced to zero, or, in some fibres, to a critical value close to zero. The form of break is now different, as illustrated in **11A(1),(2)**, and shown schematically in Fig. 11.1(c). A long tail on one end has stripped off the other end. This is a true fatigue situation: failure after cycling between zero and half to three-quarters of the normal breaking load, with break happening between 10 000 and 1 000 000 cycles, with a new characteristic form of fracture.

The sequence of events leading to the tensile fatigue is illustrated in **11A(3)–(6)**, and shown diagrammatically in Fig. 11.2. Fatigue starts with a transverse crack on the fibre surface, **11A(3),(4)** and Fig. 11.2(b). At the tip of such a crack there will be an axial shear stress, and this then becomes the dominant cause of rupture. The crack turns and runs along the fibre at a slight angle to the fibre axis, Fig. 11.2(c), as can just be seen in **11A(4)** and is clearly visible in **11A(5)**. As it proceeds, the crack gets wider and deeper, Fig. 11.2(d), until eventually the tensile stress on the reduced cross-section is large enough to cause a ductile tensile break to become the final stage of fibre failure, Fig. 11.2(e) and **11A(6)**.

The tip of the tail, which is one side of the initial transverse crack, is shown in **11B(1)**. In some tests, a number of separate fatigue cracks develop along the fibre, as shown in **11B(2)**, with one failing first. There are variants in the form of the fatigue. Sometimes, probably owing to surface damage, the initial transverse crack is at an oblique angle, **11B(3)**, and sometimes the shear stresses cause cracks to run in both directions, **11B(4)** and Fig. 11.2(f).

Tensile fatigue in another type of nylon fibre is shown in **11B(5),(6)**. Both types break in a similar way with the axial cracks running across the fibre at an appreciable angle, as indicated in Fig. 11.2(c,d,e), so that the length of the tail is typically about five fibre diameters.

In a polyester fibre, **11C(1)**, the tensile fatigue break looks very different, although the test conditions for failure are similar. The tail of a fatigued polyester fibre is extremely long. This is a consequence of the fact that the axial crack in polyester runs much more closely parallel to the fibre axis, as indicated in Fig. 11.2(g). It must therefore proceed further along the fibre before the cross-section has reduced sufficiently for tensile break to happen. Indeed, as seen in **11C(1)**, the crack has progressed beyond the point at which the final break occurs. This implies that a period of time under load is necessary in order to cause the tensile break to occur, probably at a weak place in the material.

In another set of tests, it was found that the tails sometimes split into two or more parts, **11C(2)**. Details of the fatigue are shown in **11C(3)–(6)**. In this example the final failure started from an internal flaw, which probably weakened the fibre. In other examples the break starts from the surface of the fibre. Particularly interesting are the striations visible on the crack surfaces in **11C(6)**: these may well be steps of growth of the fatigue crack in each cycle.

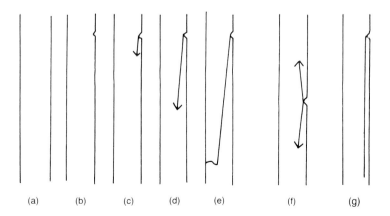

(a)   (b)   (c)   (d)   (e)   (f)   (g)

Fig. 11.2 — Sequence of tensile fatigue failure in nylon. (a) Fibre before cyclic loading. (b) After cyclic loading, transverse crack appears. (c) Shear stress at tip of crack causes axial split to start. (d) Axial crack continues getting deeper and wider. (e) Final failure by tensile break over reduced cross-section. (f) Variation with axial cracks running in both directions. (g) In polyester, the axial crack is almost parallel to the fibre axis.

Some other fibres show similar forms of tensile fatigue breaks. The tail of a fatigued Nomex (meta- aramid) fibre is shown in **11D(1)**. However, the detail of initiation in **11D(2),(3)** shows no evidence of a transverse crack: the axial splits seem to shave away directly from the surface. Possibly, surface roughness leads to the necessary shear stresses being present. Polypropylene, shown in **11D(4),(5)**, has a form similar to nylon, although with some distortions.

In the acrylic fibre, Courtelle, the fatigue failure is associated with axial splitting, **11D(6)**, but the criterion that the minimum load must be zero does not apply. An oscillating load causes failure at a reduced maximum load, whatever the value of the minimum load. There may be multiple splitting, **11D(7)**, and a tendency for a sharp separation of a surface skin layer, **11D(8)**.

Experiments on the high-strength para-aramid fibre, Kevlar, have shown no evidence of any appreciable weakening under tensile fatigue conditions. It is necessary to impose a peak load of a value within the range of experimental error for the normal tensile breaking load in order to cause break to occur. However, the form of break does show even more pronounced axial splitting, with breaks extending over very long lengths, **11E(1),(2)**, and showing complicated multiple splitting, **11E(3),(4)**.

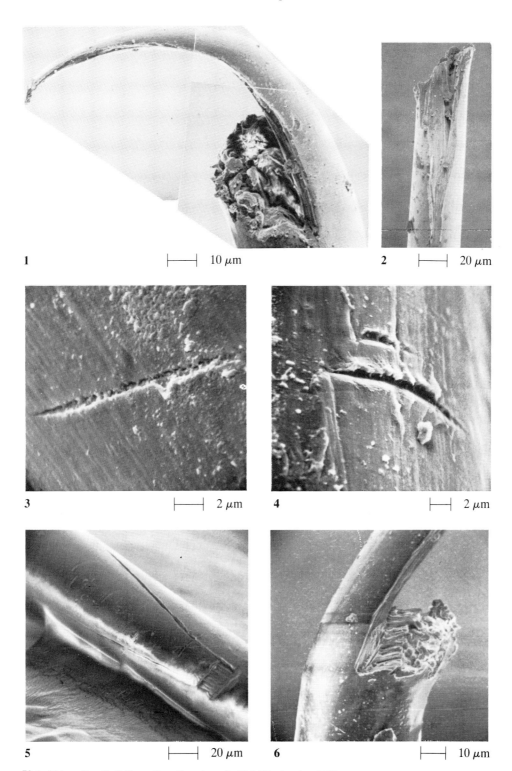

1 ├──────┤ 10 μm   **2** ├──────┤ 20 μm

3 ├──────┤ 2 μm   **4** ├──────┤ 2 μm

5 ├──────┤ 20 μm   **6** ├──────┤ 10 μm

Plate 11A — Tensile fatigue of medium-tenacity (0.4 N/tex) nylon 66 fibre.
(1),(2) Opposite ends of break after 62 000 cycles at 50 Hz between zero load and 71% normal breaking load. (3) Initiation of fatigue by a transverse crack. (4) Two transverse cracks with beginning of axial splitting. (5) Well-developed axial split. Note also other transverse cracks on fibre surface. (6) Final failure, with a ductile tensile break over the reduced load-bearing area.

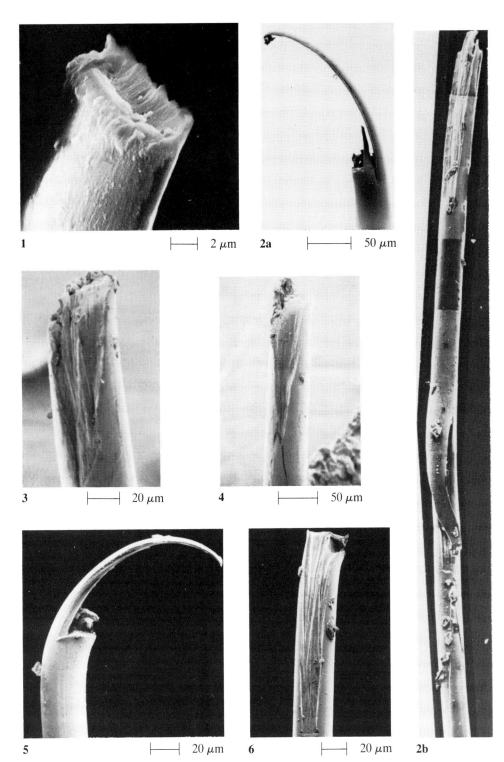

Plate 11B — Tensile fatigue of medium-tenacity nylon 66 fibre.
(1) Detail of the tip of the tail, resulting from the original transverse crack. (2a) One end of a break after 65 000 cycles at 50 Hz between zero load and 80% of normal breaking load. (2b) Opposite end of same break. Note the other developing fatigue cracks. (3) Variant form with angled transverse crack. (4) Variant with fatigue cracks running in both directions along fibre from the initial transverse crack. **High-tenacity (0.6 N/tex) nylon 66 fibre.**
(5),(6) Opposite ends of break after 58 000 cycles at 50 Hz between zero load and 66% of normal breaking load.

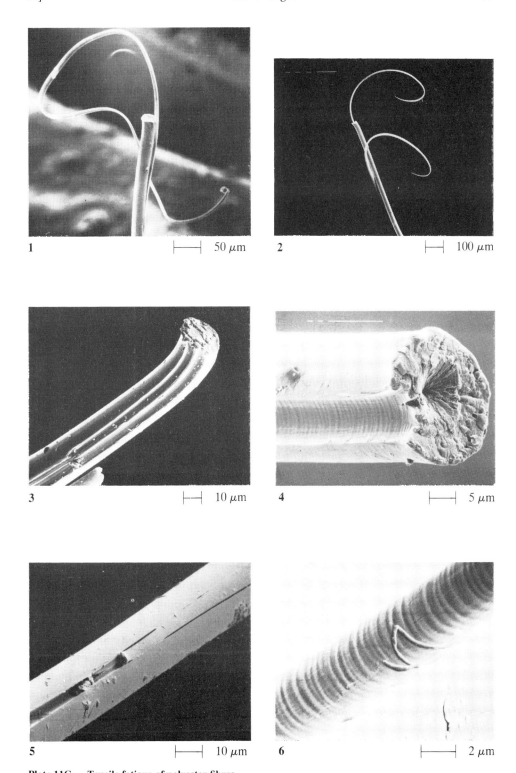

Plate 11C — Tensile fatigue of polyester fibres.
 (1) Break after 83 000 cycles at 50 Hz between zero load and 65% of the normal breaking load.
Tensile fatigue of another type of polyester fibre.
(2) With multiple splitting. (3) Detail of failure. (4) Final break, with tensile failure starting from an internal flaw, which is probably a weak place. (5) Detail of crack splitting. (6) Axial crack surface showing striations. Note that this is an enlarged view of the concave cavity in the fibre surface.

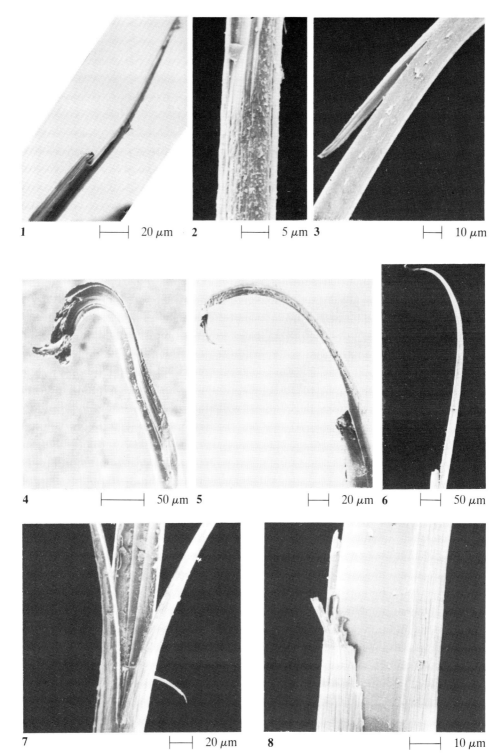

Plate 11D — Tensile fatigue of Nomex (meta-aramid) fibre.
(1) Break after 18 000 cycles at 50 Hz between zero load and 60% of the normal breaking load.
(2),(3) Initiation regions.
Tensile fatigue of polypropylene fibre.
    (4),(5) Opposite ends of break after 522 000 cycles at 86% of normal breaking load.
Tensile fatigue of Courtelle (acrylic) fibre.
    (6) Break after 157 000 cycles at 66% of normal breaking load. (7),(8) Detail of splitting.

Plate 11E — Tensile fatigue of Kevlar (para-aramid) fibre.
(1a,b) Break after 285 000 cycles at over 90% of average normal breaking load. Montage of pictures of both broken ends taken in an optical microscope. (2) Demonstrating the number of SEM pictures, at a reasonable magnification, needed in a montage to show the whole breakage region. (3),(4) Detail of multiple splitting.

12

FLEX FATIGUE

Polyester, nylon, acrylic, polypropylene, rayon

When fibres are bent, they often develop kinkbands on the inside of the bend. These can be shown up by light microscopy, sometimes with appropriate use of polarized light, **12A(1)**, or in severe examples by SEM, **12A(2)**. Whether or not visible kinkbands appear depends on the fibre and the conditions. Thus polyester develops kinkbands in a single bend at 20°C but not at 100°C, whereas nylon 66 shows the reverse combination. The kinkband is a localized buckling of the oriented structure within the fibre, as indicated schematically in Fig. 12.1, and is the mechanism of mechanical yielding under the compression on the inside of the bend. When the kinkbands are not found, even though there is yielding, this implies that microbuckling occurs at many places throughout the fibre, rather than being concentrated in a band. After repeated cycling, the kinkbands appear.

Although the deformation after a single bend, **12A(1),(2)**, looks severe, it is not mechanically damaging. The bands pull out under tension, and there is no loss of strength. But repeated cycling, by pulling a fibre backwards and forwards over a pin, leads to an intensification of the disturbance at the kinkband, with a break-up into fibrillar strands across the band, **12A(3)**. Eventually, the kinkband fails completely and becomes an angular split through the compression zone from the surface of the fibre to the neutral plane, **12A(4),(5)**. An axial split usually develops at the centre of the fibre. The other side of the fibre then suffers damage, and the final break shows characteristic angular faces, **12A(6)**, which are the original kinkband locations. This sequence of events is illustrated schematically in Fig. 12.2.

If the fibre is allowed to rotate as it is pulled across the pin, the kinkbands develop throughout the fibre. An example of a fibre which has broken in this way, before the test equipment was modified to prevent rotation, is shown in **12A(7)**. The light microscope picture shows up the kinkbands, with the final break along the same angle.

As explained in Chapter 10, flexing by pulling a fibre backwards and forwards over a pin under some tension is not simple cyclic bending. There are normal and frictional forces at the pin surface and shear stresses, which are present at places where the bending curvature is changing. More detailed investigations have shown that the form of failure depends on the type of fibre and the test environment.

The examples of nylon 6 and 66, **12B(1)–(6)**, all show a strong effect of kinkbands combined with varying degrees of axial splitting. An axial split must result from shear stress, but this can

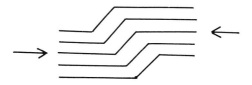

Fig. 12.1 — Schematic view of kinkband formation, due to compression of an oriented structure.

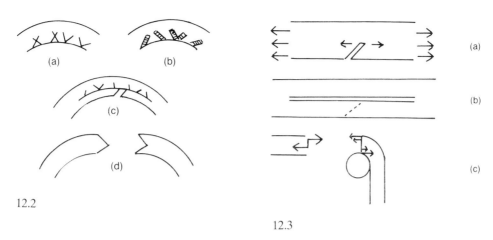

12.2

12.3

Fig. 12.2 — Sequence of events in flex failure by kinkband mode. (a) Kinkbands develop on bending. (b) They become more disturbed on cycling. (c) A kinkband fails, axial split is present, and the failure mechanism starts to operate in the other side of the fibre. (d) Final break shows angular faces.

Fig. 12.3 — (a) Shear stress at the tip of a kinkband crack. (b) Axial split occurring first, followed by break at a kinkband. (c) Shear stress in straight part balancing bending moment in bent part.

develop in two different ways. Following the rupture at a kinkband, there will be a shear stress at the tip of the crack, as shown in Fig. 12.3(a), because the fibre is under tension; and this may lead subsequently to an axial split. The other mechanism is independent of the presence of kinkbands, and could lead to axial splitting occurring first, with the kinkband later breaking through to the split, as in Fig. 12.3(b). At the place where the fibre leaves the pin, the bending moment in the curved part has to be balanced by a shear stress in the straight part, as indicated in Fig. 12.3(c). The shear stress is a maximum at the centre of the fibre. As the fibre moves across the pin, the location of the shear stress will move along the fibre and so can cause a long axial split. In reality, there is not an absolutely sharp change from curved to straight, but a small zone of varying curvature, over which the shear stresses are present.

A split of the type shown in **12B(3)** appears to have been caused by the first mechanism, but the long split in nylon, **12C(1)**, occurring before appreciable kinkband damage, must result from the second mechanism. The shear stresses can lead to multiple splitting, as in the polyester fibre, **12C(2)**. Other examples of multiple splitting, either before or after the fibre has broken, are **12D(1)–(4)**, and detail of an axial split is shown in **12D(5)**. In some situations, the axial splitting proliferates into many fine strands, **12D(6)**, although this may be an action of forces of surface wear. Snap-back after break can cause the splits to retract into spiral coils, **12E(1)**.

Further evidence for the early development of an axial split is found in studies of flexing for a certain number of cycles followed by tensile testing to break the fibre. The break can develop in two places, since the two halves of the fibre, which are separated by the axial split, act independently, **12E(2)**, or the break may show very clear axial splits, **12E(3)**. Another phenomenon observed on tensile testing of a flexed fibre is the occurrence of many micro-cracks along the fibre, **12E(4)**.

All the results reported so far in this chapter are from tests in which the combination of tension and curvature (pin diameter) were selected so as to minimize wear on the fibre surface, and cause failure to result from kinkbands and splits within the body of the fibre, as a consequence of flexing. However, if some other conditions are used, wearing away of the surface can become the dominant mechanism, **12E(5),(6)**. This tends to occur when the curvature is too high, and the tension too low. It is interesting to observe the pattern of the kinkbands showing in the worn surface in **12E(6)**. As shown in Chapter 14, surface wear becomes the dominant form of damage when flexing to and fro over a pin is applied to some other fibre types, such as aramid (Kevlar), wool and hair.

A limited study of flexural fatigue on acrylic fibre has shown that splitting and peeling can occur, **12F(1),(2)**, but there are also indications of breakage along a kinkband under compression, **12F(3)**, and of partial granular failure similar to tension breaks, **12F(4)**. Another limited study of polypropylene showed evidence of splitting and surface wear, **12F(5),(6)**.

When a standard viscose rayon fibre is repeatedly flexed, there is none of the long axial splitting or fracture along kinkbands found in other fibres. The break runs perpendicularly across the fibre, **12G(1)**, but in a very rough and uneven manner. There are kinkbands on the inside of the bend, **12G(2)**, and it is likely that the cyclic compressive stress disturbs and weakens the structure, and so leads to this unusual form of failure under the tensile load. Close examination of the fracture surface, **12G(3)**, shows bands of parallel striations and some indication of axial cracks. There is also wear of the fibre surface, **12G(4)**, due to rubbing on the pin. The kinkbands can also be seen in **12G(4)**. The flex fatigue failure of high-tenacity rayon, **12G(5),(6)**, is generally similar to that of standard rayon, although the surfaces are not as rough.

Flex fatigue tests over a pin have been extended to Dyneema HMPE (high modulus polyethylene) fibres, also referred to as HPPE (high performance polyethylene) fibres, by Sengonul and Wilding (1994,1996). At an early stage of flexing over the pin at room temperature, there is some surface abrasion and particles are shed from the surfaces. Below the surface layer, longitudinal striations appear. As shown in **12H(1),(2)** these become more pronounced and then lead to axial splits. The splitting will be due to the shear stresses associated with variable curvature bending, probably intensified by the disturbance of the structure at kink bands caused by axial compression, as seen in **12H(3),(4)**. Abrasion leads to a progressive thinning of the fibre and the final failure appears as in **12H(5),(6)**.

Another series of tests were carried out at elevated temperatures. The appearance of failures at temperatures from 40°C to 100°C are shown in **12I(1)–(6)**. The fibres show major splitting, but fine fibrillation becomes less at higher temperatures and wider ribbons are split off.

The other new observations of flex fatigue come from the yarn buckling test described in the addition to Chapter 10. The overall appearance of a wool carpet yarn after a period of cyclic buckling is shown in **12J(1)**. Localised sharp kinks can be seen in the centre of the picture, and, in places, these have led to fibre breaks. A major form of damage consists of the development of cracks at kink-bands on the inside of bends, as seen in **12J(2)**. These open up, become more pronounced, and lead to rupture, as shown in the sequence, **12J(3)–(5)**. Sometimes the break is divided into separate steps as in **12J(6)**.

Wool fibres have a tendency to split at cell boundaries, and this has occurred to some extent in the breaks in **12K(1),(2)**, though these are still predominantly along transverse cracks based on kink-bands. However, variable curvature can lead to substantial axial splitting, as seen in **12K(3),(4)**, which would lead to multiple splitting failures of the form shown in **12K(5),(6)**.

For the examination of yarns from the buckling tests carried out in FIBRE TETHERS 2000 (1994, 1995), optical microscopy was used. In addition to being a way of viewing a large number of fibres fairly rapidly and to correlate with the axial compression failures in ropes, as reported in Chapter 39, this technique was chosen to show the effects along the length of fibres on a large scale and, at a smaller scale, to show up kink bands within fibres in polarised light.

12L(1) is a fibre from a Technora (Teijin aramid fibre) yarn subject to 100000 buckling cycles. A kink in the fibre, which would weaken the yarn and would eventually break completely, is clearly seen. This occurs where the buckling develops a sharp kink at the centre of the clamped length. The internal kink-bands have progressed to a crack going about half-way across the fibre.

12L(2)–(6) are for a Kevlar 29 aramid fibre, which has been subject to buckling fatigue. Kink-bands can be seen on the inside of the bend in **12L(2)**, and these lead to breakage along angular cracks across the fibre, **12L(3)**, sometimes accompanied by axial splits, **12L(4)**.

The damage induced in a constrained buckling test of Kevlar 129 at 3000 cycles, when the yarn strength loss is about 20%, is shown in **12L(5)–(8)**. In some fibres, kink-bands run right across the fibre, **12L(5)**, presumably due to an overall axial compression without buckling. However the commoner feature is to see kink-bands on the inside of bends, **12L(6)–(8)**, developing into cracks.

Vectran is a high-performance, liquid-crystal, melt-spun, aromatic copolyester fibre from Hoechst-Celanese. The appearances of fibres from yarn buckling tests are shown in **12M(1)** for 30000 cycles and **12M(2)** for 100000 cycles.

The damage caused in buckling fatigue of Dyneema, HMPE fibre from DSM, after 100000 cycles, with about 30% loss in yarn strength, is shown in **12M(3)–(7)**. There are kink-bands, which are shown up both in the dark bands in the interior of the fibre and in kinks projecting from the fibre surface. Some fibres develop axial splits, **12M(5),(6)**, and in others, **12M(7)**, there are sharp kinks in the fibre as a whole.

Polyester fibres are more resistant to buckling fatigue. After 1000000 cycles, when the yarn as a whole has lost about 1/3 of its strength, there are a few kink-bands on the inside of bends, **12M(8)**. However the most noticeable features, **12M(9)**, are transverse lines and suggestions of axial cracks spread over the whole fibre.

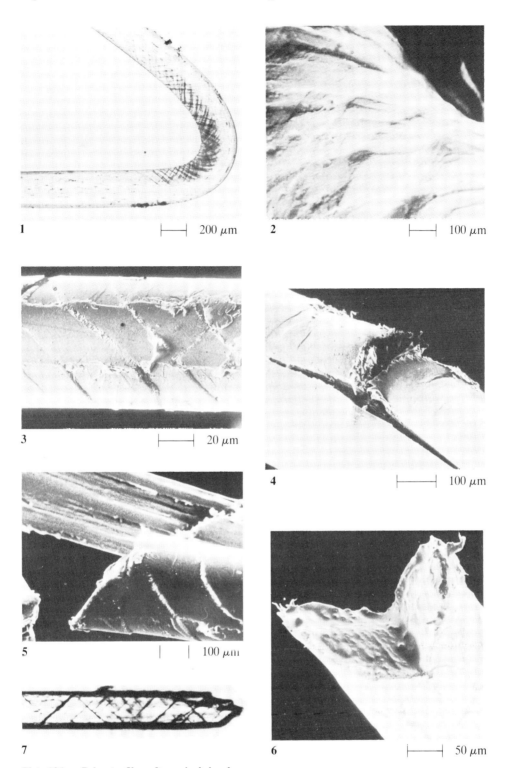

Plate 12A — Polyester fibre after a single bend.
(1) Kinkbands shown up in a bent fibre, using polarized light in a light microscope. (2) Kinkbands in a bent fibre in SEM.

Polyester fibre in flex fatigue.
(3) After some flex cycling the kinkbands have become pronounced and broken up. Note the limited wear on the surface in contact with the pin. (4) After more flex cycling a kinkband breaks completely, an axial split is present, and damage starts on the other side of the fibre. (5) Another example of kinkband failure and axial splitting. (6) Broken fibre after flex cycling, with a characteristic angular break along the planes of the kinkbands. (7) Light microscopic view of a fibre which has broken after repeated flexing, in conditions in which the fibre could rotate so that kinkbands developed all over the fibre.

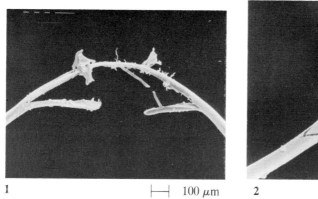

1   ⊢—⊣ 100 μm

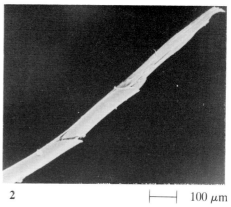

2   ⊢———⊣ 100 μm

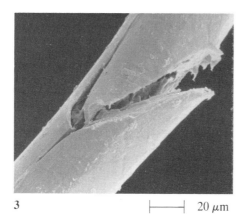

3   ⊢——⊣ 20 μm

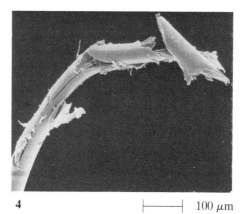

4   ⊢——⊣ 100 μm

5   ⊢——⊣ 100 μm

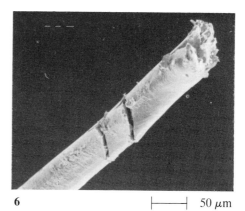

6   ⊢——⊣ 50 μm

Plate 12B — Flex fatigue of nylon 6 at 20°C, 65% r.h.
(1)–(4), Examples of kinkband failure and (4) axial splitting.
Nylon 66 at 20°C, 5% r.h.
(5) Kinkbands, with more pronounced splitting.
Nylon 6 at 20°C, 5% r.h.
(6) Clear kinkband failure.

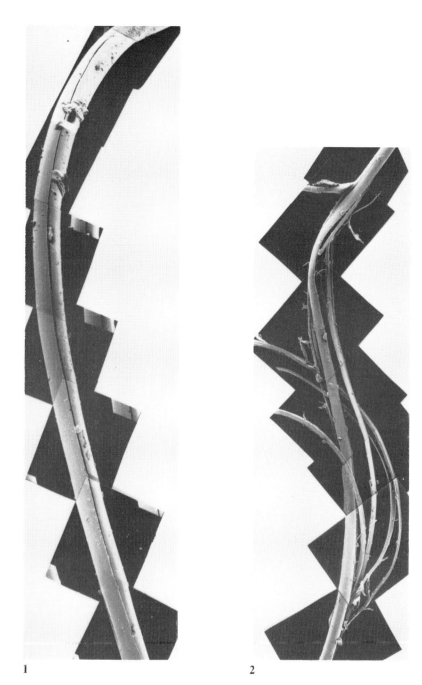

Plate 12C — Flex fatigue.
(1) Nylon 66 at 60°C, 30% r.h. (2) Polyester at 80°C, 5% r.h.

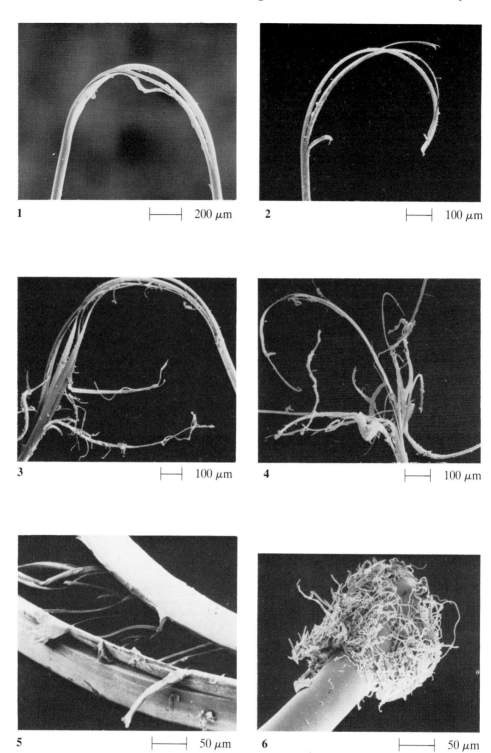

Plate 12D — Flex fatigue.
(1) Nylon 6 at 100°C, dry air. (2) Nylon 66 at 100°C, dry air. (3) Polyester at 40°C. (4) Polyester at 20°C, 65% r.h. (5) Polyester at 60°C. (6) Nylon 6 at 20°C, 95% r.h.

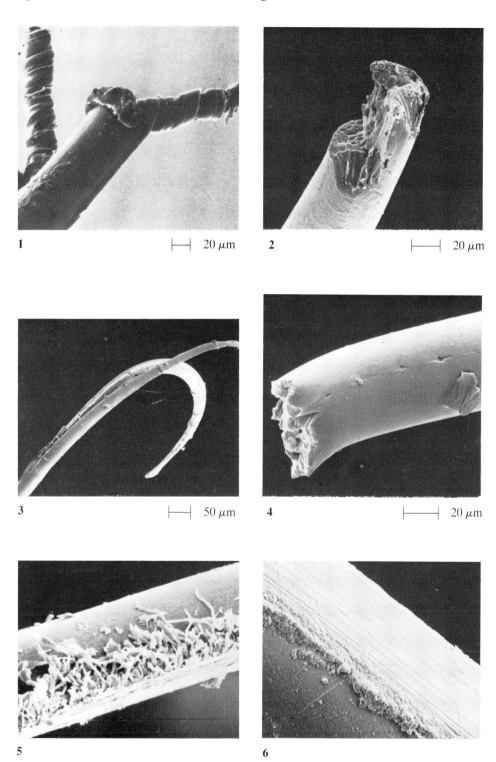

Plate 12E — Flex fatigue.
(1) Nylon 6 at 120°C, dry air: effect of snap-back after break.
Partial flexing at 20°C, 65% r.h. followed by tensile testing.
(2) Nylon 6. (3) Polyester. (4) Nylon 6.
Flexing over a pin, under conditions of surface wear.
(5) Nylon 6. (6) Polyester.

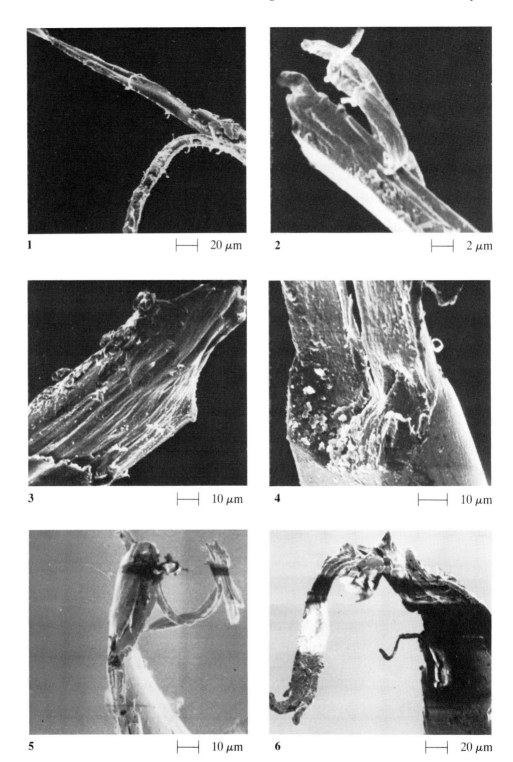

Plate 12F — Flexing over a pin: Acrilan acrylic fibre at 6.2% bending strain.
(1),(2) Tip of fibre which broke at 33 000 cycles. (3) Fibre which broke at 99 250 cycles. (4) Fibre which
broke at 125 500 cycles.
Flexing over a pin: polypropylene at 3.8% bending strain.
(5) Fibre which broke at 251 250 cycles. (6) Fibre which broke at 684 000 cycles.

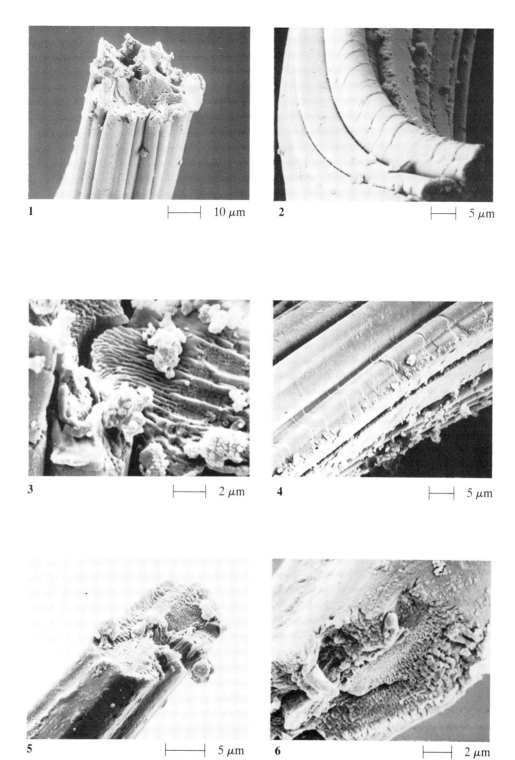

Plate 12G — Flexing over a pin: viscose rayon.
(1) Break of standard rayon. (2) Inside of bend showing kinkbands. (3) Detail of break. (4) Surface wear, due to rubbing on pin. (5),(6) Break of high-tenacity rayon.

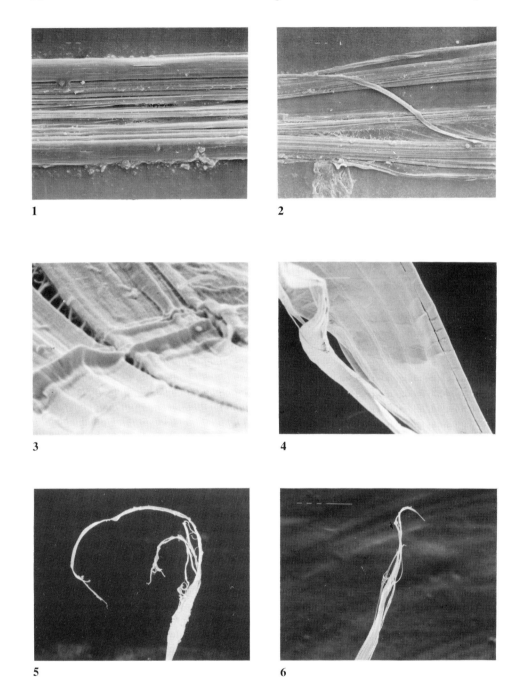

Plate 12H — Flex fatigue of Dyneema HMPE fibres at room temperature.
(1) After 20 minutes at selected test condition. (2) After 30 minutes. (3),(4) Detail of kink-bands and splitting. (5),(6) Failure after about 60 minutes.

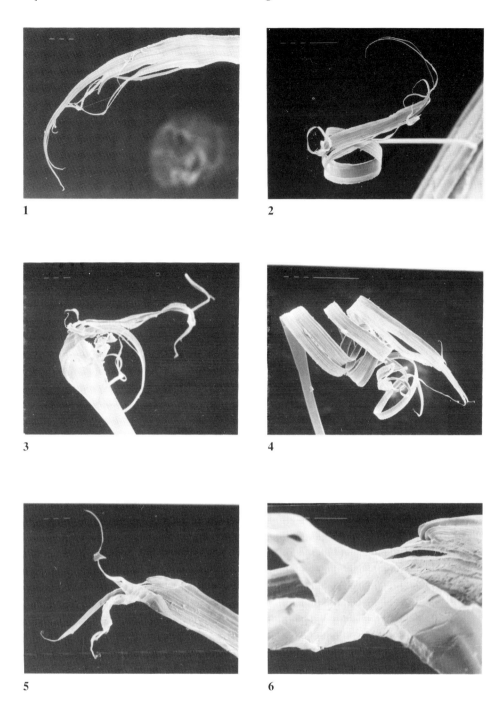

Plate 12I — Dyneema HMPE fibres failed in flex fatigue at elevated temperatures.
(1) At 40°C. (2) At 60°C. (3) At 80°C. (4)–(6) At 100°C.

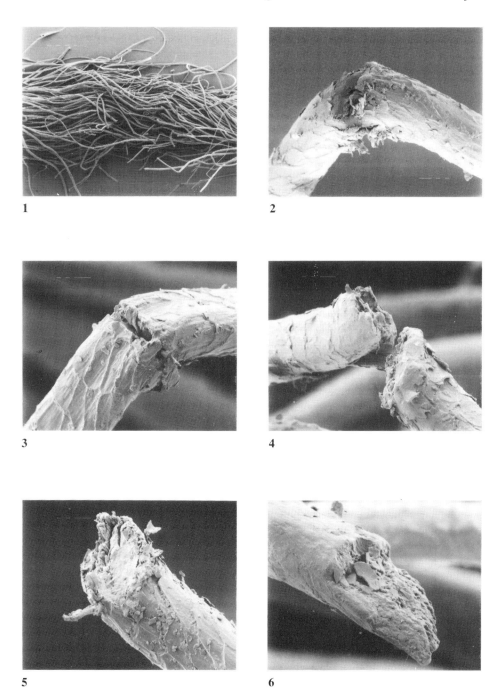

Plate 12J — Buckling fatigue of a wool carpet yarn.
(1) Overall view of failure region. (2)–(5) Progressive cracking at kink-bands leading to a sharp transverse break. (6) Break divided into two transverse cracks.

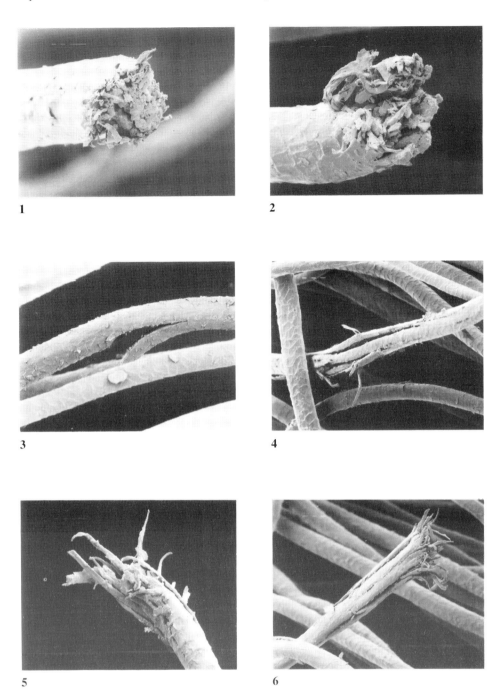

Plate 12K — Buckling fatigue of a wool carpet yarn (continued).
(1),(2) Breaks at transverse cracks with some axial splitting. (3),(4) Development of axial cracks along fibres. (5),(6) Breaks with multiple splitting.

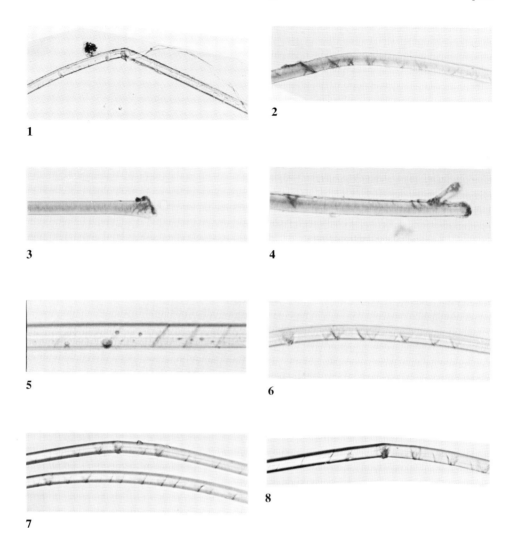

Plate 12L — Buckling fatigue of aramid yarns.
  (1) Technora after 100 000 cycles. (2)–(4) Kevlar 29. (5)–(8) Kevlar 129 after 3000 cycles.

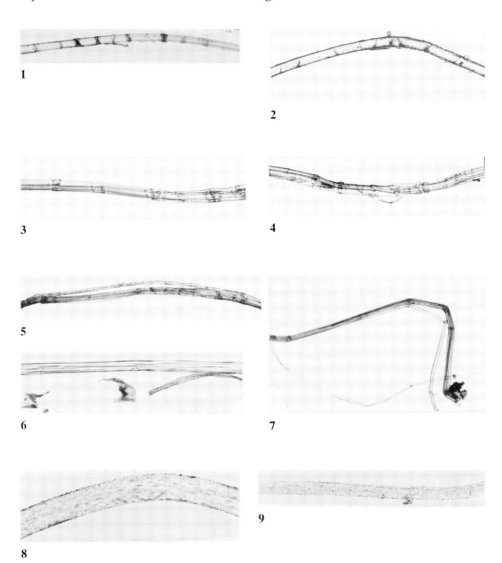

Plate 12M — Buckling fatigue.
(1) Vectran after 30 000 cycles. (2) Vectran after 100 000 cycles. (3)–(7) Dyneema SK60 after 100 000 cycles. (8),(9) Polyester after 1 000 000 cycles.

13

BIAXIAL ROTATION FATIGUE
Many fibre types

When a thick nylon monofilament is bent through about 90° and then rotated, it starts to crack axially as shown in **13A(1)**, and finally breaks to give an end with multiple splits, **13A(2)**. The principle of the test method was illustrated in Fig. 10.3(c). Note that the monofilament in **13A(1)** is twisting in opposite senses on either side of the centre point: S on the left and Z on the right.* This shows that the centre has been held back by a torsional drag, and so inserted false twist. As explained in Chapter 10, the torque results from the need to overcome the hysteresis ('internal friction') in bending.

Another example, in which a nylon monofil is just starting the final stage of break, is shown in **13A(3)**. In polyester there tend to be more splits closer together, **13A(4)**. When the bend is concentrated by holding the monofil under tension over a pin while it is rotated, the splits are usually fewer and larger, **13A(5),(6)**.

For fine textile fibres, it is essential to use a pin to get the required high curvature. When using the technique illustrated in Fig. 10.3(d), with drive from one end, the damage in nylon and polypropylene fibres shows the typical multiple splitting form, **13B(1)–(4)**. In this test method the twist is all in the same direction, since it is the free, weighted end which lags in rotation.

With a synchronous drive from both ends, as indicated in Fig. 10.3(e), there is torque on each side, and the twist develops in opposite senses in the two broken ends, as seen in the polyester fibre, **13B(5),(6)**. This test was carried out in water, but the form of break is similar to that with dry fibres.

Progressive damage to a fibre during biaxial rotation fatigue is shown in **13C**. There was a period of initiation before the first signs of damage appear at about 150 cycles, **13C(1)**. By 250 cycles, **13C(2)**, cracks are clearly apparent, and become steadily more severe, **13C(3),(4)**. Measurement of the residual strength shows that there are five stages in the test: the initiation period, when there is no loss in strength; a period when strength decreases linearly with number of cycles; another initiation period, when strength remains constant; a second period when strength falls; and final breakage. The fibre appearance during the second initiation period is shown in **13C(5)**, and indicates major damage in the outer visible part of the fibre. Shortly before final breakage, **13C(6)**, the individual splits are beginning to break. The final failure shows the typical multiple splitting, **13C(7),(8)**.

In **13C(3)–(7)**, it can be seen that the splitting divides into two regions along the fibre, and in **13C(8)** that it divides into two concentric regions in the fibre cross-section.

The reason for the axial division is that, as indicated in Fig. 13.1, the torque, which arises from friction on the pin or hysteresis, must be zero at the centre point on the pin; but the torque then builds up in opposite senses along the bent fibre to reach a maximum at the point where the fibre leaves the pin, and the torque from the drive shafts is applied through the straight fibre. So the most severe stress conditions occur where the fibre loses contact with the pin, and it as at these two places that the splitting occurs. The stresses near here will be further accentuated by the shear stresses, resulting from the change from a curved to a straight path.

---

* The terminology S and Z is used to indicate left-handed and right-handed twist respectively in textile yarns, and is adopted here to denote twist in a fibre. The definition relates to the direction of orientation on the surface of a twisted cylinder. A balanced combination of S and Z twist, so that there is no 'real twist' resulting from a rotation of one end relative to the other, is known as false twist.

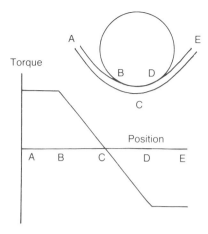

Fig. 13.1 — (a) Fibre passing over pin, bent along BCD, but straight along AB and DE. (b) Variation of torque along fibre.

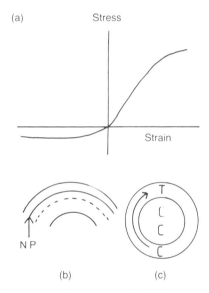

Fig. 13.2 — (a) Typical fibre stress–strain curve showing yield at a low stress in compression. (b) Location of neutral plane, NP. The dotted line is the central plane, which would be the neutral plane in an ideal Hookean material. (c) Cross-section, showing rotation of neutral plane, with outer zone in tension–compression, and inner zone always in compression.

The division across the fibre is probably caused by the fact that in a fibre which yields easily in compression, as indicated in Fig. 13.2(a), the neutral plane moves out towards the tension side, as in Fig. 13.2(b). Only the part outside this position will suffer the more severe effect of tension–compression cycling, as indicated in Fig. 13.2(c). This outer part will break up during the first stage, leaving the central zone to break up separately in the second stage.

The polyester fibre, shown in **13C(1)–(8)**, with a linear density of 4.2 tex, is rather thick for a textile fibre, although still much finer than the monofils of around 100 tex shown in **13A**. Finer polyester fibres, for example the fibre with a linear density of 0.84 tex shown in **13C(9)**, and the broken fibres in **13B(5),(6)** and **13D(1)**, do not exhibit the division of the splitting into separate zones, either axially or transversely. The reason is that the scale of the splitting relative to the fibre diameter is too coarse for separation to occur. The reduced fibre diameter not only reduces the size of the fibre cross-section, which leads to a merging of the cross-sectional zones, but also makes a smaller pin necessary to maintain the same bending strain and therefore reduces the length of fibre in contact with the pin, which causes the axial zones to merge.

Biaxial rotation fatigue with failure by multiple splitting for various types of man-made fibre is illustrated in **13D(2)–(6)**. The breaks are generally similar in character, except that the nylons show fewer and larger split portions than the other types of fibre. The same form also occurs in the natural fibres cotton, wool and hair, as shown in Chapters 18 and 19.

Multiple splitting of the form found in laboratory biaxial rotation fatigue tests is the commonest form found in the use of fibres in textiles, and many examples are included in the case studies in Parts VI and VII.

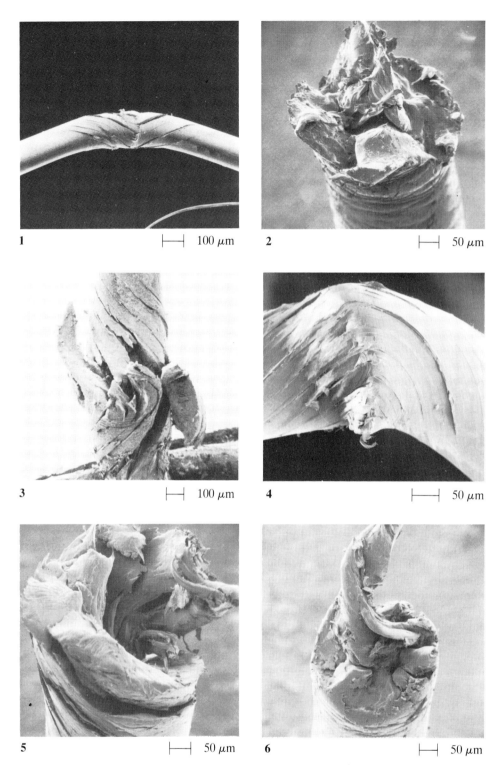

Plate 13A — Biaxial rotation fatigue of thick monofils, without a pin.
(1), (2) Nylon 66, 67 tex. (3) Nylon 66, 111 tex. (4) Polyester, 67 tex.
Rotation of thick monofil over a pin.
(5), (6) Nylon 66, 67 tex: opposite ends of break.

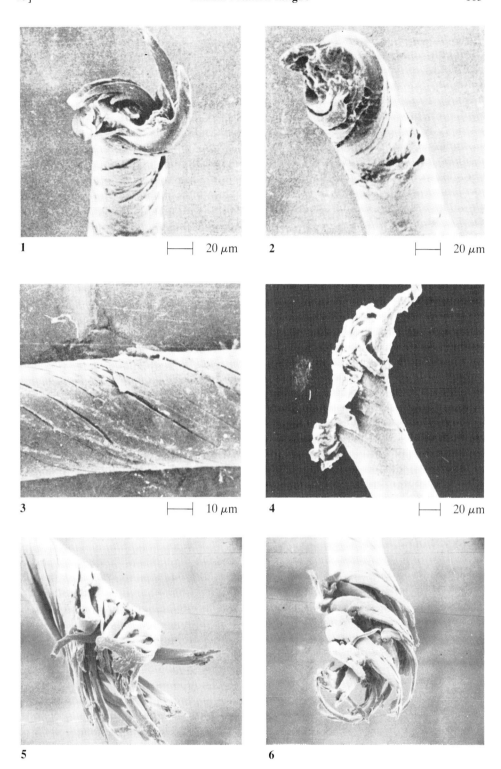

Plate 13B — Rotation over a pin (with drive from one end).
(1), (2) Nylon 66, 17 dtex, failed at 3690 cycles, opposite ends. (3) Nylon 66, 17 dtex, unbroken at 5000 cycles. (4) Polypropylene, failed at 3435 cycles.
Biaxial rotation over a pin.
(5), (6) Polyester 8.4 dtex, failed after 12 608 cycles in water, opposite ends. Note twist in opposite senses.

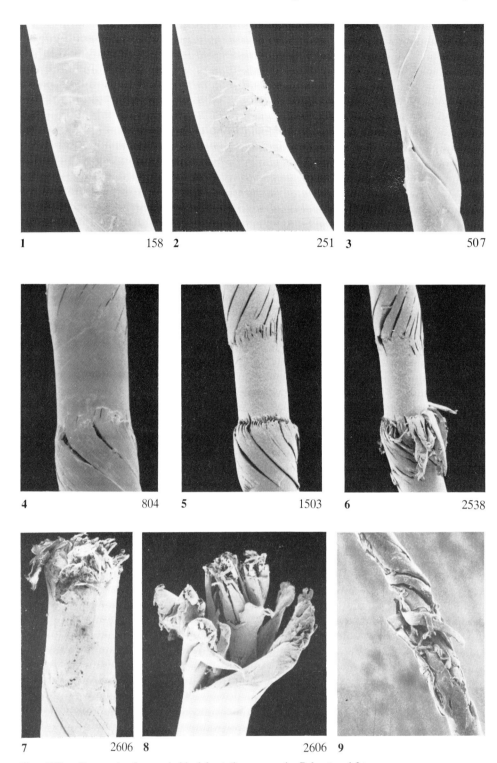

Plate 13C — Progressive damage in biaxial rotation over a pin. Polyester, 4.2 tex.
(1)–(6) Unbroken at increasing numbers of cycles. Note that these are different specimens removed from the test after different times: they are not a succession of pictures of the same fibre. The numbers below each picture indicate the number of cycles for the specimen. (7),(8) Opposite ends of failed fibre, after break at 2606 cycles.
Polyester, 0.84 tex.
(9) Close to failure, after 3687 cycles.

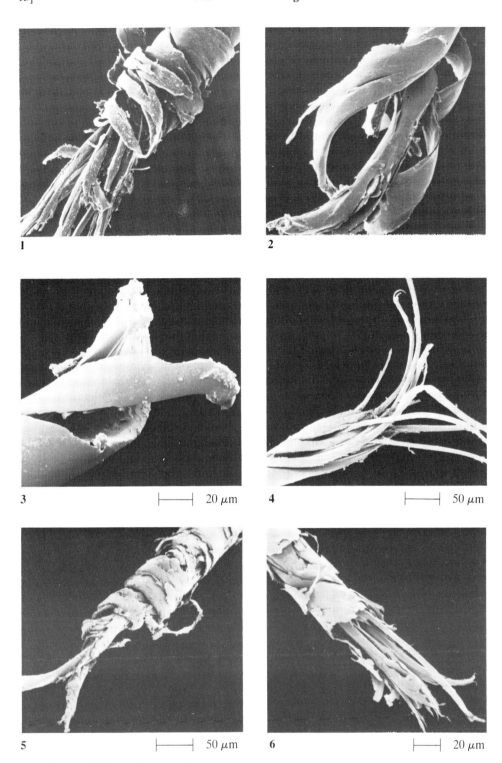

Plate 13D — Biaxial rotation over a pin.
(1) Polyester, 17 dtex, failed at 3126 cycles. (2) Nylon 6, 17 dtex, failed at 13 728 cycles. (3) Nylon 66, 17 dtex, failed at 4045 cycles. (4) Acrylic, Courtelle, 17 dtex, failed at 2885 cycles. (5) Modacrylic, Teklan, 18 dtex, failed at 8007 cycles. (6) Polyvinyl alcohol (PVA), 11 dtex, failed at 18 177 cycles.

14

SURFACE SHEAR AND WEAR
Many fibre types

As shown by the pictures of nylon and polyester fibres in **12E(5),(6)**, the test method of pulling a fibre to and fro over a pin, which is intended to demonstrate flex fatigue, can in some circumstances lead to failure by surface wear. This always happens with the highly oriented para-aramid fibre, Kevlar, and the appearance after a period of flexing over a pin is shown in **14A(1)**. A similar form of surface wear is found if the fibre is not repeatedly flexed, but is held in a fixed bent configuration over a rotating pin as shown in Fig. 10.5. The final failure of the Kevlar fibre, when the tensile stress on the reduced cross-section reaches its limiting value, is by axial splitting, **14A(2),(3)**, although evidence of the surface wear can be seen.

Examples of surface wear in wool, cotton and rayon caused by a rotating pin are shown in **14A(4)–(6)**. The picture of wool, **14A(4)**, shows very clearly how the wearing of the surface eventually reduces the area of cross-section to a size in which tensile rupture takes place. The mechanism is illustrated schematically in Fig. 14.1.

Results from another series of experiments show the detail of how the wear takes place in nylon, **14B(1)–(6)**, and polyester, **14C(1)–(4)**, fibres. In some instances, **14B(1),(2)** and **14C(1)**, fine fibrillar strands are worn away; but in other cases, **14B(3)** and **14C(2)**, ribbon-like strips peel away. Sometimes the final break tapers right across the fibre, **14B(4),(5)**, but, in other examples, **14B(6)** and **14C(4)**, a substantial part breaks transversely.

The influence of surface shear stresses in causing damage to fibres is clearly important, but basic laboratory studies on single fibres have been very limited. Speculation on possible forms of damage resulting from shear stresses on fibre surfaces is illustrated in Fig. 14.2. Other evidence of splitting and peeling under surface shear comes from studies of yarn-on-yarn abrasion (Chapter 24) and from some case studies reported in Chapters 31 (underwear), 39 (ropes) and 40 (some industrial products). In these situations the shear stresses will be cyclic and reverse in direction, which can be more damaging than the unidirectional shear stresses applied by a rotating pin. It must also be remembered that cyclic shear stresses arising indirectly at the tip of a crack in tensile fatigue testing (Chapter 11) and due to change of

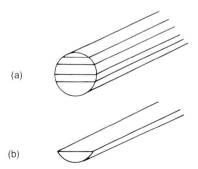

Fig. 14.1 — (a) Successive stages in wearing away of surface, (b) leading to tensile failure of reduced cross-section.

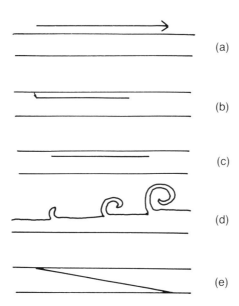

Fig. 14.2 — (a) Shear stress on fibre surface. (b) Crack penetrates into fibre and then runs along fibre. (c) Crack starts below surface. (d) From (b) or (c) multiple layers may peel off surface. (e) Alternatively split may run across fibre.

curvature in flex testing (Chapter 12) have been shown to be important failure modes. The whole subject of the effect of shear in promoting rupture of fibre structure justifies much more experimental and theoretical research.

Finally, we show an example from the more recent series of experiments of the surface wear in Kevlar, **14C(5)**, which leads to a tensile break with multiple splits, **14C(6)**.

Further studies of fibre-to-metal abrasion by the method of holding a fibre under tension for a 90° arc round a rotating pin have been made by C. Cork and M. A. Wilding at UMIST. In these tests, the geometry was changed from that shown in Fig. 10.5, so that the rotating pin was immersed in a dish of water. The fibre came down at 45° from a clamp into the water, round the pin and up again over a guide to a weighted hanging end. The abraded nylon 66 fibres, shown in **14D(1)–(4)**, were fairly thick (18.7 dtex), comparatively weak, with a tenacity of 3.8 gf/dtex, and high breaking extension (76%). The breaks show an angled wearing away of the fibre surface until there is not enough material left to support the tension and the thinned fibre cross-sections break, sometimes with axial splitting. The nylon 6 fibres, illustrated in **14D(5),(6)**, are similar in properties (16 dtex, 5.1 gf/dtex, 53%); commonly, as shown, one broken end has a simple smooth rupture but the other shows more splitting.

Fibre-to-fibre abrasion, using an arrangement like the yarn-on-yarn test shown in Fig. 10.6, of partially oxidised polyacrylonitrile fibres (PAN) was studied by Zhu. As noted in Chapter 10, the use of this method on single fibres involves appreciable bending and twisting, as well as surface abrasion. The tests of the fibres shown in **14E** were made at 65% r.h. and 20°C under a tension of 0.5 gf, which is about 15% of break load, with 1 turn of twist, a wrap angle of 35°, a stroke of 12 mm, and at a frequency of 120 Hz. The PAN fibres had been removed from various positions in the oven, so that they had been stabilized for different times; the total time in the oven for full stabilization was 90 minutes. In addition to wear of the surface, as seen in **14E(1),(3)**, there is axial splitting, **14E(2)**, before the breaks shown in **14E(4)–(6)**.

Plate 14A — After pulling to and fro over a pin.
(1) Kevlar 29 (para-aramid) after 45 000 cycles of flexing, with an apparent bending strain of 4.7% under a tensile stress of 0.2 N/tex.
After abrasion against a rotating pin.
(2) Kevlar 29 broken after abrasion under relatively high tension. (3) Kevlar 29 broken after abrasion under relatively low tension. (4) Wool. (5) Cotton. (6) Viscose rayon.

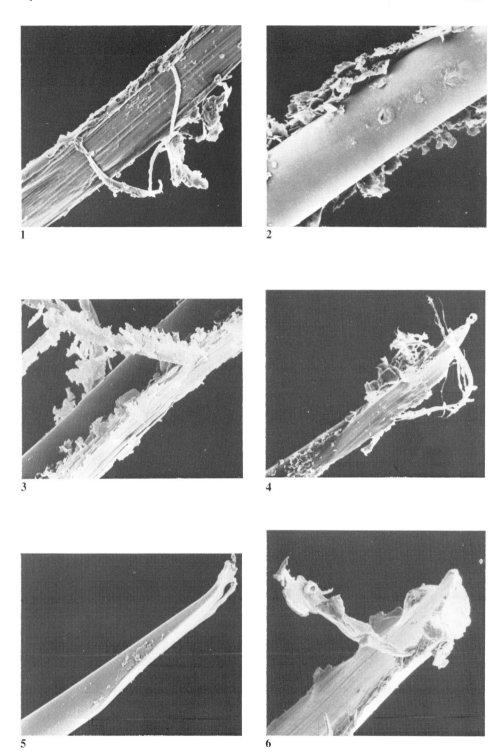

Plate 14B — Nylon fibre held under tension against a rotating pin.
(1–3) Intermediate stages of surface wear. (4–6) Final breaks.

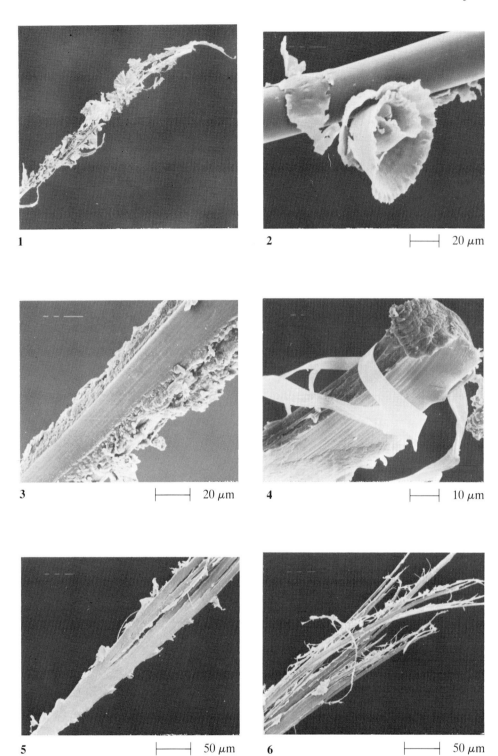

Plate 14C — Polyester fibre held under tension against a rotating pin.
(1) Final break. (2),(3) Detail of wear. (4) Final break.
Kevlar (para-aramid) fibre held under tension against a rotating pin.
(5) Surface wear. (6) Final break.

Plate 14D — Fibre-to-metal abrasion of nylon fibres.
(1),(2) and (3),(4) Opposite ends of break of nylon 66 fibres. (5),(6) Opposite ends of break of nylon 6 fibre.

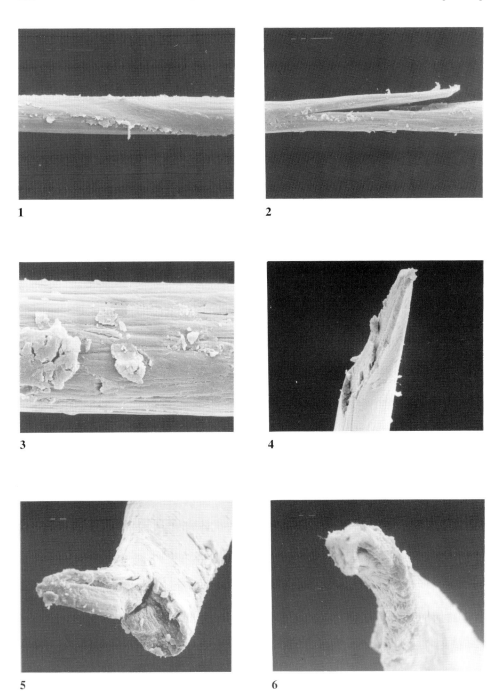

Plate 14E — Wrapped fibre-to-fibre abrasion of oxidised PAN fibres.
(1) Stabilized for 15 minutes, after 3600 cycles. (2) Stabilized for 15 minutes, after 7200 cycles. (3) Stabilized for 60 minutes, after 6000 cycles. (4) Stabilized for 45 minutes, after 13 300 cycles. (5),(6) Stabilized for 60 minutes, after 15 000 cycles.

Part IV
Other fibre studies

15

Introduction

In Parts II and III we have described six well-defined and different modes of tensile failure (classes 1–6 in Fig. 1.5) and four well-defined modes of fatigue failure (classes 8–11 in Fig. 1.5), together with surface wear and some examples of effects of shear splitting and peeling (classes 12 and 13 of Fig. 1.5). All of these results were obtained from simple laboratory tests on single fibres, which were mostly the widely used types of man-made fibre.

In this part there is a more diverse collection of breaks involving chemical degradation (including class 7 in Fig. 1.5), other forces such as twisting and cutting, effects of heat on fibres, natural fibres producing a variety of forms of break, and finally breaks in a variety of specialized fibre types.

16

DEGRADED FIBRES

Fibres are, in varying degrees, subject to attack by light, heat and chemicals. These agents may break polymer molecules or cause other changes in structure and thus weaken and degrade the fibre. In this chapter the effects of degradation on forms of breakage will be explored.

Nylon fibres lose strength on exposure to light, but even before there is any substantial reduction in tenacity the form of break changes. With exposure to moderate sunlight (a Manchester summer!) for 2 days, some voids show up in the V-notch region of a ductile fracture, **16A(1)**; after 3 weeks the simple V-notch and catastrophic region have both become strongly fragmented, **16A(2)**; and after 24 weeks the original form has almost completely disappeared, and the break has become a collection of separate turrets, **16A(3)**. Even with this complete change of the appearance of the break, the loss of strength is less than 10%.

Close examination, **16A(4)**, shows that each turret, or part of a turret, is associated with a conical cavity, where break has occurred. At the base of many of the cavities, a delustrant titanium dioxide particle can be seen. During drawing of the fibre, elongated (cigar-shaped) voids form round the particles, and these become enlarged by the photodegradation of the nylon, which is promoted by the TiO_2. Under sufficient tensile stress the voids become sites for the development of ductile failure with cracks round the edge of each void opening out to give cone-shaped cavities, similar to the single internal cones shown in **5D(1),(2)**. Eventually, catastrophic failure starts, and the multiplicity of internal conical failure regions join up. The sequence is illustrated schematically in Fig. 16.1. Since the mechanics differs only in that the crack and deformation mechanism is spread over many sites rather than being concentrated in one position, the strength is not much reduced.

Prolonged exposure to intense sunshine does reduce the strength to very low values, and although the same void mechanism can be seen to be operative the appearance of the break shows that the fibre is very heavily degraded, **16A(5),(6)**, with the surface cracking of the filaments and longitudinal splits giving staggered breaks.

In bright nylon fibres, which do not contain delustrant particles, the breaks with many turrets do not occur, although the rupture is often staggered at several levels.

Thermal degradation of nylon and polyester when heated in air, so that oxidation can occur, usually changes the form of break from the ductile fracture shown in Chapter 5 to the granular form shown in Chapter 8. Examples are **16B(1)–(3),(6)**. Sometimes, **16B(1)**, the

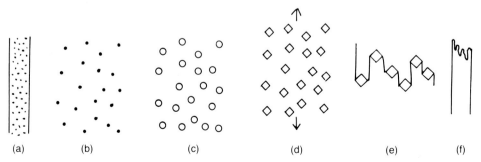

(a) (b) (c) (d) (e) (f)

Fig. 16.1 — (a) Nylon fibre containing titanium dioxide particles. (b) Enlarged view of portion of fibre. (c) After exposure to light, voids are enlarged. (d) Cracks form under tension and lead to internal cones. (e) Cracks go catastrophic, and join up to give final break. (f) Broken fibre shows series of turrets.

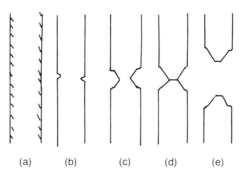

Fig. 16.2 — (a) Fibre with chemical degradation in surface layer. (b) Under tension, surface layer breaks. (c) Ductile crack develops all round fibre. (d) Catastrophic rupture at centre. (e) Broken fibre ends.

degrading action can be seen to be penetrating inwards from the fibre surface; and, in other examples, **16B(2),(3)**, there is a residual indication of an initiation region of the fracture at a point on the fibre surface. However, the granular form is not always found: in some examples, **16B(4),(5)**, the break is glass-like, suggesting a brittle failure. More research is needed to establish the circumstances which determine the form of failure, and the reasons for it.

Relatively mild oxidative action on nylon by means of a chemical environment can have an effect similar to heat, **16C(1)**. The immediate surface layer is most strongly attacked, and the form of break suggests that it fails first in a granular mode, and that this provides the initiation for crack propagation to develop as a ductile fracture in a circular notch all round the fibre, with the final catastrophic region forming the centre of the break, as illustrated in Fig. 16.2.

More severe forms of chemical attack, as illustrated in polyester in **16C(2)**, cause obviously greater damage to the fibre. The magnitude, character and effects of the reactions between a vast range of chemicals and different fibres under different conditions is a major subject which goes beyond the scope of this book, and these two pictures of tensile breaks are merely given as examples of what can happen.

Alternatively to chemical action followed by mechanical action, the two can be combined if a fatigue test, or a dead-weight loading, is carried out in a hostile environment. For example, the biaxial rotation fatigue life of nylon in hydrochloric acid or caustic soda is almost independent of pH for values greater than 2, but drops sharply below 2. Examples of the forms of break found in the biaxial rotation fatigue test under different liquid and gaseous environments are shown in **16C(3)–(6)**. The polyester fibre in nitric acid, **16C(3)**, is little affected in the appearance of break and only suffers a 10% reduction in fatigue life, but nylon in weaker nitric acid solution **16C(4)** shows a more distinct change of appearance, compared to **13D(3)**, and the fatigue life has fallen by 60%. Nitrogen dioxide also has a more severe effect on nylon, **16C(6)**, where the fatigue life has fallen from over 10 000 cycles without NO_2 to 200 cycles with NO_2, than in polyester, **16C(5)**, where the fatigue life remains over 1500 cycles.

There is one very distinctive form of fibre fracture which has only been observed after chemical attack on fibre assemblies. It was first noted by Martin Ansell at the University of Bath, and was subsequently found in a case study at UMIST, discussed in Chapter 34, **34G**. Two of Ansell's pictures are shown in **16D(1),(2)**, and others are shown in **26(B)**, where they are discussed as an example of testing a composite material. Briefly, they are tensile breaks from a PVC-coated polyester fabric which has been boiled in water for a long time. There is degradation of a surface layer, similar to that in **16C(1)**, but the break inside this layer occurs by a conical split running into the fibre, to give what has been called a stake-and-socket break, and is type 7 in Fig. 1.5. This form of axial split must be a result of shear stress, acting in the way indicated in Fig. 16.3.

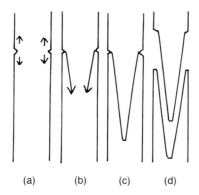

Fig. 16.3 — (a) Initial break in degraded surface layer, with shear stress at tip. (b) Shear crack propagation. (c) Completion of break. (d) Two ends of stake-and-socket break.

Finally, in this chapter, we give four examples of the effect of irradiation from work at the University of California. Moderate proton-irradiation on nylon 6 causes a partial change from ductile to granular break, **16D(3)**, but a more severe dose leads to a brittle fracture, **16D(4)**. Moderate gamma-irradiation causes most change in a surface layer, **16D(5)**, which penetrates deeper with a more severe dose, **16D(6)**.

The stake-and socket form of rupture was first found by Ansell (1983) as a result of chemical and thermal attack in PVC-coated polyester fabric, **16D(1),(2)** and **26B(3)–(6)**, then in autoclaved laboratory overalls, **34G**, in hair, **19J(1),(2)**, and in a nylon rope, **39J(1),(2)**. In none of these was the chemistry of the attack well defined. The 'cathedral spire' breaks of hair after exposure to ultra-violet radiation, **19J(1),(2)**, were an example of controlled laboratory testing, and stake-and-socket breaks have also now been found in laboratory studies of polyester fibres.

Holmes (1996) aminolyzed polyester fibres by treatment with n-butylamine in both aqueous and vapour form. The amine penetrates the fibre and breaks polymer chains, particularly extended chains in amorphous regions. Due to the action of internal stresses on the weakened material, cracks develop after prolonged exposure to amine, but these are axial cracks after vapour exposure and radial cracks after aqueous exposure. Following 192 hours in n-butylamine vapour, both molecular weight and strength had fallen to about 20% of the values in the original fibres. In aqueous amine the drop was only about half as much. When a vapour aminolyzed polyester fibre is extended, both axial and radial cracks are found in the outer layers, **16E(1)**, and then pieces break away, **16E(2)**. A tensile break of a fibre exposed to vapour for 144 hours shows the characteristic stake-and-socket form, **16E(3),(4)**, though this is less well developed in a fibre exposed to aqueous amine, **16E(5),(6)**, which also does not show any long axial cracks.

A variety of complicated forms of failure of polyester fibres, free of UV stabilizer, after exposure to ultra-violet radiation alternating with humidification have been found in studies by Salem and Ruetsch (1997). The form of break depends on the spinning conditions and consequently the resulting fibre structure. In a conventionally spun and drawn fibre, radial microcracks develop in the outer layer, **16F(1)**, but then axial and slightly off-axis cracks develop and lead to rupture, as seen in **16F(2)**. The detail may show even more complex shattering, **16F(3),(4)**.

An undrawn fibre, spun at 1500 m/min, also shows radial cracks, but these lead to smooth brittle fractures, **16F(5)**. An oriented, but high extension, fibre, spun at 5500 m/min shows shear-induced splits, **16F(6)**, and may also have some transverse cracks, giving a stepped break, **16G(1)**. Skin-core fractures, with multiple splitting in the core, are found in fibres spun at 7000 m/min, **16G(2),(3)**; similar fibres after heat-setting have the forms of damage seen in **16G(4)–(6)**. Since these fibres do not contain delustrant particles, they do not show the turreted forms found in light-degraded nylon, which were shown in **16A**.

The degradation of wool by ultra-violet light has been examined by Diane Jones at UMIST. **16H(1)** shows the surface of an unexposed wool fibre. After exposure to UV for 208 Xenotest hours, **16H(2),(3)**, there is considerable damage to the fibre surface, including loss of scale definition, pitting and accumulation of debris. The tensile break of unexposed wool, **16H(4)**, is simpler than that of wool exposed to UV for 52 hours, **16H(5),(6)**, which shows an apparent loss of intercellular adhesion.

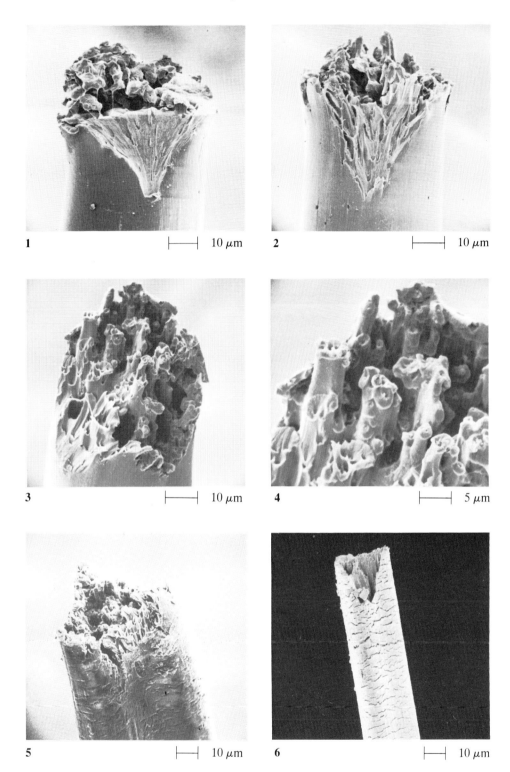

1 |—————| 10 μm **2** |—————| 10 μm

3 |—————| 10 μm **4** |—————| 5 μm

5 |—————| 10 μm **6** |—————| 10 μm

Plate 16A — Tensile breaks of light-degraded nylon 66, 17dtex, exposed in summer in Manchester, facing WSW.
(1) 2 days exposure. (2) 3 weeks exposure. (3),(4) 24 weeks exposure.
Light-degraded trilobal nylon 66, exposed in Florida.
(5) 6 months exposure (600 hours of sunlight). (6) 12 months exposure (1000 hours of sunlight).

Plate 16B — Tensile breaks of heat-treated nylon and polyester.
(1) Nylon 66, heated slack for 4 hours at 150°C. (2) Nylon 66, heated slack for 6 hours at 230°C. (3),(4) Nylon 66, heated slack for 4 hours at 225°C. (5) Polyester, heated slack for 6 hours at 230°C. (6) A different type of polyester fibre, heated slack for 6 hours at 230°C.

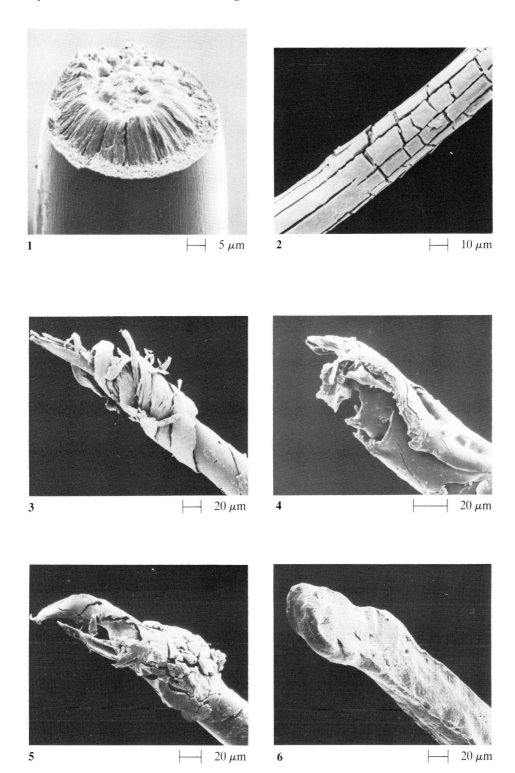

Plate 16C — Tensile breaks of chemically degraded fibres.
(1) Nylon 66, exposed to 100 vol. H_2O_2 for 2 hours. (2) Polyester treated with *n*-propylamine for 24 hours at 25°C plus 7 hours at 40°C.
Biaxial rotation fatigue in a hostile environment.
(3) Polyester in 1M HNO_3. (4) Nylon in 0.1M HNO_3. (5) Polyester in air + NO_2. (6) Nylon in air + NO_2.

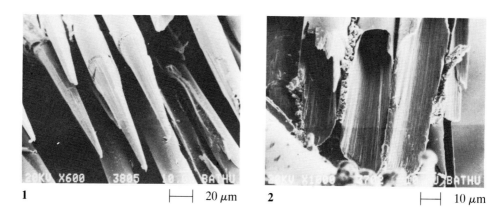

1   ├──┤ 20 μm   2   ├──┤ 10 μm

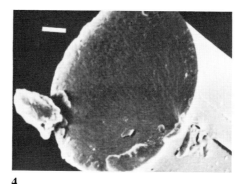

3   4

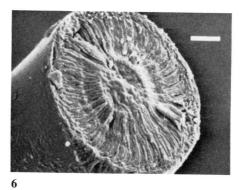

5   6

Plate 16D — PVC-coated polyester fabric, boiled for 6 weeks, and broken in tension (from Ansell, 1983).
(1) Conical ends of stake-and-socket fibre fractures. (2) Fibre sockets on other end.
Tensile breaks of nylon 6 fibres after exposure to radiation (from Ellison, Zeronian and Fujiwara, 1984).
(3) After 12 Ci proton-irradiation dose. (4) After 76 Ci proton-irradiation dose. (5) After 10 Mrad gamma-irradiation. (6) After 50 Mrad gamma-irradiation.

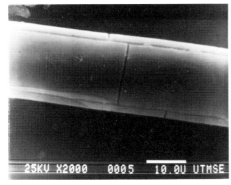

1

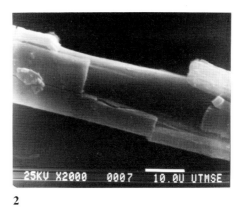

2

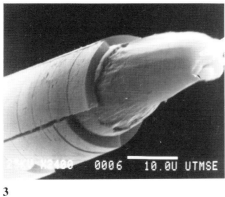

3

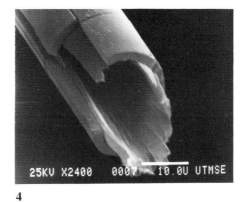

4

5

6

Plate 16E — Aminolysis of polyester fibres, Holmes (1996).
(1) Fibre exposed to *n*-butylamine vapour for 144 hours and extended to 60% of break load. (2) Taken to 75% of break load. (3),(4) Opposite ends of tensile failure after 144 hours exposure to vapour. (5),(6) Opposite ends of break of fibre exposed to aqueous amine for 144 hours.

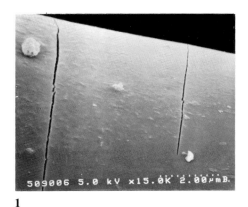

1

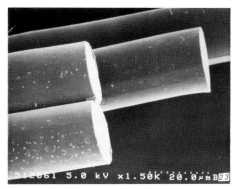

2

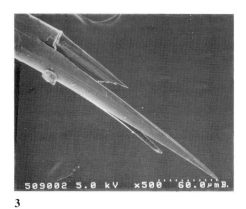

3

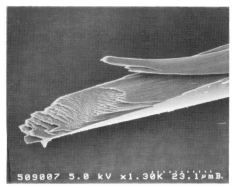

4

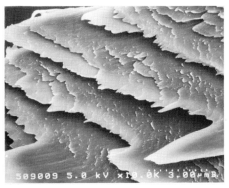

5

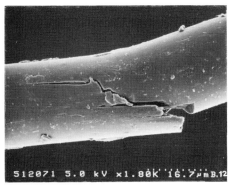

6

Plate 16F — Effects of ultra-violet exposure on tensile fractures of polyester fibres, courtesy of D. Salem and S. B. Ruetsch, TRI, Princeton.
(1)–(4) High molecular weight fibre, conventionally spun and drawn. (5) Fracture of undrawn textile molecular weight fibre spun at 1500 m/min. (6) Shear fracture of textile molecular weight fibre spun at 5500 m/min. All after alternating exposure to UV and humidification.

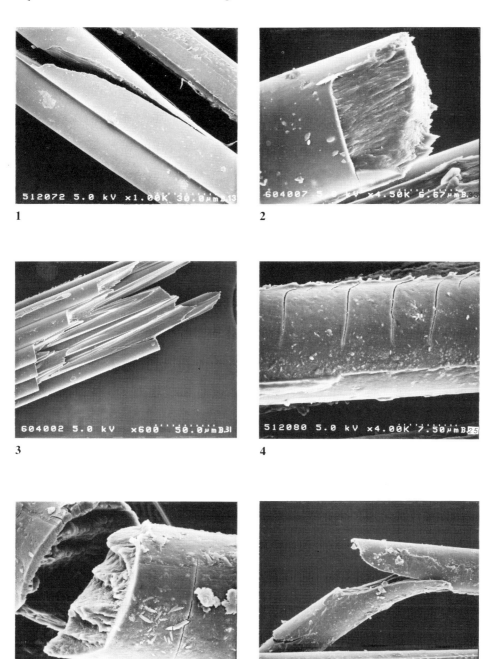

Plate 16G — Effects of ultra-violet exposure of polyester fibres, courtesy of D. Salem and S.B. Ruetsch, TRI, Princeton (continued).
(1) Step fracture of textile molecular weight fibre spun at 5500 m/min. (2),(3) Skin-core effects in textile molecular weight fibre spun at 7000 m/min. (4)–(6) Damage in heat-set textile molecular weight fibres spun at 7000 m/min. All after alternating UV exposure and humidification.

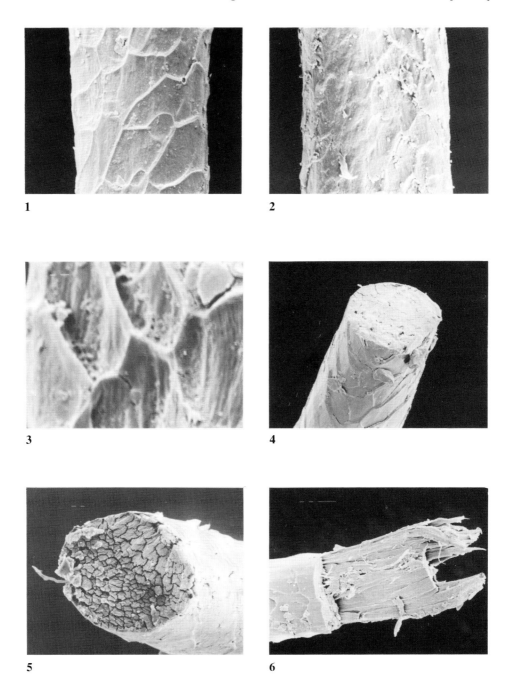

Plate 16H — Effect of ultra-violet exposure on wool, from Diane Jones, UMIST
(1) Surface of unexposed wool. (2),(3) Surface of UV-exposed wool. (4) Tensile break of unexposed wool. (5),(6) Tensile breaks of UV-exposed wool.

17

TWIST BREAKS

Although the overwhelming majority of laboratory fibre breakage tests are carried out by axial extension to the breaking point, it is also possible to twist fibres until they break. In many cases, the essential failure mechanisms are the same, although the geometry of the rupture is changed by the different direction of deformation. For example, **17A(1)** shows a brittle fracture in a glass fibre, with cracks following the line of twist.

In nylon, twist breaks usually show a ductile V-notch, angled to the fibre axis in the twist direction, followed by a catastrophic region, **17A(2)**, but the fibre can also split in two, **17A(3),(4)**.

In addition to simple twisting to break, fibres can be cyclically twisted alternately in either direction and fail by torsional fatigue. In nylon, this leads to coarse multiple splitting, **17A(5),(6)**. Other studies of torsional fatigue have been reported by Van der Vegt (1962) and Goswami, Duckett and Vigo (1980).

The twist breaks of acrylic fibres range from simple twist distortions of the granular fracture, **17B(1),(2)**, influenced also by the dog-bone cross-section of Orlon, through some degree of splitting, **17B(3)**, to breaks in which splitting is a predominant effect, **17B(4)–(6)**.

Twist breaks of rayon fibre also show a typical granular break, **17C(1)**, and the occurrence of transverse cracks, **17C(2)**, similar to those found in tensile breaks, suggests that the tension generated by the prevention of contraction on twisting is an important factor. The acetate twist breaks, **17C(3),(4)**, show similar effects.

In the high-strength, para-aramid fibre, Kevlar, twisting at constant length generates tensile stress, and the rupture comes from axial splitting, **17C(5)**; but if the twisting is carried out at a constant small tension, much higher twist levels can be reached and the fibre is more sharply ruptured, **17C(6)**.

The most extensive studies of torsional fatigue have been made at the University of Tennessee. Using a single station tester, Goswami *et al* (1980) investigated the torsional fatigue of a medium-tenacity commercial polyester fibre. The applied tension of about 0.1 mN/dtex, which is about 0.2% of break load, was low. At the high torsional strain amplitude of about 10°, failure occurred in about 500 cycles and the breaks were similar to the twist breaks of nylon fibres in **17A**. However, at a lower strain amplitude, about 3°, failure took about 6000 cycles and the breaks were different. **17D(1),(2)** show two ends of a fibre, which has formed a helical snarl and failed due to torsional fatigue under these conditions. Along most of the fibre there are skin-deep cracks, but near the point of break the cracks have penetrated into the core of the fibre. The damage and deformation are similar to that found in a pill extracted from a polyester/cotton shirt.

Using a multi-station tester, developed by Duckett and Goswami (1984), studies of the effect of structure of polyester fibres on torsional fatigue were carried out by Fu-Min *et al* (1985). The fibres were fairly thick and had been spun at different sizes to have almost the same diameters, about 50 μm, after drawing. When cycled with a strain amplitude of 15° under a nominal tension of 0.2 mN/dtex, a partly drawn fibre, with a draw ratio of 1.65 and a break extension of 157%, showed no loss of strength up to 1000 cycles. It then rapidly lost 70% of its strength, followed by a slower drop until catastrophic failure occurred at about 25 000 cycles. A highly drawn fibre, with a draw ratio of 5.08 and a break extension of 20%, had lost 10% of its strength by 1000 cycles; then a further 70% loss took place gradually to about 100 000 cycles, followed by a rapid drop.

17D(3)–(6) show the progressive change during cycling of the partly drawn fibre. This starts with transverse cracks, followed by axial cracks and finally more plastic deformation. A tensile break after cycling, **17D(7)**, shows coarse axial splitting. The damage in the highly drawn fibre at two levels of cycling is shown in **17D(8),(9)**. Tensile rupture after torsional fatigue, **17D(10)**, shows an extremely high degree of axial splitting.

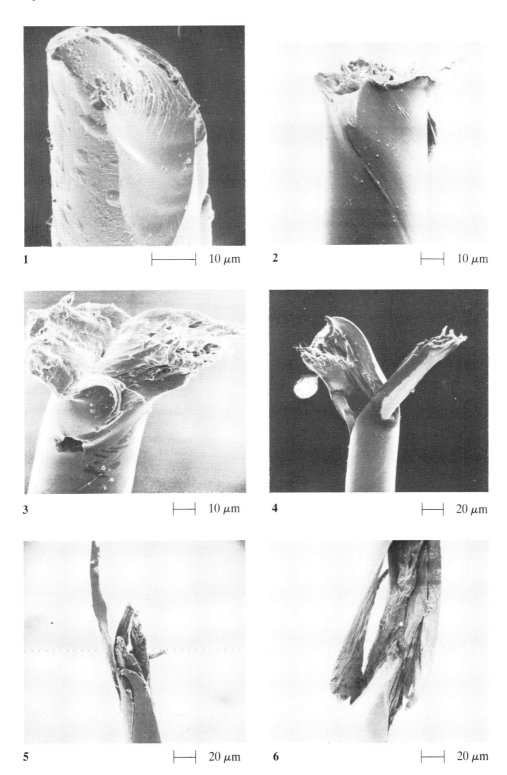

Plate 17A — Twist breaks.
(1) Glass fibre. (2–4) Nylon 66, 17 dtex, twisted to break at constant length, failing at a twist factor of about
100 tex$^{\frac{1}{2}}$cm^{-1}.
Torsional fatigue.
(5),(6) Nylon 66, 17 dtex, cycled to ±47° at 5.2 Hz for 13 hours.

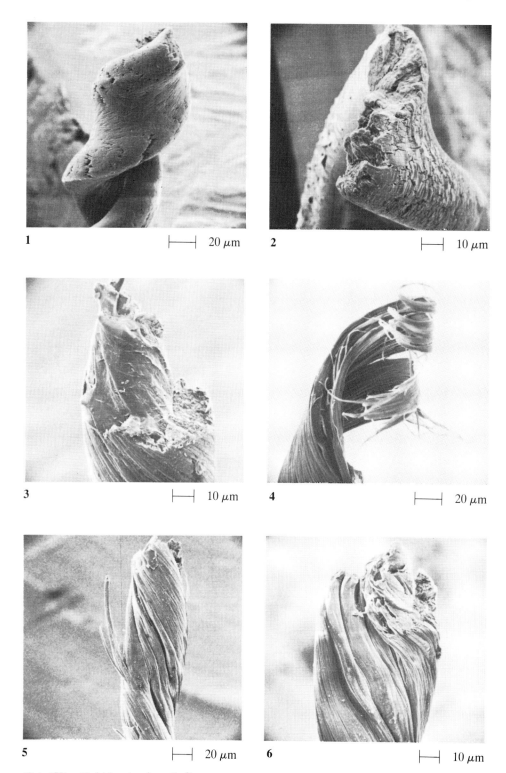

Plate 17B — Twist breaks of acrylic fibres.
(1),(2) Orlon, 17 dtex. (3),(4) Courtelle, 17 dtex. (5),(6) Acrilan, 17 dtex.

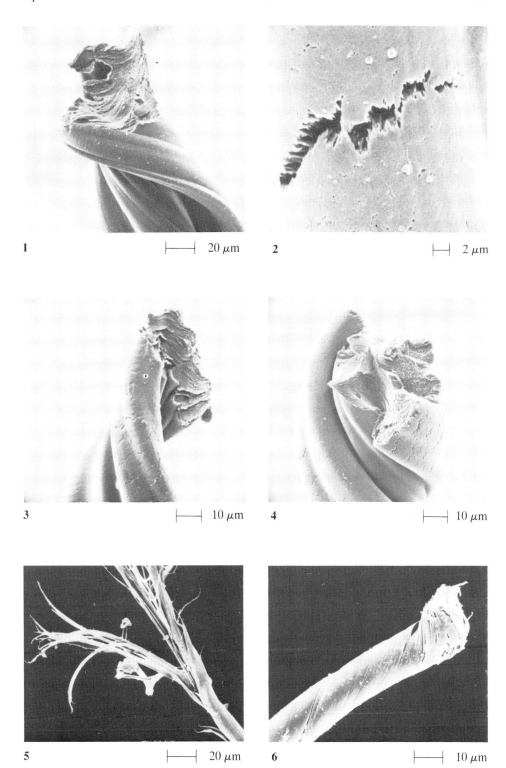

1 $\vdash\!\!-\!\!-\!\!-\!\dashv$ 20 μm   2 $\vdash\!\!-\dashv$ 2 μm

3 $\vdash\!\!-\!\!-\!\dashv$ 10 μm   4 $\vdash\!\!-\!\!-\!\dashv$ 10 μm

5 $\vdash\!\!-\!\!-\!\!-\!\dashv$ 20 μm   6 $\vdash\!\!-\!\!-\!\dashv$ 10 μm

Plate 17C — Twist breaks of cellulosic fibres.
(1),(2) Viscose rayon, Fibro, 20 dtex. (3) Secondary acetate, Dicel, 6 dtex. (4) Triacetate, Tricel, 14 dtex.
Twist breaks of para-aramid fibre (Kevlar).
(5) Twisted to break at constant length, failed at 30 tex$^{\frac{1}{2}}$cm^{-1} (12°). (6) Twisted to break at constant
tension of 27 mN/tex, failed at 73 tex$^{\frac{1}{2}}$cm^{-1} (45°).

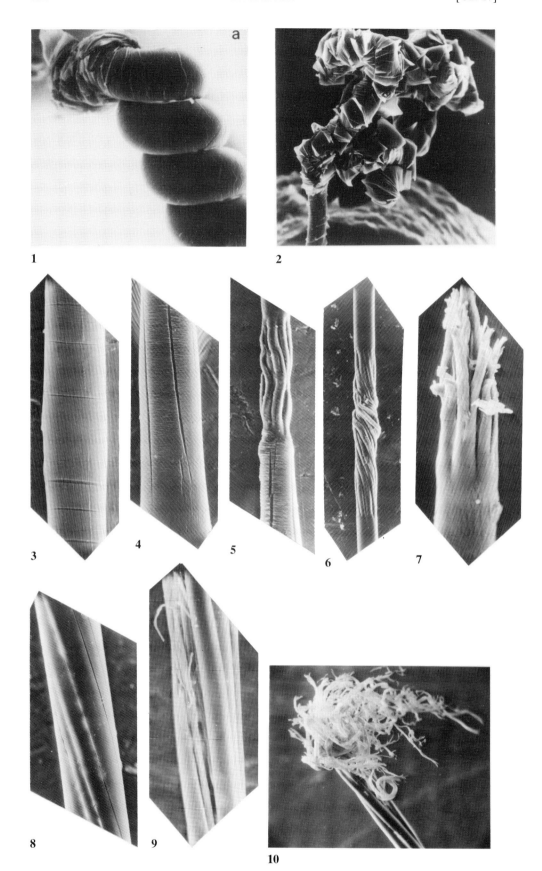

Plate 17D — Torsional fatigue of polyester fibres, courtesy of B.C. Goswami, Clemson University.
(1),(2) Opposite ends of a commercial fibre, broken after torsional cycling with an amplitude of about 3°.
(3)–(6) Progressive damage of a partly drawn fibre after cycling at 15° amplitude, 1000, 3000, 6000 and
12 000 cycles. (7) Tensile break of partly drawn fibre after 3000 torsional cycles. (8),(9) Damage in a
highly drawn fibre after 3000 and 15 000 cycles. (10) Tensile break of highly drawn fibre after 150 000
cycles.

18

COTTON

Cotton shows different forms of tensile break, depending on the environmental conditions and on any chemical modification of the fibre. The common form of rupture of cotton at 65% r.h., **18A(1)**, shows an axial split which runs round the fibre and then tears back along the fibre, as indicated schematically in Fig. 18.1. This reflects the structure of the cotton fibre, which is formed as a hollow tube of helically wound microfibrils. On drying, there is a collapse to a kidney-shape. Zone A is the most tightly packed structure and zone C is the most disturbed and internally buckled. A line of weakness runs along the fibre between zones A and C. Furthermore, at intervals the fibrils reverse direction from a left-handed to a right-handed helix. When the fibre is put under tension, the reversal point will tend to untwist in order to elongate the fibre, as shown in Fig. 18.2, and this generates shear stresses which cause the axial splitting between the fibrils. The split runs round the fibre until it reaches the line of weakness, and then tears back.

At the reversal itself, the fibrils (and the cellulose molecules) will be axially oriented, so that this is a zone of high strength. But close to the reversal the change of orientation generates additional stress concentrations, and so the break normally starts adjacent to a reversal, as in **18A(1)**.

Although the form described above is the commonest form for untreated cotton at 65% r.h., there is considerable diversity, with some breaks being shorter, **18A(2)**, and others longer **18A(3)**. Similar forms of break are found in mercerized cotton **18A(4)** at 65% r.h.

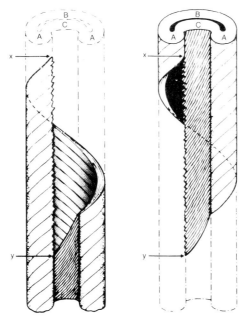

Fig. 18.1 — Typical, but somewhat idealized, form of break of cotton fibre. The full lines show the two opposite broken ends, with the dotted lines being the 'missing parts'. The break starts at x and runs as a crack along the spiral angle through zones A-B-A-C to y on the AC boundary on which it started. It then tears, somewhat irregularly, along the line xy.

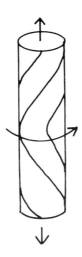

Fig. 18.2 — Untwisting under tension at a reversal point.

In other conditions, other forms are found. The break of untreated cotton when wet was described in Chapter 9, as a particular form of rupture, namely independent fibrillar failure.

If the cotton fibre has been chemically cross-linked by resin treatment, the cellulose fibrils hold more firmly together and axial splitting is hindered. The breaks at 65% r.h. show distorted forms of granular break across the fibre, possibly with a short length of axial separation linking different zones of transverse fracture, as in **18B(1)**. Detail of the granular breakage is seen in **18B(2)**. In other examples, such as **18B(3)**, the break, which is adjacent to a reversal, runs in a single transverse fracture round the fibre. Similar transverse breaks are found in untreated cotton at 0% r.h., when the fibrils are hydrogen-bonded together, and at 65% r.h., if a very short test length is selected with the exclusion of any reversal, so that the shear stresses due to untwisting are not present.

In wet conditions, the greater freedom allows a long split to develop in cross-linked, resin-treated cotton, **18B(4)**.

The selection of the type of break depends on the degree of chemical attraction between the fibrils, and the changes in the form of breakage may be summarized as follows:

(a)   *strong interaction*   granular break, across
      raw cotton at 0% r.h.;   fibre, characteristic
      cross-linked at 65% r.h.   of bonded fibrillar elements

(b)   *weak interaction*   axial split between
      raw cotton at 65% r.h.;   fibrils
      cross-linked, wet

(c)   *very weak interaction*   independent fibrillar
      raw cotton, wet   break

Twist breaks of cotton show axial splitting, with a rather sharp tearing off at the end of the broken fibre, **18C(1),(2)**. The break is similar to a twist break of Kevlar, **17C(6)**.

Tensile fatigue testing of cotton leads to marked separation of fibrils before failure, **18C(3)**.

Fatigue testing of cotton by biaxial rotation over a pin is complicated by the shape of the fibre, which gives uneven movement, and by the variability of fibre diameter, which makes it difficult to standardize strain levels. The scatter of results is large, and no significance should be attached to the differences in lifetimes in the captions to **18C(4)–(9)**. Another complication is that the torque developed in the test alternately increases and decreases the natural reversing twist in the fibre, **18C(4)**. A typical raw cotton break in air shows multiple splitting, similar to that found with other fibres, **18C(5)**. After testing in water, **18C(6)**, the broken end tapers to a point, but this may be a result of the split portions being twisted together, and then bonded on drying before examination.

The chemically treated fibres also fail by multiple splitting, **18C(7)–(9)**.

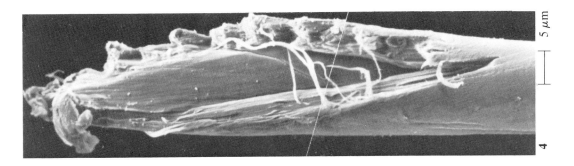

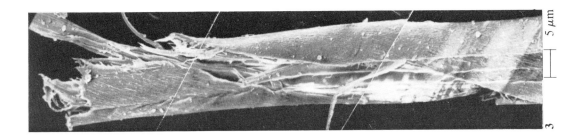

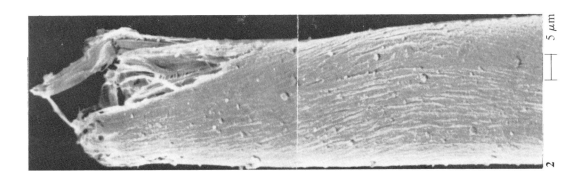

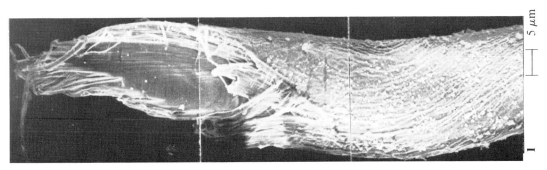

Plate 18A — Tensile breaks of cotton at 65% r.h.
(1)–(3) Raw cotton. (4) Mercerized cotton.

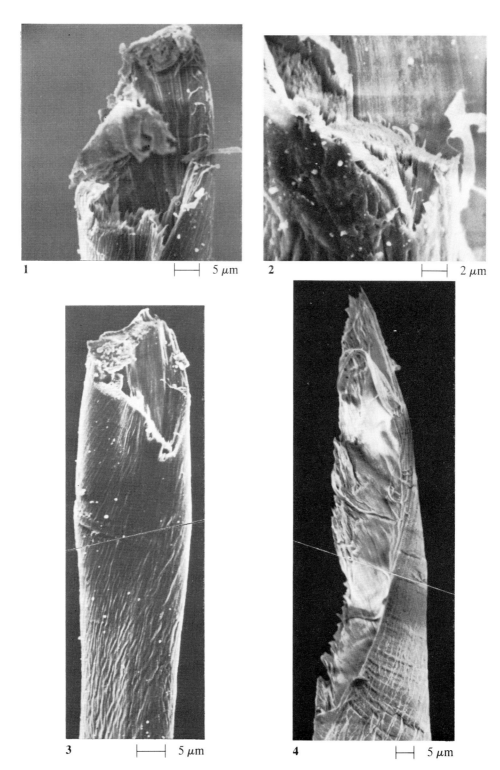

Plate 18B — Tensile breaks of resin-treated (cross-linked) cotton.
(1)–(3) At 65% r.h. (4) In water.

Plate 18C — Twist breaks of cotton.

(1) Raw cotton. (2) Mercerized cotton.

Tensile fatigue.

(3) Raw cotton.

Biaxial rotation fatigue.

(4) Raw cotton at 65% r.h., failed at 1898 cycles. (5) Raw cotton at 65% r.h., failed at 903 cycles. (6) Raw cotton in water, failed at 3499 cycles. (7) Prograde-treated cotton at 65% r.h., failed at 7762 cycles. (8) Mercerised cotton at 65% r.h., failed at 5979 cycles. (9) Resin-treated cotton (10% Fixapret) at 65% r.h., failed at 1334 cycles.

19

WOOL AND HUMAN HAIR

Wool and hair are natural fibres with structural features at many levels. Chemically, they are made of a mixture of complex proteins. In physical fine structure they consist of crystalline fibrils, with the molecules in helical coils, embedded in an amorphous matrix. In biological structure they are composed of separate cells, differing in composition from one part (ortho-) of the central cortex to another part (para-), and with a special form of cuticle cells or scales on the surface.

The tensile break of wool is commonly a granular fracture running across the fibre, **19A(1),(2)**, but it often splits into two separate breaks linked by an axial split, **19A(3),(4)**. It is not certain whether the central split is present before the transverse cracks appear, and thus divides the fibre into two parts, which break independently, giving the form shown in Fig. 19.1(a); or whether the transverse cracks form first, and then join up by an axial split as in Fig. 19.1(b). Both the length of the split in **19A(3)** and the continuation of the split in **19A(4)** suggest that the axial split is there first. Sometimes, there are more than two steps, **19A(5)**.

Human hair, perhaps because it suffers a greater variety of chemical and mechanical treatment, shows an even greater diversity of form than wool. The break of dry hair may be a single plane running perpendicularly across the fibre, **19B(1)**, with a granular surface similar to the breaks of solution-spun fibres in Chapter 8. However, this simple form may be complicated by steps, axial splits, separate breakage of the cuticle and fibrillation.

Most of these features can be seen in the montages of two matching ends, shown over the whole failure length in **19B(2)** and in part in **19B(3).** The rupture is in three main steps. The break appears to start at X in **19B(3)**, and radiates from a point on the fibre surface until it reaches the first short axial step. The second transverse surface at Y reaches the major split at the centre of the fibre. This split is inclined at a small angle to the fibre axis, and thus causes a tapering over the long length (many fibre diameters) that reaches to the third transverse fracture at Z, which completes the break. Both the second and third fracture surfaces are smaller in area compared with the single planar surface near X. Two contrasting aspects of axial splitting can be seen in **19B(3)**: the central split continues for a long distance past the

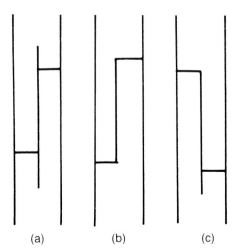

(a)   (b)   (c)

Fig. 19.1 — Three possible combinations of transverse and axial cracks.

transverse break at Y, but the other short axial split stops at the transverse fracture, X. On the other end, as seen in **19B(2)**, with corresponding locations marked X′, Y′, Z′, the central split Y′ Z′ does not continue past the transverse break at Z′. Examination of breaks of many fibres suggests the occurrence of all three possible combinations, shown in Fig. 19.1, of splits stopping or continuing.

It must be emphasized that the suggestion that the break starts at X is speculative. It is possible, but unlikely, that the break starts at the other end, Z, turns into an axial split, ZY, and that a subsidiary final stage of break occurs at X and Y. Another feature visible in **19B(2),(3)**, is some short fibrillation on the surface of the axial splits.

It is difficult to know whether the complicated forms found in breaks of hair are due to the interaction between the mechanics of stress distribution and some random variations in structure or in damage within the fibre, or whether they derive from the major structural features of wool and hair fibres, namely the divisions of the cortex into two halves, with the cuticle being a third region. The break shown in **19B(4)** could easily form the basis for a schematic drawing of the fibre structure, with the central split dividing the ortho-cortex from the para-cortex, and the scale cells of the cuticle clearly seen at the edge. However, study of the other end of the break, **19B(5)**, does make clear that the sharp fracture of the scales also includes a portion of the cortex. Another feature clearly seen in **19B(5)** is a plug of material, somewhat out of focus, which runs down the length of the axial split and extends for some distance into the other end. Re-examination of **19B(4)** shows where this plug fits.

Examples of variant forms of splits and plugs in broken dry hair fibres are shown in **19C(1),(2)**.

Breaks of wet hair, **19C(3)–(5)**, are usually similar to breaks of dry hair. However, the fibrillation is often more pronounced, with longer strips of material pulled away from fracture surfaces. An extreme example is shown in **19C(6)**.

Tensile fatigue of wool and hair intensifies the tendency to axial splitting, and very long breaks are common, **19D(1)**. In one example of fatigue failure in human hair, the break had the stake-and-socket form, **19D(2)–(4)**, which is illustrated in **16D(1),(2)** and is type 7 in Fig. 1.5. In this situation it could be regarded as a variant of the form of tensile fatigue failure found in nylon and illustrated in Fig. 11.2, except that the initiating transverse crack encircled the fibre and caused the axial split to run conically into the centre of the fibre. The tendency of wool and hair to develop axial splits under oscillating tensile stress may well be a cause of split fibre ends, which can be a problem in human hair.

Biaxial rotation fatigue of human hair in air leads to the usual mode of failure by multiple splitting, **19D(5)**. Some earlier studies, using the method of rotation over a pin with the drive from one end, showed similar breaks with multiple splits in air, and sometimes in water. Other tests in water gave a sharper form of splitting, together with some surface damage, as shown in **19D(6)**.

Typical multiple split breaks result from biaxial rotation fatigue tests of wool, **19E**. Two pairs of broken ends, **19E(1)–(4)**, show the splitting occurring over the whole break zone, with the twist in opposite directions on the two ends, due to the drag resulting from friction on the pin, and internal hysteresis. Some of the splits in **19E(1)**, and to a lesser extent in **19E(3)**, appear to be axial rather than helical, but this may be due to elastic recovery after rupture. It is also possible that the selection of lines of splitting is influenced by the multicellular morphology of wool fibres. In the third pair, **19E(5),(6)**, there are two zones of splitting, similar to the polyester fibre in **13C(7),(8)**, with considerable surface wear in the region between zones caused by rubbing on the pin. The effects are thus similar to those discussed in Chapter 13. There is very high statistical variability in the test results on wool fibres, due to variability in diameter, which influences the level of strain in bending, and in structure or selective damage.

Flex fatigue of wool by pulling backwards and forwards over a pin can lead to both surface wear and axial splitting, **19F(1)**, which are two of the forms discussed in Chapter 12. In some tests, **19F(2)**, surface wear was dominant, with the final failure being a tensile break over a reduced cross-section, similar to effects in surface wear from a rotating pin, as was shown in **14A(4)**. In other samples, **19F(3),(4)**, the multiple splitting was dominant.

Similar tests on human hair also showed clear evidence of surface wear, **19F(5)**; but there were also very complicated final breaks, **19F(6)**, reflecting a strong influence of the cellular structure of hair.

Another form of damage, which occurs in wool, results from attack by the larvae of moths and beetles, which are able to digest cross-linked proteins as a source of food. A bite mark of the larva of the furniture carpet beetle, *Anthrenus flavipes* var. *seminiveus* Casey, on wool is shown in **19G(1)**. Other examples of insect damage are given by Anderson and Hoskinson (1970) and by Cooke (1989).

An extensive investigation of larval excreta, in order to show how the breakdown of the keratinous material occurs in wool pests, has been reported by Hammers, Arns and Zahn (1987). Generally, the excreta are a mass of small granules of the same colour as the wool fabric fed to the larvae, but some fragments at various stages of digestion are found. Initially, the cuticle swells and the cells begin to lift off the cortex, as seen in the fibre in **19G(2)**, excreted by a larva of the webbing clothes moth, *Tineola bisselliella*. More severe breakdown, with partially digested cortex, fragments of cortical cells, and cortical and membrane fragments, are shown in **19G(3),(4)**, from *Anthrenus* Casey. A collection of fibrous fragments occurs in the excreta of the black carpet beetle, *Altagenus piceus*, **19G(5)**. All the above examples are from larvae fed on untreated wool. The exceta of larvae fed only on wool treated with a low concentration of a moth-proofing agent showed similar forms of breakdown: undigested and partly digested treated fibres from *Anthrenus* Le Conte can be seen in **19G(6)**.

Studies of human hair made during the first decade of commercial scanning electron microscopy were described in a paper by Brown and Swift (1975). Referring to studies of mechanical properties and of the appearance of deformed fibres, they say:

A logical development is to combine these two techniques so that both physical and structural data could be collected simultaneously, thereby enabling a more detailed and accurate assessment of the breakdown of structural components to be made. The SEM, because of its great depth of focus, wide range of magnification and large area for specimen manipulation, has been adapted for conducting dynamic experiments in situ. In addition, the manner in which the visual information was processed made direct recording of the results on to videotape possible.

Unfortunately, stills from video-recordings do not give a good impression of the insights to be obtained by watching the moving pictures. However **19H(1)–(4)** illustrate the studies made by Swift over a number of years. The paper mentioned above includes studies made on the hair of six young women. Samples of hair, over 50 cm long, which fell out naturally with intact roots during brushing and combing, were collected, and 4 cm portions from root to tip were broken in an Instron tester. Brown and Swift (1975) noted five main types of fracture. Type 1 at the root end (similar to **19B(1)** and to **19H(5)** from the TRI studies in the next section) was 'a clean transverse fracture . . . almost as if cut with a knife . . . the cuticle has split circumferentially about the transverse fracture through the cortex'. Type 2, also near the root and similar to **19B(4)** and **19I(1)**, had 'the cortex stepped and with some disturbance of the cuticle behind the point of fracture, either in the form of a longitudinal split back from the main fracture or a narrow circumferential split some distance from the point of primary fracture.' In type 3, similar to **19C(1),(2)**, 'part of the primary fracture is transverse but the remainder tails off with segments of cortex pulled out.' In types 4 and 5, occurring near the end of weathered hair, the fracture becomes 'ragged with the cortex splitting into fibrillar elements.' This is beginning to be seen in **19C9(2)**, and is clear in the extreme example of type 5, **19H(1)**. In going from root to tip, the stepped and fibrillated forms are reached more quickly for badly weathered hair, and the most extreme forms are not found in hair which has not been much weathered.

In addition to changes in the fracture of the cortex, effects are also seen in the cuticle, with some cracking and lifting of scales. A particularly severe example of scale lifting is shown in **19H(2)**. This is a picture of a human hair that had been stretched 20% in water and then steam set. The delamination and circumferential fracture of the cuticle, which results from the differences in extensibility of cortex and cuticle, have exposed the surface of the cortex.

A classic split end is shown in **19H(3)** in woman's hair, 35 cm long. The subject used no toiletry treatments except for twice-weekly shampooing, and brushing and combing about twice per day. Another hair, **19H(4)**, from the same person shows *trichorrhexis nodosa*, which is an intense focal splitting of the hair shaft revealing cortical fragments. This damage is visible as a bright node. The hair is intact on either side of the node and a droop at this point of weakness is particularly apparent with long hair styles. The problem is found with people who spend excessive amounts of time in the sun, since the mechanism is embrittlement due to crosslinking of the plasticising components of the hair by free radicals induced by ultraviolet exposure.

RESEARCH ON HUMAN HAIR AT TRI PRINCETON
by H-D. Weigmann and S. B. Ruetsch

Keratin fibres have a rather complex cellular morphology consisting basically of an assembly of closely packed cortical cells, which are surrounded by single or multiple layers of cuticular cells. In human hair these can amount to up to ten layers at the root end of the fibre. Since grooming, weathering and cosmetic treatments impact on the cuticle, progressively fewer cuticle cells are found along the hair shaft towards the tip of the fibre. Total ablation of the cuticle cells in very long and heavily stressed fibres leads to the phenomenon of split ends, where the cortical cells lose cohesion, separate and sometimes fibrillate on a macrofibrillar level. Each of the morphological components contains various structural elements which affect its tensile, torsional, bending and shear properties. Intercellular adhesion is provided by the so-called cell membrane complex. In some hair fibres, especially those of large diameter, a loosely packed porous region called medulla is located near the centre of the fibre. In view of the complex morphology of hair fibres, it is not surprising that their fracture behaviour yields a number of interesting patterns, which sometimes permit conclusions to be made regarding the cause of damage experienced by the hair.

The application of tensile stresses results in a variety of fracture patterns, depending on molecular cohesion within the fibre. Smooth radial fractures occur most frequently when the fibre is wet and the fully swollen cortical cells press against the cuticular envelope. Initiation of fracture occurs almost exclusively at a point between cuticle and cortex — possibly at a pre-existing flaw. Only in very dry conditions have we observed crack initiation in the centre of the cortex. From its initiation, the crack propagates radially until it reaches critical size and catastrophic failure occurs, **19H(5)**. While there is obviously some unevenness in the catastrophic region of the fracture surface cortex, it appears that the cuticle fails as a unit, **19H(6)**.

A wide variety of step fractures is the most common type of fracture pattern, which reflects the presence of weak points in the intercellular cohesion within the cortex. Step fractures are most frequently observed during fracture under ambient conditions, namely 65% relative humidity. The fracture always starts as a smooth radial crack, which suddenly changes direction when it encounters a region of weak intercellular adhesion. The crack travels axially along cortical cell boundaries until it encounters another radial crack and the fibre fails. The step length can be much longer than that shown in **19I(1)**, can be angled relative to the fibre axis, and can end in a smooth, jagged or fibrillated end.

Chemical treatments such as bleaching (6% alkaline hydrogen peroxide) affect intercellular cohesion in the cortex as well as in the cuticle. This can manifest itself in long step fractures as shown in **19I(2)**, which also shows poor adhesion between cortex and cuticle resulting in a hollow cuticular sheath. This lack of adhesion is more clearly seen at higher magnification in **19I(3)**, where a loose, undulating cuticle is observed. **19I(4)** shows that failure in the cuticle also occurs in a stepwise manner with the individual cuticle cell failure, reflecting poor intercuticular adhesion.

Tensile fatigue, 100 000 cycle at 1 Hz at stresses below the yield point of the fibres, causes considerable damage to intercellular cohesion both in the cortex and the cuticle. An abundance of scale lifting is observed in the cuticle, **19I(5)**, while failure along the cortical cells shows fibrillation on the macro fibrillar level with possibly some fragments of intercellular cement also sticking out of the cell surfaces, **19I(6)**.

Exposure of human hair fibres to 700 hours of alternating cycles of ultra-violet radiation and humidification at 95% relative humidity results in modification gradients within the fibres, which lead to rather interesting fracture patterns, **19J(1),(2)**. High levels of photo-oxidation occur in the fibre periphery with a steep gradient to lower levels in the interior. The highly photo-oxidised, fused periphery, which includes the cuticle and outer cortical cells, has become brittle and rigid, losing its elastic properties and failing as a unit all around the fibre during extension. It would appear that radical reactions result in these outer domains. Failure in the not yet fused interior occurs in the form of a 'cathedral spire' fracture pattern (referred to elsewhere as 'stake-and-socket') at individual sites alternatively along the cortical cell boundaries and across the cortical cells tapering off to the least modified centre of the fibre. The corresponding opposite site of the cathedral spire is a hollow opening.

A shorter exposure of the hair fibres to only three hours of ultra-violet radiation and humidification preferentially results in smooth, radial fractures. The photo-oxidised periphery of the fibre has lost its original extensibility and develops deep radial cracks, **19J(3)**. As stresses increase during fibre extension, fracture is initiated between cortex and cuticle on the side opposite to the radial cracks and propagates from there in the form of a smooth fracture surface until catastrophic failure occurs, **19J(4)**.

As pointed out above, the medulla seen in certain higher diameter fibres consists of an array of empty cells, which can constitute a significant part of the centre of the fibre. While the medulla does not contribute in terms of mechanical performance of the fibre, it does appear to have significant effect on the optical or light scattering of the fibres. Fibrous elements in the differentiated medullar cells apparently fuse with the cell membranes. These reinforced cell membranes cannot collapse during desiccation and thus form large intracellular cavities, which are rather stiff. During tensile failure, these medullar cells tear apart and form part of the fracture surface of the fibre, **19J(5),(6)**.

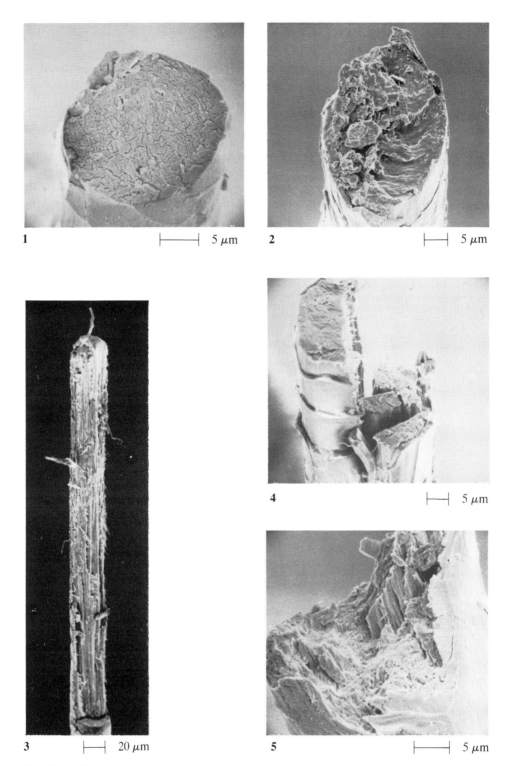

1   ⊢——⊣ 5 μm   2   ⊢——⊣ 5 μm

4   ⊢——⊣ 5 μm

3   ⊢—⊣ 20 μm   5   ⊢——⊣ 5 μm

Plate 19A — Tensile breaks of wool.
  (1) Untreated wool, 64s. (2) Untreated wool, 56s. (3)–(5) Untreated wool, 64s.

Plate 19B — Tensile breaks of dry human hair.
(1) Simplest form, from female subject A. (2), (3) Opposite ends of another break, showing complex
features, from subject A. (4), (5) Opposite ends of break, from another female subject B.

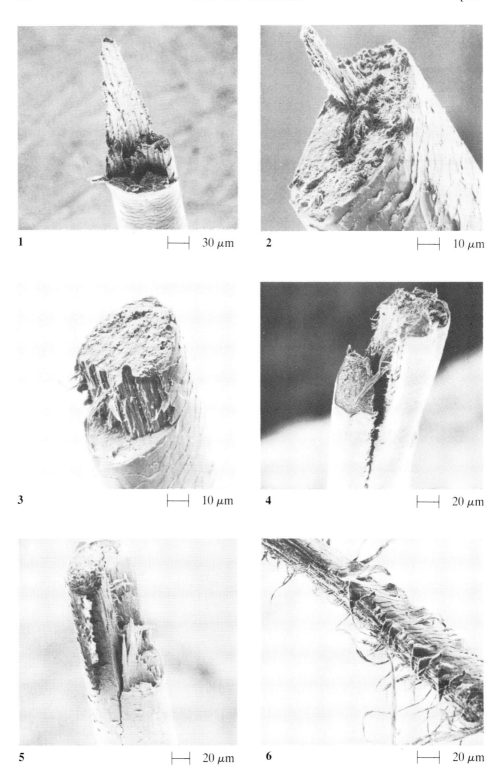

Plate 19C — Tensile breaks of dry human hair.
(1) From female subject A. (2) From female subject B.

Tensile breaks of wet human hair.
(3) From subject A. (4),(5) From subject B. (6) Extreme fibrillation near broken end, from subject A.

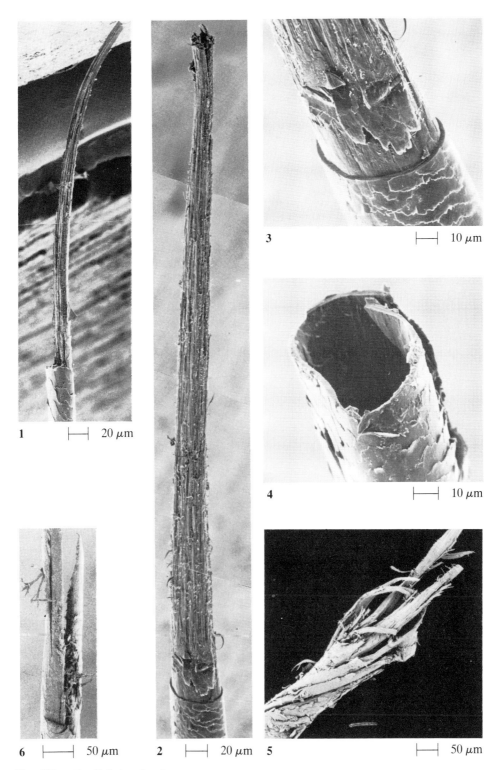

Plate 19D — Tensile fatigue breaks.
(1) 56s untreated wool. (2) Human hair. (3) Detail of break (2). (4) Opposite end of the break shown in
(2), (3).

Biaxial rotation fatigue over a pin in air.

(5) Human hair.

Rotation over a pin in water.

(6) Human hair.

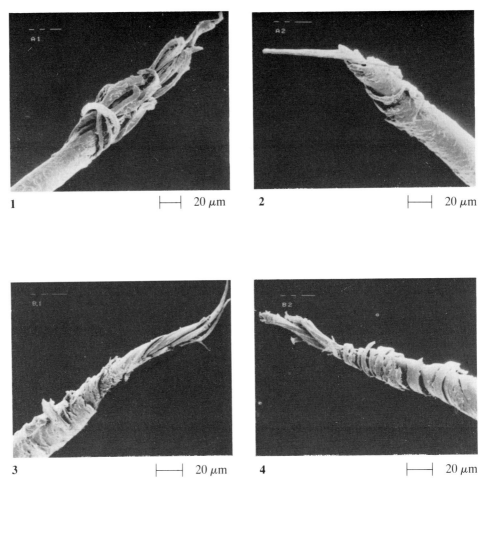

1 |—| 20 μm

2 |—| 20 μm

3 |—| 20 μm

4 |—| 20 μm

5

6

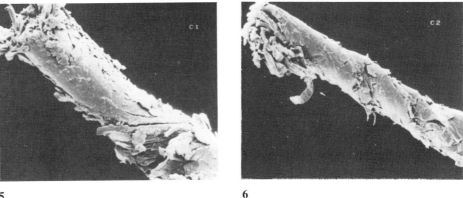

Plate 19E — Biaxial rotation fatigue of wool fibres extracted from bleached fabric tested at 120°C, 65% r.h., approximately 10% strain amplitude, and tension of about 10 mN.
(1), (2) Opposite ends of fibre, which broke at 7108 cycles. (3), (4) Opposite ends of fibre, which broke at 3537 cycles. (5), (6) Opposite ends of fibre, which broke at 1550 cycles.

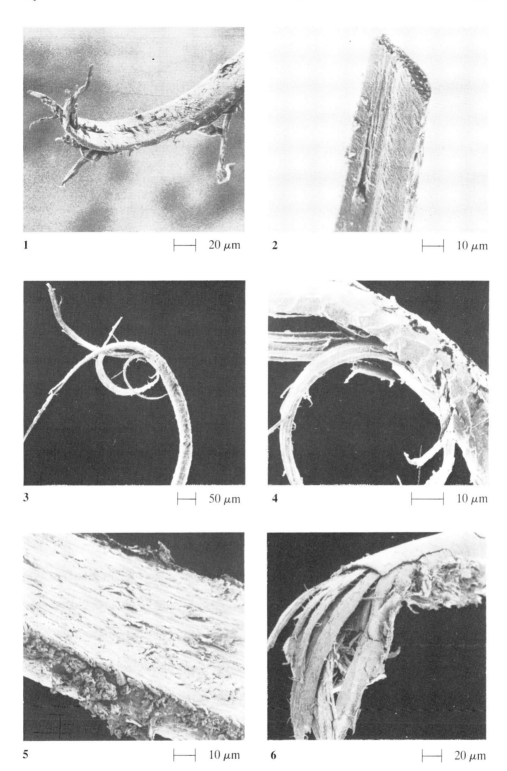

Plate 19F — Flex fatigue by pulling to and fro over a pin.
(1) Wool fibre broken after 14 750 cycles. (2) Wool fibre broken after 43 250 cycles. (3), (4) Wool fibre broken after 1750 cycles. (5) Human hair after 5000 cycles, before failure. (6) Human hair broken after 11 250 cycles.

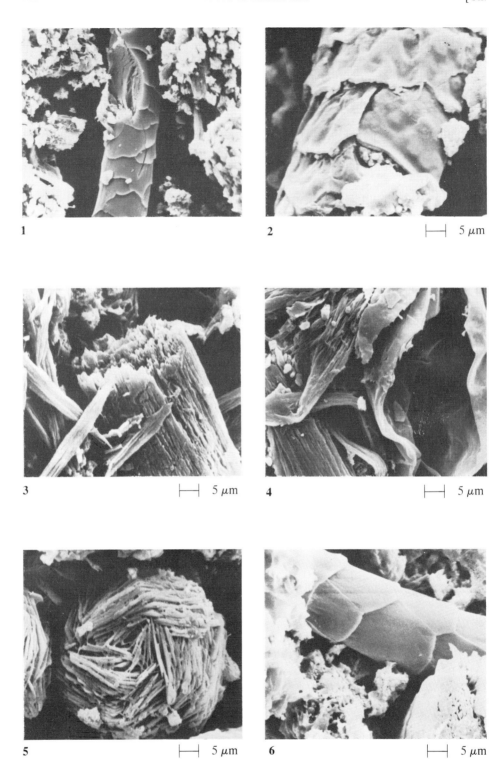

Plate 19G — Larval bite (from I. Hammers and W. Arns, private communication).
(1) Bite-mark of larva of *Anthrenus* Casey on wool fibre.

Larval excreta (from Hammers, Arns and Zahn, 1987).
(2) *Tineola bisselliella*, fed on untreated wool. (3), (4) *Anthrenus* Casey, fed on untreated wool. (5) *Altagenus piceus*, fed on untreated wool. (6) *Anthrenus* Le Conte, fed on moth-proofed wool.

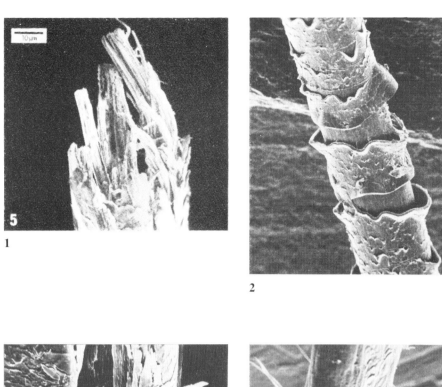

1   **2**

3   **4**

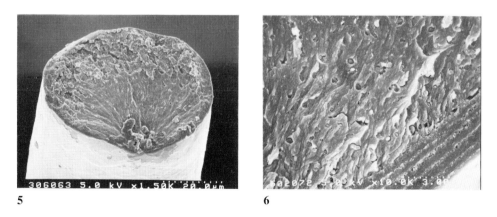

5   **6**

Plate 19H — Studies of damage in human hair by J. Alan Swift.
(1) Multiple splitting at the end of highly weathered hair. (2) Hair stretched in water 20% and then steam set. (3) Classic split end. (4) *Trichorrhexis nodosa.*
Studies of human hair at TRI
(5) A common form of smooth radial fracture of normal hair. (6) Detail near the cortex/cuticle boundary.

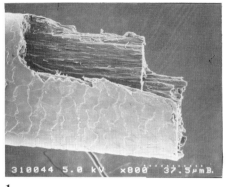

1

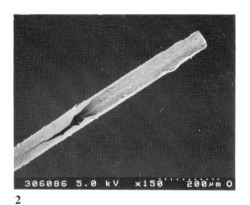

2

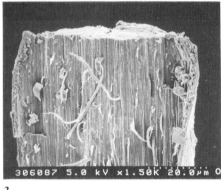

3

4

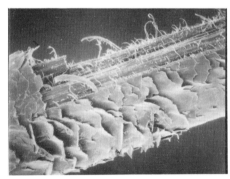

5

6

Plate 19I — Studies of human hair at TRI (continued).
(1) Stepped tensile fracture of normal hair. (2) Long stepped fracture of hair bleached with 6% alkaline hydrogen peroxide. (3) Detail of split surface. (4) Poor intercellular cohesion in cuticle and between cuticle and cortex. (5) Tensile failure of hair after 100 000 cycles of fatigue. (6) Detail of fibrillation of cortical cells.

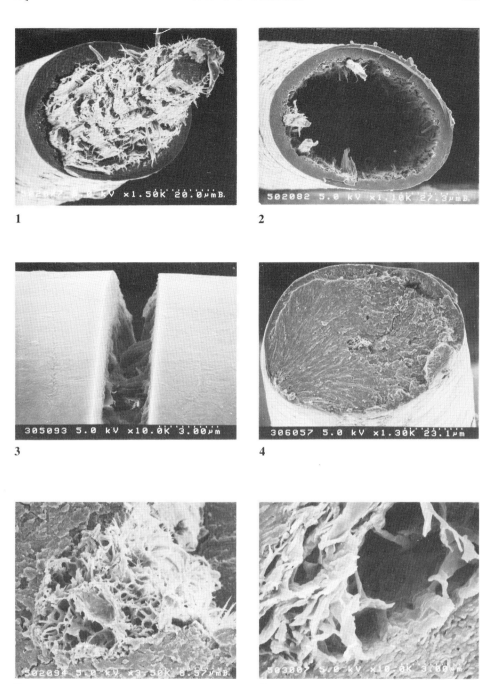

Plate 19J — Studies of human hair at TRI (continued).
(1) 'Cathedral spire' tensile fracture of hair exposed to 700 hours of alternating ultra-violet irradiation and humidification. (2) The hollow opening in the opposite end, alternatively called the socket in a 'stake-and-socket' break. (3),(4) Tensile failure of hair exposed for 300 hours of ultra-violet radiation and humidification. (5) Fracture of hair including a medulla. (6) Detail of damage in medulla.

20

OTHER FORMS OF SEVERANCE

Fibres can be severed in other ways than by the application of tensile, torsional, flexural or shear stresses. Cutting is a method which may be deliberate or accidental, and the appearance of the break depends not only on the fibre type but also on the instrument. With a blunt knife, cutting may be little more than a means of applying tension, so that the break is a distorted form of tensile failure. But with sharper instruments distinct forms occur.

A sharp razor gives a clean cut across a fibre as seen in nylon, **20A(1)**, with the only features being some grooves in the direction of cutting and a small lip at the edge of the fibre. In contrast to this, cutting with a knife shows more spreading of the break, **20A(2)**. There is some similarity to the high-speed breaks shown in Chapter 6, and a common feature may be heating of the fibre, which could occur as the knife is drawn across it.

The grooves in the end of a nylon fibre cut with a razor can be quite deep, **20A(3)**, and show interesting surface detail at high magnification, **20A(4)**. In polyester, perhaps because there is a higher resistance to cutting, there is more distortion of the fibre end in a razor cut, **20A(5)**.

Scissor cuts of nylon or polyester show a characteristic form, which is caused by the two blades pressing together, **20A(6)**.

The effects of different forms of cutting on most other fibres are broadly similar in form. A razor cut of cotton is clean, with a few grooves on the surface, **20B(1)**, and perhaps some tearing, **20B(2)**. Tearing and squashing are much more apparent in a knife cut, **20B(3)**: melting does not occur, because cotton chars before it melts. The scissor cut of cotton is somewhat sharper, **20B(4)**. The razor cut of viscose rayon, **20B(5)**, shows much less distortion of the fibre end than the knife cut, **20B(6)**.

Differences between the clean cuts with a razor and the greater distortion of knife and scissor cuts are shown by acetate fibres, **20C(1),(2)**, acrylic fibres, **20C(3),(4)**, and wool, **20C(5),(6)**.

Melting is another deliberate or accidental action, which changes the appearance of a fibre end. The melting of nylon gives a bulbous end to the fibre, but this may be elongated, **20D(1)**, or spherical, **20D(2)**, depending on the distribution of heat. Sometimes, as in the polyester, **20D(3)**, there may be a combination of the two shapes. An acrylic fibre, **20D(4)**, shows a similar form.

When wool is heated it undergoes a combination of chemical change, by decomposition and burning, and physical change, by contraction and melting. The effects are seen as a change to the fibre surface, **20D(5)**, which becomes a more drastic transformation in very severe conditions, **20D(6)**. A bulbous end also forms.

Cellulose fibres, although they normally burn or char when heated, can show some softening similar to melting, as in the viscose rayon fibre, **20E(1)**, which has been held near a flame, although the effects can be more complicated, **20E(2)**, when the fibre is dipped into the flame.

A loose tuft of cotton fibres which has been quickly singed in a spirit lamp is shown in **20E(3)–(7)**. The fibres can take on a quasi-molten appearance, **20E(4)**, or be more severely burnt away, **20E(5).** Where the fibre is free to contract, bulbous ends can form, **20E(6)**, and these may also become more burnt away, **20E(7)**.

The study of cuts in fibres and fabrics is important for forensic scientists as described in Chapters 44–46. Foos (1993), from the Bayerischen Landeskriminalamtes, has examined cuts in several different fibre types. Polyester fibres cut with a scalpel are shown in **20F(1)**. On the side where the cutting edge makes contact the end of the fibre is slightly rounded, whereas on the opposite side material is drawn out. An oblique cut of a polyester fibre, **20F(2)**, clearly shows the marks of the scalpel. The fibre moved as it was being cut. Similar effects are seen in the cutting of acrylic fibres with a scalpel, **20F(3)**.

Piercing of a fabric with a knife gives polyester fibres with more irregularly torn ends, **20F(4)**. A blunt knife can cause acrylic fibres to weld together, **20F(5)**. This test was carried out for comparison with damage in an actual crime, as shown in **44A(5)**. The suspect knife was used to produce the specimen shown in **20F(5)**. The irregular break-up and welding of acrylic fibres in **20F(6)** was produced by an unknown tool, probably a screwdriver. Finally, **20F(7)** shows acrylic fibres cut with scissors, which also causes fibres to weld together.

Aramid fibres are extremely difficult to cut. However, in order to obtain cross-sections, they can be cut with a razor after dipping in liquid nitrogen ($-196°C$). Kevlar 29 is shown in **20G(1)** and Kevlar 49 in **20G(2)**.

In connection with the studies of papermakers' felts, described in Chapter 40, C. Cork and M. A. Wilding of UMIST developed an impact test. A fibre under tension is held in contact with a flat glass anvil and is then impacted by a cylindrical surface of a glass hammer driven by a vibrator. **20G(3)–(4)** show the appearance of a 19 dtex nylon fibre subject to the impact test. There is initial flattening followed by axial splitting and rupture.

For thermogravimetric analysis, it is necessary to break fibres down into small pieces. As an incidental result, this gives a way of studying the forms of breakdown. Sharma *et al* (1996) have tested unretted flax fibres, which were scutched and hackled from sulphur dioxide treated green straw, and water and enzyme retted fibres from Belgium. The samples were prepared in three ways: by cutting to less than 1 mm length with serrated scissors; by milling in a cyclotec mill to pass through a 0.5 mm sieve; and by freezing in liquid nitrogen for three minutes, followed by grinding in the mill. As seen in **20H(1)–(6)**, there are appreciable differences in the size of particles depending on the method of preparation. The mean fibre length ranged from 2.5 mm for the cut unretted sample to 10.7 mm for the freeze-ground water-retted material. Freeze-grinding causes shearing of the fibres and produces more fine fibre dust. The fibre ends, **20I(1)–(6)**, also differ. Scissor marks can be seen in the cut fibres, whereas the milled fibres had rounded ends. The freeze-ground fibres had deep axial splits.

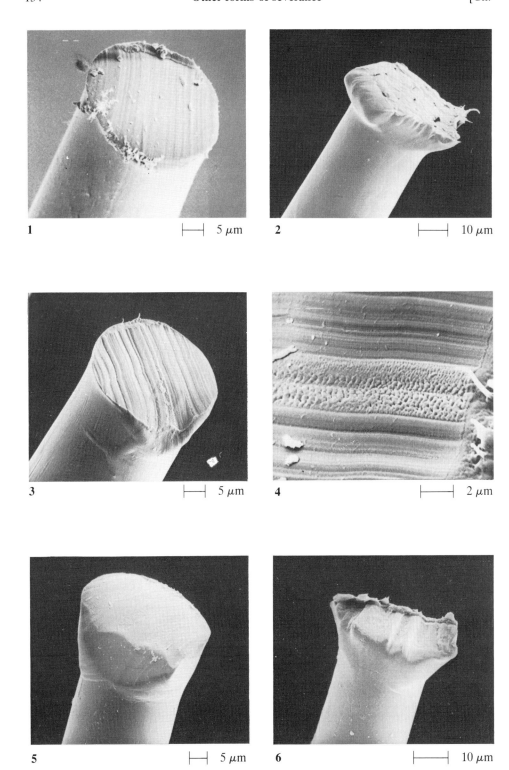

Plate 20A — Cutting of nylon and polyester fibres.
(1) Nylon cut with a razor. (2) Nylon cut with a knife. (3),(4) Nylon cut with a razor. (5) Polyester cut with a razor. (6) Polyester cut with scissors.

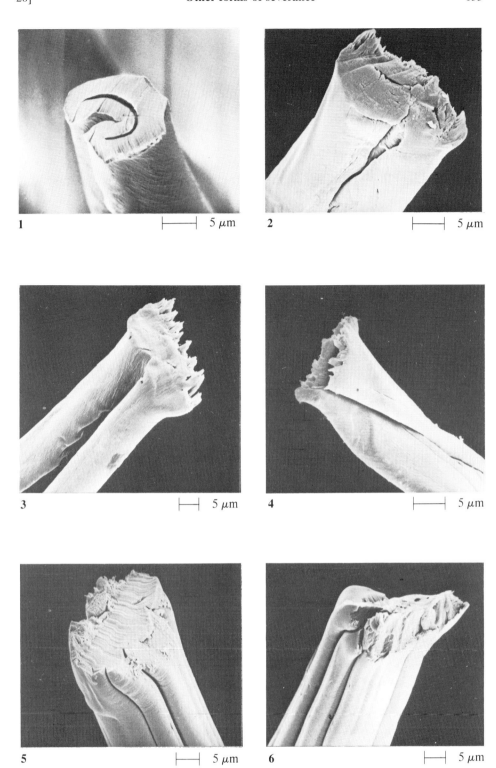

Plate 20B — Cutting of cellulose fibres.
(1), (2) Cotton cut with a razor. (3) Cotton cut with a knife. (4) Cotton cut with scissors. (5) Viscose rayon cut with a razor. (6) Viscose rayon cut with a knife.

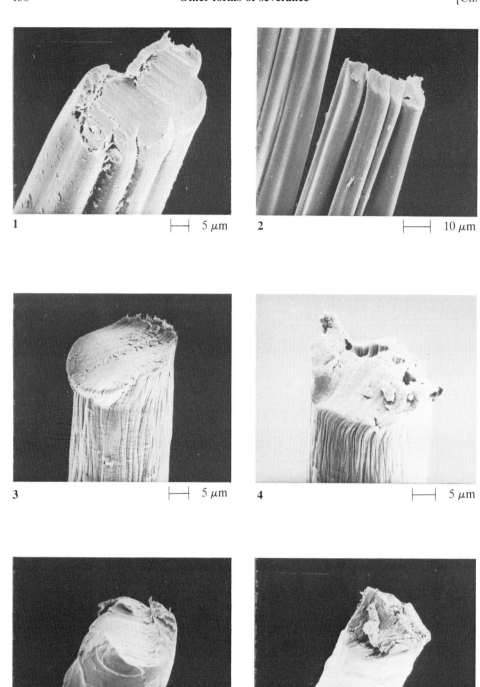

Plate 20C — Cutting of acetate, acrylic and wool fibres.
(1) Secondary acetate cut with a razor. (2) Secondary acetate cut with a knife. (3) Courtelle acrylic fibre cut with a razor. (4) Courtelle acrylic fibre cut with scissors. (5) Wool cut with a razor. (6) Wool cut with scissors.

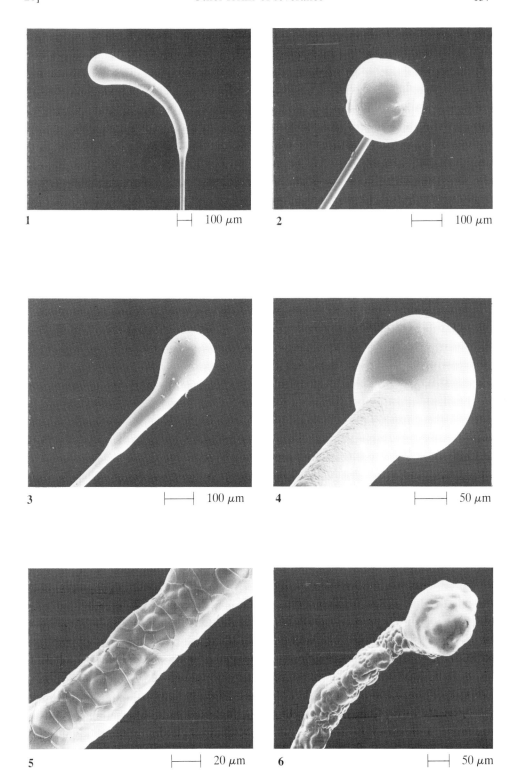

Plate 20D — Melting of fibre ends.
(1) Nylon held near a flame. (2) Nylon quickly dipped into a flame. (3) Polyester held near a flame. (4) Courtelle acrylic fibre held near a flame. (5), (6) Wool drawn through a flame.

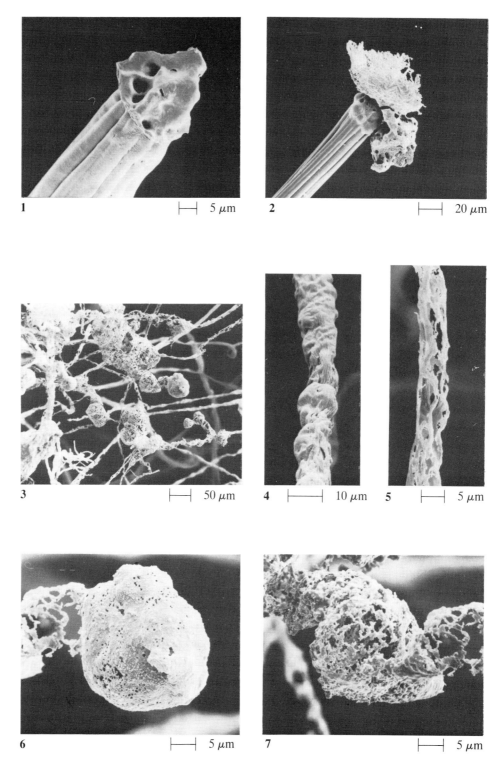

Plate 20E — Effect of heat on cellulose fibres.
(1), (2) Viscose rayon fibre dipped in a flame. (3)–(7) Tuft of cotton singed in spirit lamp flame.

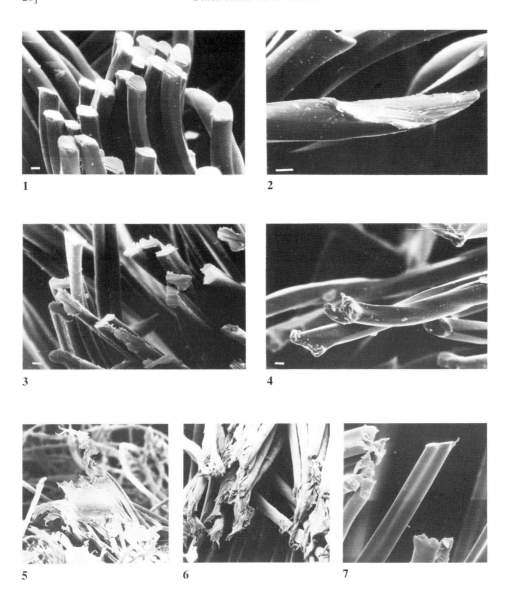

Plate 20F — Cutting of fibres, courtesy of Karlheinz Foos, Bayerischen Kriminalamtes.
(1) Polyester fibres cut with a scalpel. (2) Oblique cut of polyester fibre with a scalpel. (3) Acrylic fibres cut with a scalpel. (4) Polyester fibres pierced with a knife. (5) Acrylic fibres welded together by a blunt knife. (6) Welding of acrylic fibres by an unknown tool. (7) Acrylic fibres cut with scissors.

Plate 20G — Cuts of Kevlar fibres after cooling to −196°C.
(1) Kevlar 29. (2) Kevlar 49.
Impact on 18 dtex nylon fibre.
(3) Flattening. (4),(5) Axial splitting. (6) Rupture.

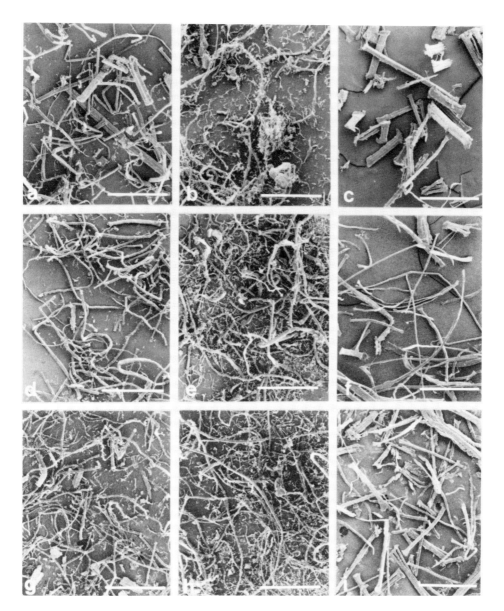

Plate 20H — Flax fibres prepared for thermogravimetric analysis, Sharma et al (1996).
[a] Unretted milled. [b] Unretted freeze-ground. [c] Unretted cut. [d] Enzyme-retted milled. [e] Enzyme-retted freeze-ground. [[f] Enzyme-retted cut. [g] Water-retted milled [h] Water-retted freeze-ground. [i] Water-retted cut.

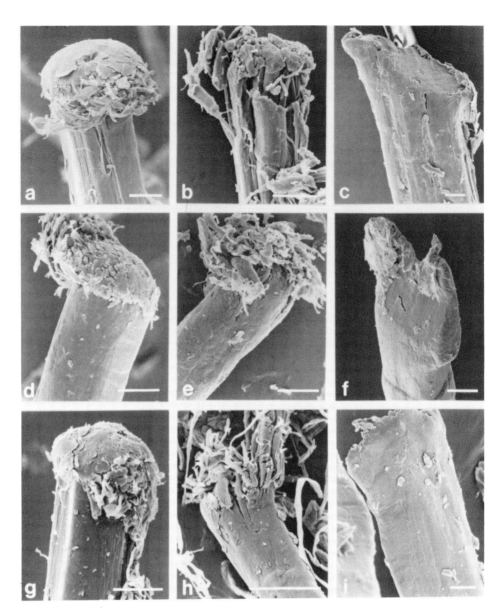

Plate 20I — Ends of flax fibres prepared for thermogravimetric analysis, Sharma et al (1996).
[a] Unretted milled. [b] Unretted freeze-ground. [c] Unretted cut. [d] Enzyme-retted milled. [e] Enzyme-retted freeze-ground. [f] Enzyme-retted cut. [g] Water-retted milled. [h] Water-retted freeze-ground. [i] Water-retted cut.

21

MISCELLANY

This chapter contains a miscellaneous collection of breaks which have been found in laboratory tests either of fibres of specialized interest or in special conditions.

Apart from cotton, most plant fibres in commercial use as textile or cordage fibres have a more highly oriented molecular arrangement, with fibrils following helical paths at an angle of about 10° to the axis, and they are multicellular. Their tensile breaks tend to be granular across the fibre, but show evidence of the separate cells. This is illustrated for jute, **21A(1), (2),** and flax, **21A(3)**. A hollow, unicellular seed fibre shows a granular break straight across the fibre, **21A(4)**.

Silk is a natural fibre which is spun from a solution extruded by the silkworm. The tensile breaks, **21A(5),(6)**, are granular and similar to those of the man-made fibres spun from solution and shown in Chapter 8.

Hollow polyester fibres may show a tensile break, **21B(1)**, which is a modified form of the normal ductile failure; but we also find examples in which two separate fractures are joined by an axial split, **21B(2)**.

The thermally resistant meta-aramid fibre, Nomex, shows rather unusual breaks, which appear to be intermediate between granular and ductile, **21B(3),(4)**.

PVC fibres are made in France and used to a limited extent: they are solution-spun fibres, similar to other vinyl fibres such as the acrylics, and the form of break is similar, **21B(5)**.

Polytetrafluorethylene is better known as a bulk plastic or as a coating, but it is made into fibres for specialized industrial purposes. It is an extremely inert material; there are no solvents for solution-spinning; and it decomposes at a high temperature as it begins to soften, prior to the melting which would occur if the decomposition had not occurred. These difficulties make it necessary to spin by extrusion of a dispersion, followed by sintering to hold the fibre together. The tensile break shows many separate fibrillar ruptures, **21B(6)**. This may be another example of independent fibrillar failure, type 6 of Fig. 1.2, otherwise only found in wet cotton, as described in Chapter 9. Weak cohesion of separate elements of the dispersion, even after sintering, would explain this.

Monvelle is a bicomponent fibre in which one half is nylon and the other half is an elastomeric segmented polyurethane. Differential contraction makes this a helically crimped fibre. The snap-back into a helical tangle after break, **21C(1)**, makes detailed examination of the broken end difficult; but it is possible to see that each half has broken with its own characteristic form, **21C(2),(3)**. The nylon shows the usual ductile failure, described in Chapter 5, while the elastomer shows a brittle failure, as found in single-component elastomeric fibres, **4C(4)–(6)**. The breaks of the two parts are sometimes separated along the fibre.

A different sort of bicomponent fibre, used as a high-performance fibre in composites, consists of silicon carbide which has been vacuum deposited on a tungsten core. The diameter of the fibre is 100 μm, and of the core is 20 μm. A single-fibre tension test in air causes brittle failure, as shown in **21C(4)**, initiated at a surface defect at a strain of about 0.6%, corresponding to a stress of 2.1 GPa. However, when the fibre was embedded in epoxy resin, the break propagates radially outwards, following initiation at the tungsten core, as shown in **21C(5)** for a fibre which broke at 1.0% strain or about 3.5 GPa stress. In some instances, as seen in **21C(6)**, the break originated at a defect between the silicon carbide and the tungsten core; this was a source of weakness, with the failure occurring at 0.8% strain, 2.8 GPa stress.

The effect of bending on the highly oriented, para-aramid fibre, Kevlar, is shown in **21D**. In a loop test around a wire, giving an apparent bending strain of 4.7%, the strength had fallen to 47% of the value for a straight fibre, but the broken fibre, **21D(1)**, was very similar to the tensile breaks, shown in Chapter 7, with long axial splits.

In one set of bending fatigue experiments, a short length (2.5 mm) of the Kevlar fibre was buckled by decreasing the distance between the grips by 30%. It was then oscillated about this mean value, between 20% and 40% contraction. The smooth buckling curve soon changes to a sharp kink, but even after 1 week at 50 Hz (over 30 million cycles) the fibre had not broken, **21D(2)**. A subsequent tensile break of such a fibre shows that it has split into fibrils in the kinked region, **21D(3),(4)** thus relieving the strain.

Flex fatigue of Kevlar, by pulling the fibre backwards and forwards over a pin which is free to rotate, so that the surface wear reported in Chapter 14 does not occur, leads to failure with thinning, fibrillation and some flattening, **21D(5)**. Rotation over a pin gives a typical multiple splitting failure, **21D(6)**.

In the biaxial rotation fatigue test of nylon or polyester fibres, discussed in Chapter 13, it sometimes happens that break occurs after relatively few cycles, before the twisting has reached a steady state. These breaks are called direct breaks, and they occur when the angle of wrap round the pin is too large, so that a high torque is needed to overcome the external and internal friction. One might expect a simple twist break to occur, but this is not so. Complicated forms of deformation occur, and these are probably associated with the influence of torque on a fibre which is weakened and softened by the repeated bending and perhaps by frictional heating. The motion in the pin at the beginning of a test can be jerky, as a fibre rolls along and then slips.

The first effect is the formation of bulbous zones at positions of high torque where the fibre leaves the pin, **21E(1)**. Sometimes these appear at intervals along the fibre, **21E(2)**, but the reasons for this are not understood. The structural changes within the fibre can be seen by examination in polarized light microscopy, **21E(3)**. The fringes at A and B are typical of undamaged fibres, and are due to increasing optical path difference with thickness in a birefringent fibre. The way in which the fringes swing round at C and D, to indicate a lower optical path difference in a thicker fibre cross-section, means that the birefringence must be much reduced. The fibre has softened, disoriented and contracted.

The final failure can show at least three different forms: 'wagon-wheel' cavitation, **21E(4)**; pinching off under torque, **21E(5)**; and spreading out, **21E(6)**.

Calcium alginate fibres are used in medical dressings. Their breaks, seen in **21F(1)–(3)**, have a granular form, with evidence of failure at internal faults in the fibre. Thistledown is a natural fibre, made up of a collection of tubular structures, **21F(4)**. These easily fibrillate, **21F(5)**. Rupture is shown in **21F(6)**.

Tencel from Courtaulds is a new lyotropic cellulose fibre, which is having a significant market impact. Its tensile breaks were shown in **8E**. As shown in **21F(7),(8)**, it easily fibrillates. Although this might be a disadvantage in some circumstances, it is a property that can be exploited favourably, as described in Chapter 23.

BACTERIAL FIBRES by J. J. Thwaites and N. H. Mendelson

Bacterial thread is a multifilament fibre formed from cultures of a cell-separation-suppressed mutant of the Gram-positive bacterium *Bacillus subtilis*. The bacteria, which are normally rod-shaped, $0.8\,\mu$m in diameter and up to $4\,\mu$m in length, are grown as cellular filaments up to 1 mm in length, producing in a Petri dish an aggregation that resembles a random textile web. Thread is produced by lifting part of the web from the culture by means of a wire hook. The surface tension has the effect of a die and other filaments are drawn radially into a fibre as the hook is raised. The filaments are close-packed and highly aligned parallel to the fibre axis. The fibre has a circular cross section and, given a uniform web, is of constant diameter over lengths up to about 100 mm. For a fibre of diameter $100\,\mu$m the cross section contains about 15 000 filaments; a specimen of length 100 mm contains therefore almost 10^9 cells, Thwaites and Mendelson (1985).

Bacterial thread can be tested in tension in the same way as other fibres. No interfilament slippage is observed; it is clear that the cells adhere strongly to each other, even in very humid atmospheres. The material shows typical polymer behaviour of relaxation and recovery; it is stiff and brittle when dry but ductile at high relative humidity, with initial modulus smaller by a factor of about 1000. Bacterial cells have no internal structure so that the material involved is the cell wall, which occupies about one-fifth of the fibre cross section. The load bearing polymer of the wall, peptidoglycan, is a polysaccharide with short peptide side chains. It accounts for half the wall mass. There is no evidence of crystallinity. The measured initial modulus is about 20 GPa when dry, its tensile strength is about 300 MPa and its extensibility $<0.5\%$. At high relative humidity the extensibility can be as high as 70% and the tensile strength falls to about 15 MPa, Thwaites and Surana (1991a).

21G(1), a scanning electron micrograph, shows the fracture cross section of a bacterial thread as first drawn. The filaments are aligned and appear not to have collapsed during drying but have deformed into approximately hexagonal shape. **21G(2),(3)** show similar cross sections for bacterial thread from which the residual culture medium has been removed by washing in water, **21G(2)**, and for which the walls have been attacked by lysozyme, **21G(3)**, Thwaites *et al* (1991b). The effect of washing is merely to raise the 'transition' relative humidity by 18% without changing the measured parameters. Surprisingly the properties are not changed by lysozyme attack, even though it produces substantial circumferential cracks in the fibre surface. Lysozyme preferentially cleaves the cross walls between adjacent cells in a filament. This accounts for the cracks. It breaks the glycosidic bonds of the peptidoglycan backbone. The unchanged tensile properties indicate respectively how strong the lateral cell adhesion is, and that, although the cell wall is amorphous, there is order in the peptidoglycan, with its backbone lying on average in a circumferential direction in the cell wall.

Both peptidoglycan and the other major cell wall polymer in *Bacillus subtilis*, teichoic acid, carry ionizable groups that collectively result in a highly electro-negative material. Counter-ions must neutralize the majority of these charges to maintain the integrity of cell wall. Many different ions can serve to do so, thus cell wall can act as a complex ion exchanger. Ion binding can lead to nucleation of crystallization within the interior and on the cell wall surface. Such processes are thought to be responsible for the geochemical deposition of some minerals.

Mineralization of cell walls can also be achieved under laboratory conditions and large fibre-like structures called bionites have been made, Mendelson (1992). Chloride salts of iron, copper and calcium when added directly to the culture medium containing a *Bacillus subtilis* web of filaments give rise to mineralized fibres when the web is later drawn from the solution. Bionites have also been produced by hydrating bacterial thread in ion solutions and then redrawing, and by transferring a filament web from its culture medium into an ion solution prior to drawing. The structure and material properties of bionites differ depending upon the composition of the inorganic solid. All of them resemble bacterial thread in having their bacterial filaments aligned along the fibre axis. Bionites are generally shorter and of increased diameter compared to threads produced from similar cultures.

21G(4) shows a SEM image of a crack edge cross section obtained from a bionite produced by addition of $FeCl_3$ to the growth medium. The inorganic solid present is Fe_2O_3. Mineralization throughout is evident. Bundles of cell filaments can be resolved in the bionite interior as well as on its surface. Ferric bionites are brittle structures that spontaneously crack when stored at low relative humidity (<30% rh), Mendelson (1992). **21G(5)** shows the surface of a KDP bionite obtained by suspension of a filament web in a 1 M solution of potassium dihydrophosphate (KDP). The fibre axis is aligned with that of the long hollow tube crystals. All the crystals were produced during the drying process after drawing. Their upper ends begin in the bionite interior. The open ends point downwards in the direction of fluid drainage during drying. The crystal composition (KDP) was determined by comparison of its X-ray diffraction powder pattern with a known KDP spectrum, Mendelson (1994). **21G(6)** shows the fractured surface of a bionite produced by first suspending a filament web in 1 M $CaCl_2$, drawing the structure and immediately resuspending it in a solution of 1 M KDP before final drawing. Individual cell filaments that lie along the fibre axis are coated with mineral. The composition of the inorganic solid has not yet been determined, Mendelson (unpublished).

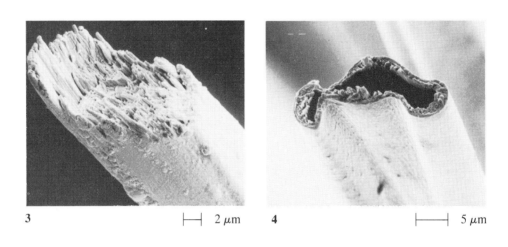

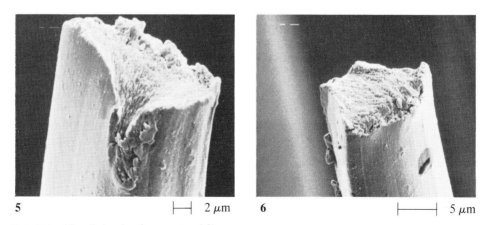

Plate 21A — Tensile breaks of some natural fibres.
　　(1), (2) Jute. (3) Flax. (4) A seed fibre from the Iranian desert. (5), (6) Cultivated silk.

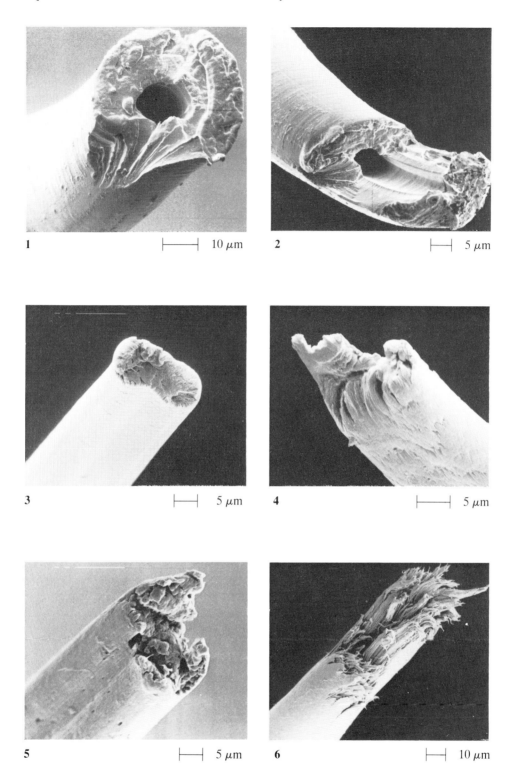

Plate 21B — Tensile breaks of some man-made fibres.
(1), (2) Hollow polyester fibres. (3), (4) Nomex meta-aramid fibre. (5) Rhovyl PVC fibre. (6) Polytetra-fluorethylene (PTFE), Teflon fibre.

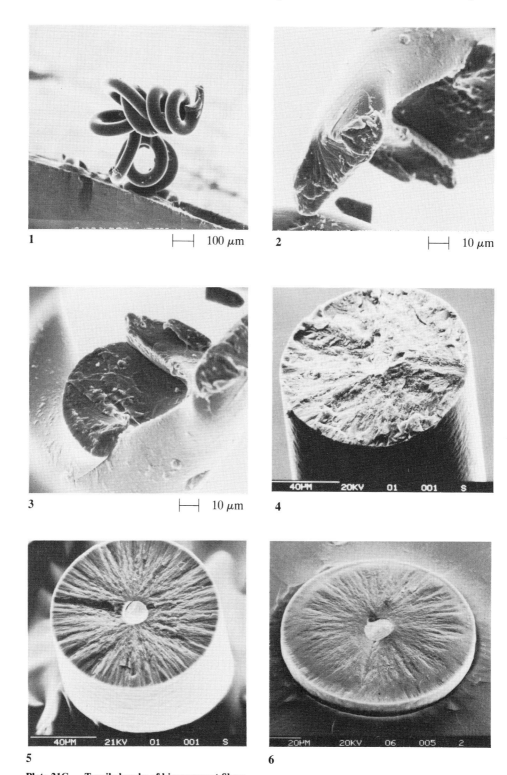

Plate 21C — Tensile breaks of bicomponent fibres.
(1)–(3) Monvelle bicomponent fibre. (4)–(6) Silicon carbide fibre (by courtesy of M. G. Bader and D. A. Clarke, University of Surrey).

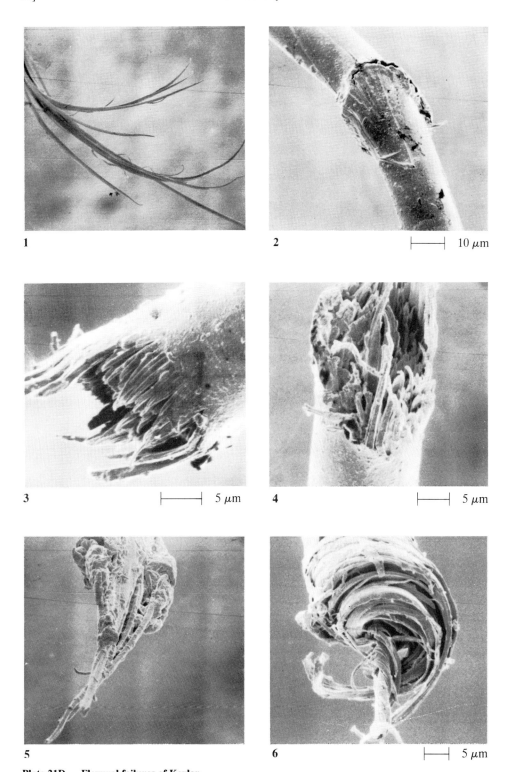

Plate 21D — Flexural failures of Kevlar.
(1) Broken in a loop tensile test around a wire with an apparent bending strain of 4.7%, with 47% loss of strength. (2) After 30×10^6 cycles of buckling between 20% and 40% contraction. (3), (4) Opposite ends of tensile break of Kevlar after repeated buckling, as in (2). (5) Flex fatigue over a rotating pin, failed after 111 000 cycles with a nominal bending strain of 2.14% and a tension of 0.5 N/tex. (6) Fatigue by rotation over a pin with a weight attached to the free end, failed after 11 000 cycles, a nominal bending strain of 7.7% and a tension of 0.04 N/tex.

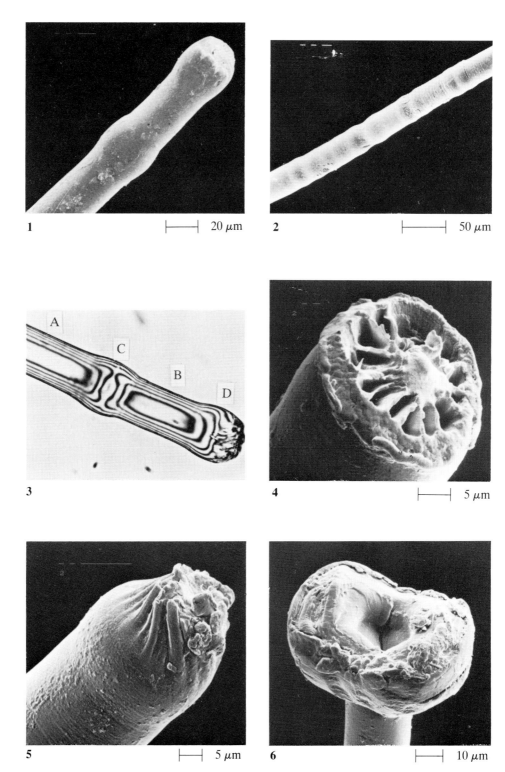

Plate 21E — 'Direct' breaks of nylon in biaxial rotation fatigue tests.
(1) Bulbous forms seen in SEM. (2) Bulbous forms appearing at intervals along a fibre. (3) Polarization fringes in optical microscopy. (4)–(6) Forms of break.

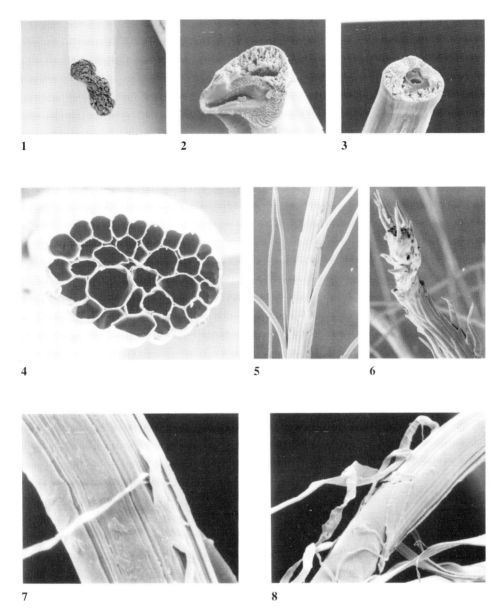

Plate 21F — New and unusual fibres.
(1)–(3) Breaks of calcium alginate fibres. (4)–(6) Thistledown. (7),(8) Fibrillation of Tencel

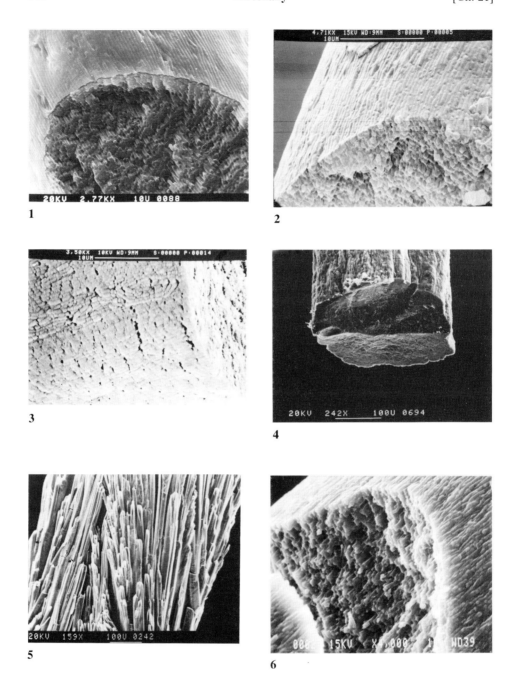

Plate 21G — Bacterial threads.
(1) Fracture of drawn bacterial thread. (2) Fracture of thread washed in water. (3) Fracture of thread attacked by lysozyme. [Bars are 10 μm.] (4) Crack edge of ferric bionite [Bar = 100 μm.] (5) Surface of KDP bionite [Bar = 50 μm.] (6) Fracture of a bionite [Bar = 1 μm.] Micrographs reproduced by permission of American Society for Microbiology, American Association for Advancement of Science and Material Research Society.

Part V
Textile processing and testing

22

INTRODUCTION

Parts II, III and IV have dealt with fibre breaks in the informative but artificial situation of laboratory studies of single fibres. In Part V we come closer to the reality of fibre utilization by industry and the consumer.

We examine first fibre ends which result from textile processing. Apart from the inherent interest in these breaks, it is important to be able to recognize them as not being a result of wear itself, when they occur in case studies of wear of textile materials. We then consider laboratory tests on textile materials: yarns, fabrics and composites. Sometimes such tests are used as easier ways of studying single-fibre properties, because the test specimens are larger, but more often they are used as a guide to performance of materials in use.

23

PROCESSED AND NATURAL FIBRE ENDS

When the raw material for the textile industry is supplied in the form of baled fibres, each bale will contain at least 10^{10} fibre ends, which will have either developed naturally or have been produced during harvesting or shearing of natural fibres or cutting of man-made fibre tow. Another way of converting tow into staple fibres is to break the filaments on a stretch-breaking machine as a first stage of textile processing. All these fibre ends will show up in textile materials.

Cutting of staple fibre gives a distorted and complicated fibre end, **23A(1)–(3)**, which may be a mixture of true cutting and pulling the fibre apart by tension. Stretch-breaking gives a tensile break, **23A(4)**, although in nylon and polyester this will have the mushroom form of a high-speed break. Cotton shows a difference between the natural tip of the fibre, **23A(5)**, and the root, **23A(6)**, in fibres taken from a bale. Cotton fibres damaged by processing may show a more ragged appearance, **23A(7)**, possibly associated with multiple splitting due to bending and twisting.

Apart from accidental and unwanted fibre breakage during textile processing, there is some deliberate formation of fibre ends. For example, the hairiness of a textile fabric may be reduced by singeing, and this leads to the bulbous polyester fibre ends in **23B(1)**. In some very open and hairy fabrics, singeing can cause problems if fibres get melted in two places and break off, as shown in the light microscope view in **23B(2)**. Other examples of singeing are in **29A(1)** and **34D(1)**.

The contrary situation is found in fastener fabrics, like Velcro, where fibre ends sticking out of the fabric have a positive role to play. The fibres are fairly thick monofilaments, and the form of the ends depends on the nature of the fibre and how it has been cut, **23B(3a,b)**. It is necessary to have a loop or a mushroom end, in order to cause the fastening to occur. The mushroom end probably results from cutting with a hot wire between two layers of a double fabric linked by the monofilaments, but the loops are formed by cutting a loop pile.

Finally there are situations in which fibres are cut in order to produce a pile fabric. Velvet can be woven or knitted as a double fabric, and then cut between the two layers by a sharp knife, to give fibre ends, **23B(4)**, similar to fibres cut with a razor. Another example in a cotton cord fabric is shown in **28F(3)**. In cut pile carpets the loops which are formed in tufting or weaving are more crudely cut, and the ends look more like blunter knife cuts, **23B(5),(6)**. Other examples are in **33F(1),(2)**.

Fibres are subject to damage as a result of the severe forces that can be imposed on them during textile processing. One example consists of the actions in the early stages of wool processing, which are described below. However the 'damage' is not always harmful; it can be at least partly beneficial if not wholly necessary. In this chapter both categories are illustrated.

Davis (1989) describes the way in which multiple carding increases the dyeability of polyester fibres. The staining technique for transmission electron microscopy, which was described in Chapter 1, provides pictures to explain the reasons for this. A low magnification view, **23C(1)**, of a trilobal polyester fibre, which had been subject to three laboratory cardings, shows three kinds of damage: 'first, the surface is highly distorted; second, a large number of cracks are observed in the fibers; third, much of the lobe area appears to absorb substantially more stain than would normally be expected.' Higher magnification, **23C(2)**, shows that 'the internal boundary structure has opened up in the vicinity of the cracks and in the dark stained regions', and therefore 'would be expected to diffuse and absorb dye rapidly, just as it does with the stain'.

The next two pictures show deliberate rupture of fibres in the stretch-breaking of a continuous filament acrylic tow into staple fibres. **23C(3)** has the characteristic granular tensile fracture of acrylic fibres, as seen, for example, in **8B(2)**, but there is splitting in **23C(4)**, which is more like **8C(2b)**.

As an example of controlled attack, **23C(5),(6)** show the effects of treating polyester fibres with caustic soda, so as to reduce their diameter and change their surface character. This can give an enhanced appearance and feel to fabrics.

In order to make paper, wood-pulp fibres must be caused to fibrillate by beating in the wet state. **23D(1),(2)**, from Hamad and Provan (1995), shows how the outer layers of the fibre are removed and inner layers break up into fibrils.

In many other processes, damage is undesirable and should be avoided as far as possible. Studies by Greenwood and Cork (1984) show the intense fibrillation which results from abrasion during the weaving of Kevlar aramid fabrics, **23D(3),(4)**. The failure of filaments causes a substantial reduction in yarn strength. There are fibrils in unwoven yarn, **23D(5),(6)**, but to a much smaller extent.

The appearance of fabrics of the new lyocell fibre, Tencel from Courtaulds, at various stages of processing is shown in **23E** and **23F**. The fibre, which consists of cellulose spun from an organic solvent, fibrillates easily. Properly controlled, this can give a very attractive hand to fabrics. As woven, **23E(1),(2)**, the fibres appear circular with hardly any fibrillation. They are even cleaner after singeing and desizing, **23E(3),(4)**. However, after steam-tumbling, **23E(5),(6)**, there is extensive fibrillation. Other stages of processing are shown in **23F(1)–(4)**. Finally, enzyme treatment, **23F(5),(6)**, enhances the appeal of the fabrics.

WOOL FIBRE RUPTURE AND MICRODAMAGE IN OPENING PROCESSES
by Nigel Johnson and Ali Akbar Gharehaghaji

Fibres are mechanically damaged during their conversion from raw fibre to finished product. Opening processes, which separate clumps of fibres into smaller clumps and even into individual fibres, can be particularly damaging, often evidenced as a significant shortening of the mean fibre length. In loose fibre opening processes, such as those found in the blowroom and carding, opening elements (e.g., pins or sawtooth wire mounted on rollers) tear through fibre clumps in order to straighten and disentangle them. Similarly, sliver opening involves pin or wire opening elements passing through more aligned assemblies of fibres; specially severe is the sliver opening necessary to create the fibre flow for open-end spinning.

Wool fibres are particularly prone to damage in opening processes, often getting broken because of their relatively low tenacity. Their crimp and long length make them difficult to separate into individual fibres, while debris from removed surface scales leads to rotor deposits in conventional rotor spinning.

A series of investigations has attempted to uncover the forms of damage that occur in opening processes. These investigations were conducted with pristine fibres to ensure that the observed damage did not occur in some earlier process. Even with unprocessed wool, the fibre tips are weathered (degraded by sunlight) and the root ends severed by the shearer s cutting blades, so experimental fibres are prepared by cutting off both ends with scissors. Good fibre uniformity can be achieved by selecting fibres from hand-fed sheep kept indoors. An original fibre end, with the weathered tip removed by a scissor cut, is shown in **23G(1)**.

Following preliminary work by Young and Johnson (1988, 1990), Yan (1991) studied the fracture morphology of ruptured fibre ends. Using the opening unit of an Investa BDA12 long staple rotor spinner, he explored the effect of opening roller speed. The BD A12 opening rollers were unusual in that they consisted of widely-spaced rows of sawtooth wire set across the roller surface, parallel to the axis. At slow speeds of both opening roller and feed roller, the wire elements travelling at 1000 m/min contact the fibres relatively gently so that no fibres were actually ruptured, but the high number of contacts caused wear. Scale edges were broken and removed and the fibre end was worn down and rounded, **23G(2)**. At a slightly higher element speed of 1100 m/min, there was still no fibre breakage, but with a much faster feed speed (giving fewer pin contacts on the fibre) there was much less wear of the fibre end, although scale lifting was still seen and fibre splitting was sometimes observed, **23G(3)**.

At substantially higher speeds of the opening element (1860 m/min) many of the fibres were broken. The new ends created by this breakage, but still held by the feed rollers, can be struck many further times by the opening elements. These usually developed severe axial splitting, **23G(4)**, or other damage, such as scale lifting, **23G(5)**. Young and Johnson (1988) had observed such fibrillation developing in a section of the fibre prior to rupture, **23G(6)**.

Following Yan's (1991) investigation of damaged fibre ends, Gharehaghaji (1994) searched for the more subtle damage that might be caused to the body of the fibre without actually breaking it. He termed this *microdamage*. He looked for such microdamage in fibres subjected to sliver opening via the opening units of short staple rotor spinners, using pin and sawtooth clothed opening rollers. While both pin and sawtooth elements caused microdamage, that due to the sawtooth was far more severe. It should also be remembered that the fibres could inflict damage on each other as they are pulled out of entanglements or tightly into knots. He found instances of the following features:

- Wearing away of the surface scales, **23H(1)**.
- Transverse cracks, **23H(1)**.
- Transverse cracks which often coincided with the tip edge of the overlapping scale, **23H(2)**.
- Longitudinal cracks, usually together with transverse cracks, **23H(3),(4)**.
- Cracks which were quite large and deep, **23H(5)**.
- Holes, often surprisingly circular and sometimes in clusters giving a honeycomb effect, **23H(6)**. These holes were more prevalent in pin-damaged samples than wire-damaged samples.
- Plastic compressional deformation where the fibre had been partly crushed by the opening element or another fibre, **23H(7)**. The visco-elastic wool fibre might be expected to behave elastically at the impact speeds likely in these experiments, but it appears that the time of contact can be sufficiently long for permanent deformation to occur. This may happen as a fibre is detached from the feed roller and carried forward by the element, or if a fibre is bent sharply around another which is then dragged forward with it.

The interactions between fibres and opening elements are very complex and highly variable in an actual opening process. To get a better understanding of the individual contact process, Gharehaghaji looped fibres around the leading edge of opening elements and pulled the fibre loop against the element at a constant rate of extension in an extensometer; strain rates from 5 to 1000% per minute were investigated. Fibres were extended to two-thirds of their average breaking extension, and immediately released. The element contact regions of those fibres which did not break were then examined. Again, the damage features were more severe when the fibre loop was pulled against the sharp corners of sawtooth wire than against the more gentle rounded shape of the pin.

Seven distinctive features were created in these experiments:

- Longitudinal cracks (axial splits), **23I(1)**, were found only with wire damaged samples. Such cracks are probably caused by the high shear stress associated with the strong variable curvature bending at the square edges of the sawtooth.
- Transverse cracks occurred at the outside of the bend in both wire and pin-damaged samples. In **23I(2)**, some fibre debris lies towards the camera; the compressed region, which contacted the sawtooth edge, is on the left; and a transverse crack can be seen to the right. Transverse cracks became more frequent at higher strain rates.
- Permanent compressive deformation was commonly caused by the sharp corners of the sawtooth wire, **23I(3)–(6)**, but only at the lower rates of strain. At the faster rates, the fibre behaved more elastically and was able to recover from the deformation. In some cases, the concentration of axial and lateral compressive stresses led to platelet buckling.
- Scale lifting, **23I(7)**, was more common at the higher rates of strain, on the outside of the bend.
- Surface wear occurred where the fibre slipped around the element, even though efforts were made to prevent this happening. The sliding action of the sawtooth edge sometimes caused a rippling effect, **23I(8)**.
- Cuts were sometimes created by the sharp edges of the sawtooth wire, **23I(8)**.
- Holes (circular voids) were also created on some of these samples, **23I(9)**.

Some forms of microdamage were also induced at places away from the region of contact with the element. The transverse cracks, including those at the tipline of overlapping scales, scale lifting, longitudinal micro-cracks and holes must have been caused purely by the axial extension.

The effects of opening elements contacting fibres can be examined in even greater detail by using a tensile stage in the scanning electron microscope. The stage allows regions of the fibre to be examined as the fibre is slowly extended, and the progressive changes in the fibre surface can be recorded on video. Using tensile stages, Hepworth *et al* (1969) and Yang *et al* (1988) slowly extended undamaged wool fibres and observed scales lifting from the fibre surface, then transverse microcracks developing, some of which grow into larger cracks. In some instances, necking of the fibre can occur prior to final rupture. The microcracks initiate at critical points, such as natural micro-voids within the fibre and the weak region at the root of overlapping scales where stress concentrates. Another characteristic feature is the craze, a special form of transverse crack which eventually grows rapidly and becomes the rupture site. White lines in the SEM record, see **23J(5)**, are indicative of crazes.

After confirming these observations for simple extension of undamaged fibres, Gharehaghaji extended fibres that had already suffered damage from opening elements. Controlled damage was first inflicted by pressing the edge of the opening elements into a fibre resting on a flat surface. He then observed the damaged region as the fibre was extended slowly in the SEM by a simple tensile stage.

Of course, this does not precisely simulate the action in an opening process, because the damage occurs by slow compression against a flat surface rather than by high speed impact in free space, and the extension is necessarily extremely slow so that changes can be observed and recorded. Nonetheless, it does simulate the mechanical behaviour of a damaged fibre under tensile load in a subsequent process. Very interesting phenomena have been observed which give insights into the forms of microdamage which can weaken the fibre and lead to eventual tensile failure.

23J(1) shows an extended fibre which was previously cut by a sawtooth element being pressed into it. As the bulk of the fibre was uniformly strained, the two faces of the cut separated, leading to a shear stress which formed a longitudinal split, initiated at the base of the cut and running in both directions. With further extension, one of these longitudinal splits developed a transverse component, so that the crack propagated at an angle across the fibre, leading to rupture and leaving a tapered ruptured end, similar to that shown in **23J(3)**. Pushing the sawtooth element deeper into the fibre induced permanent deformation, sometimes accompanied by a cut at the sharp edge, **23J(2)**. When this fibre was stretched, an angled longitudinal split developed from the base of the cut and propagated down through the deformed region, leading to rupture, **23J(3)**. Because the deformed region was so much weaker than the fibre above the cut, the longitudinal crack did not propagate into this undamaged side.

The rounded surface of a pin causes much less dramatic damage and in most cases, it is not possible to see any damage on the surface of the compressed fibre before it is stretched. As it is stretched, features similar to those seen in the stretching of undamaged fibre are apparent (lifted scales and transverse cracks), except that they arise preferentially at the damaged site. This is probably because the area is weak and so takes more of the strain energy. In **23J(4)**, a wide crack has opened, allowing a clear view of the stretched interior of the fibre, which has the appearance of an aligned fibrillar material. Some charging is evident on this exposed interior surface because it is not coated. At the left side of this crack and in line with its centre, a new craze (seen as a faint, thin white-lipped crack) can be seen developing, **23J(5)**. This craze grew rapidly and was the site of final rupture.

The effects of the sharp cutting edge of the sawtooth wire were simulated in a more controlled fashion using a sharp blade, pressed into the fibre at a 60° angle to the fibre axis. When such blade-damaged fibres were slowly stretched, a series of transverse crazes developed, **23J(6)**; one of these eventually became dominant, growing into a larger crack which finally burst rapidly through the fibre and ruptured it.

In a second series of dynamic SEM experiments, the tensile stage was used to pull loops of undamaged fibres against opening elements mounted inside the SEM, **23J(7)**. A key difference between these experiments and the previous ones is that the compressive effects are created by the fibre tension, so that the fibre tension may initiate failure before any severe crushing or compressive effects have been induced. In fact, as has already been noted by others, the majority of the fibres pulled around a round pin broke at weak places (natural flaws?) away from the pin contact area. The sharp edges of the sawtooth wire have a more

Table 23.1 — Experimental materials and threads
a) Experimental materials

Substrate	Construction	Mass, g/m^2	Fibre content, %
Knitted wool	1 × 1 knitted rib 7.0 wales/cm 10.0 courses/cm	408.2	100 wool
Knitted cotton	Knitted interlock 12.5 wales/cm 14.5 courses/cm	170.4	100 cotton
Woven wool	Woven twill Z 2/1steps 22 warp yarns/cm 22 weft yarns/cm	318.5	67.4 wool 31.8 acrylic 0.8 elastomer

b) Threads

Thread	Construction	Turns/m	Breaking load, N	Breaking extension, %	Count, tex	Diameter, mm
Mercerised cotton	3-fold yarn	636	12.6	5.61	R49/3	0.21
Spun polyester	3-fold yarn Folding Z twist singles S twist	733	11.9	18.29	R34/3	0.18
Continuous filament polyester	3-fold multi-filament	634	23.5	26.37	R50/3	0.21

dramatic effect because of the high curvature induced in the fibre, **23J(8)**. Scales lift at the outer edge of the bend, and fracture commonly initiates at this outer edge and propagates inwards. Examination of the fractured ends showed that the crack had travelled across the fibre from the high tension zone on the outer side of the bend to the inner compression zone. Once in the compression zone, the crack tended to deviate along the fibre, leaving one side of the fractured end with a tapered end. Sometimes, this crack would deviate into a longitudinal (axial) split, due to the high shear stresses induced by these high curvatures.

DAMAGE IN STITCHED SEAMS by Janet Webster

The final example of processing damage comes at the end of the garment manufacturing chain, and is based on the PhD thesis, Webster (1996).

Damage, which results from sewing and from subsequent mechanical action, was examined in the materials listed in Table 23.1. The seams of stitch type ISO-301 were made with a Singer Centurion 210B needle feed machine at 120 stitches/minute. After preparing the test pieces, the effects of subsequent wear were simulated by up to 50 000 extension and recovery cycles at a rate of 150 mm/min, applying 5000 cycles a day for 10 days in a direction parallel to the stitch line. The test pieces remained in the tensile test machine at zero load between each successive day. After completion of extension cycling, the fabrics were allowed to relax at 65% r.h. and 20°C, and were then extended to the point of first stitch break at 50 mm/min.

The first two pictures are of sewing threads in the knitted cotton fabric immediately after seaming. Polyester filaments, **23K(1)**, show surface damage, and the mercerised cotton thread, **23K(2)**, has surface splintering. The fabric can also be damaged by the action of the sewing needles: **23K(3)** shows breaks caused in a woven wool fabric after sewing with the continuous filament polyester thread. Such damage was found only in the immediate vicinity of the seam.

The next two pictures are after the extension cycling. The surface damage on the polyester filaments, **23K(4)**, from a seam in the woven wool fabric is more severe. Splintering of fibres from the spun polyester thread stitched into the knitted wool fabric is seen in **23K(5)**. The last picture, **23K(6)**, is a break of a fibre from the mercerised cotton thread after the extension to the first stitch break.

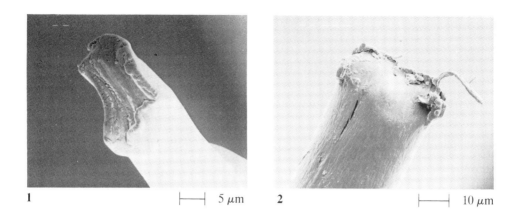

1   ⊢——⊣   5 μm   2   ⊢——⊣   10 μm

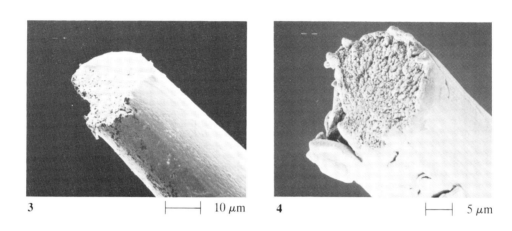

3   ⊢——⊣   10 μm   4   ⊢——⊣   5 μm

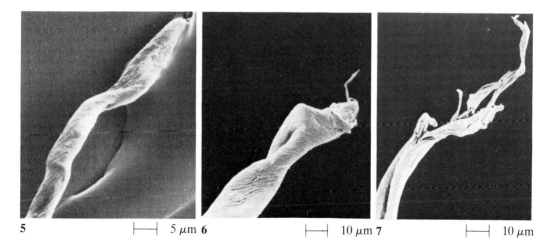

5   ⊢——⊣   5 μm   6   ⊢——⊣   10 μm   7   ⊢——⊣   10 μm

Plate 23A — Staple fibre ends.
(1) Cut polyester. (2), (3) Cut acrylic. (4) Stretch-broken acrylic. (5) Tip end of cotton. (6) Root end of cotton. (7) End of cotton fibre damaged in processing.

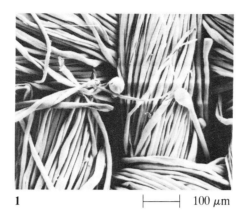

1 ├─────┤ 100 μm   2

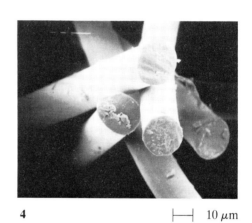

3a ├─────┤ 500 μm   3b ├───┤ 200 μm   4   ├───┤ 10 μm

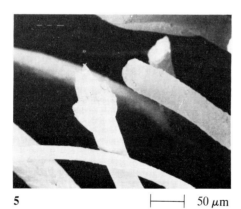

5   ├────┤ 50 μm   6   ├───┤ 10 μm

Plate 23B — Fabric singeing.
(1) Tightly woven polyester fabric, made from ring-spun yarn. (2) Optical microscope view of short
   lengths of polyester fibres, singed at both ends, from a loose fabric made from open-end spun yarn.

Fastener fabrics.
                    (3a, b) Forms of projecting ends in different types of fastener fabric.

Pile fabrics.
(4) Fibre ends in polyester velvet. (5) From cut pile carpet (nylon and wool). (6) From cut pile carpet
                    (wool), showing medulla.

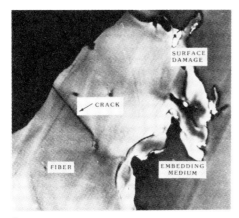

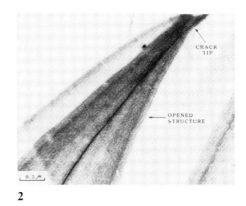

1

2

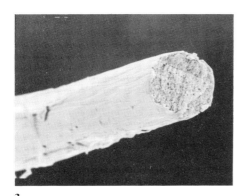

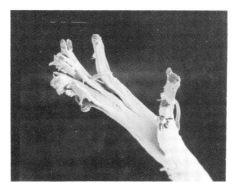

3

4

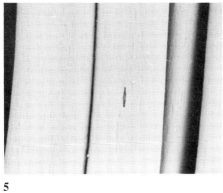

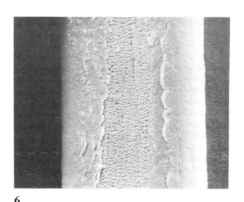

5

6

Plate 23C — Transmission electron micrographs of stained sections of polyester fibres after carding, Davis (1989).
(1) Low magnification view showing damage. (2) High magnification view showing opened structure round crack.
Acrylic fibres from stretch-broken tow.
(3) Granular fracture over single cross-section. (4) Break with multiple splitting.
Polyester fibres treated with caustic soda.
(5) Initial development of surface cracks. (6) More severe attack.

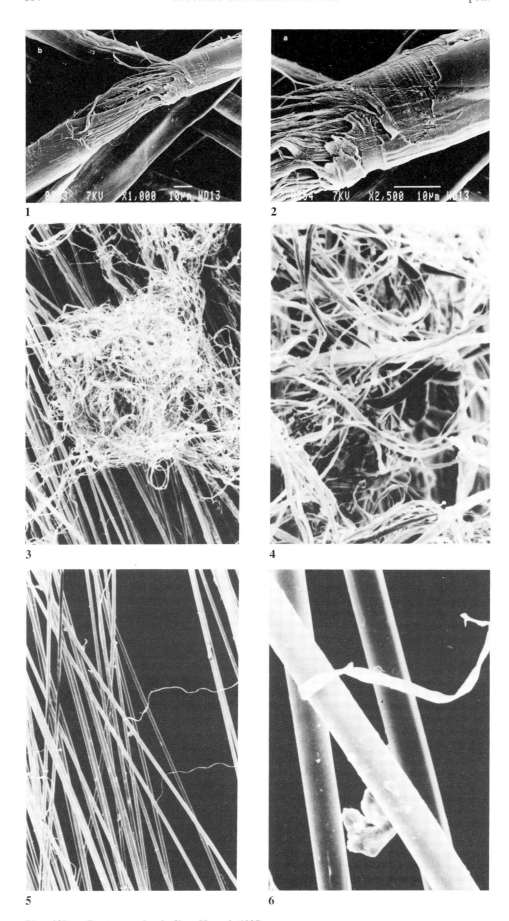

Plate 23D — Beaten wood-pulp fibre, Hamad (1995).
(1),(2) From pulp refined at 7.0 GJ/t, with partial removal of P and S_1 layers and exposure and disruption
of S_2 layer (courtesy of A. Karnis, Pulp and Paper research Institute of Canada).
Weaving damage in Kevlar aramid fabric, Greenwood and Cork (1984).
(3),(4) Fibrillation in woven fabric. (5),(6) Unwoven yarn.

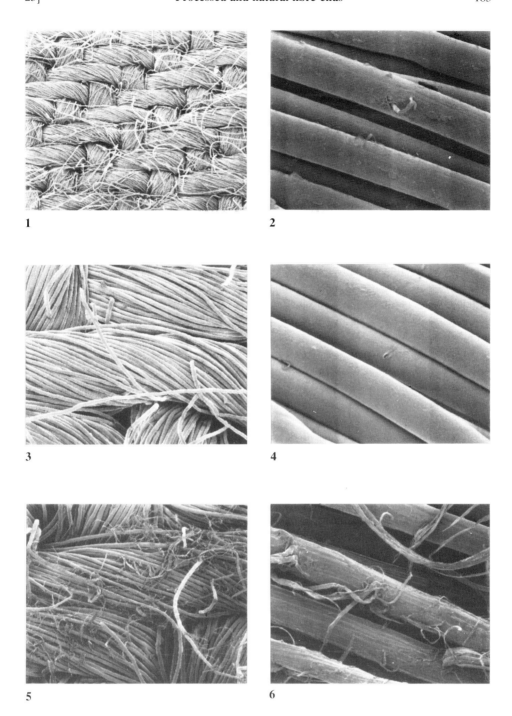

Plate 23E — Courtaulds Tencel fibres in fabrics at various stages of processing, courtesy of T.R. Burrow, Courtaulds Fibres Tencel.
          (1),(2) Loomstate fabric. (3),(4) Singed and desized. (5),(6) Steam tumbled.

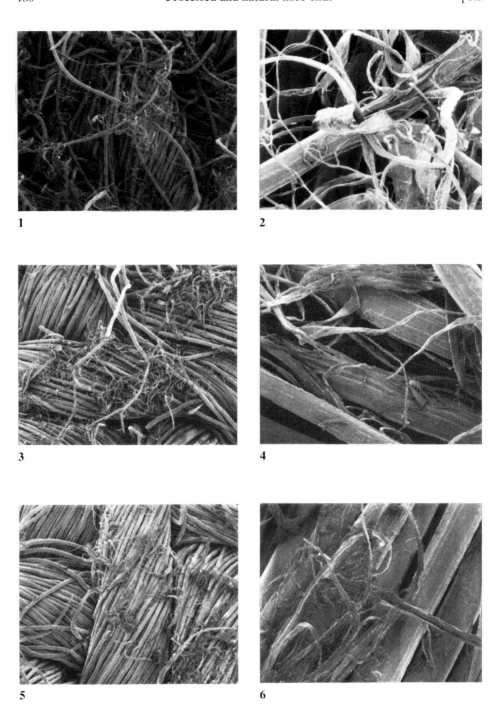

Plate 23F — Courtaulds Tencel fibres in fabrics at various stages of processing, courtesy of T.R. Burrow, Courtaulds Fibres Tencel (continued).
(1),(2) Prepared for beam dye. (3),(4) Set. (5),(6) Enzyme treated.

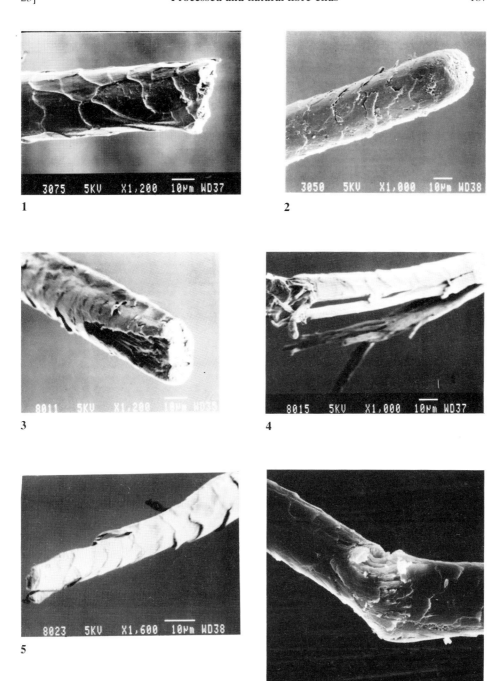

Plate 23G — Studies of opening of wool.
(1) A cut end for experimental studies. (2) Rounded end due to wear at 1000 m/min. (3) At higher speed.
(4),(5) At 1860 m/min. (6) Incipient fibrillation.

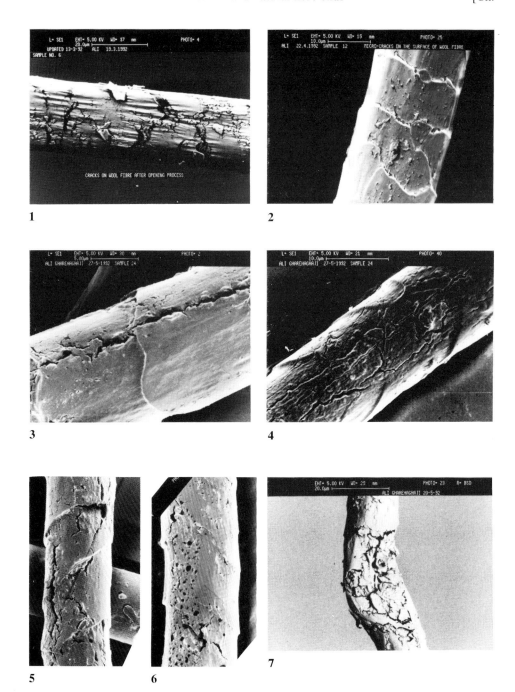

Plate 23H — Studies of opening of wool (continued).
(1) Surface wear and transverse cracks. (2) Transverse cracks at edges of scales. (3) Longitudinal cracks.
(4) Joining of transverse and longitudinal cracks. (5) Deep cracks. (6) Clusters of holes. (7) Plastic
deformation.

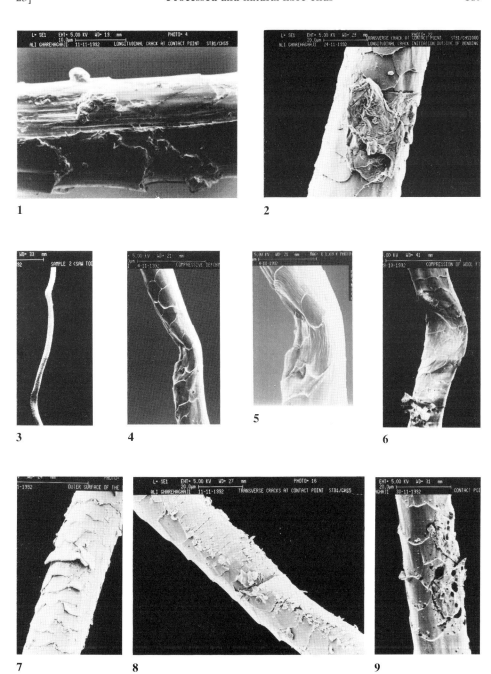

Plate 23I — Wool pulled against opening elements in a tensile tester.
(1) Axial splits, against wire. (2) Transverse cracks. (3)–(6) Plastic deformation. (7) Scale lifting. (8)
Rippling and a cut. (9) Holes on fibre surface.

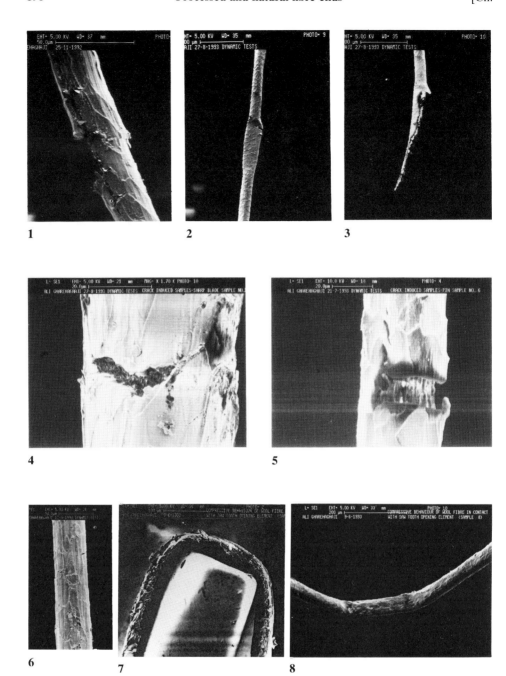

Plate 23J — Fibres extended in SEM after opening damage.
(1)–(3) Extension of a fibres that had been previously pressed against a sawtooth element. (4),(5) Crack and crazes. (6) Transverse crazes. (7) Wool fibre pulled against a sawtooth element in a live test in SEM. (8) Resulting plastic deformation.

Plate 23K — Damage from sewing and subsequent cycling, Webster (1996).
(1) Continuous filament polyester sewing thread following seaming of knitted cotton fabric. (2) Mercerised cotton sewing thread following seaming. (3) Fibres from woven wool fabric in the vicinity of a seam. (4) Continuous filament polyester thread, sewn into woven wool fabric, after 50 000 cycles between 20 and 25% extension. (5) Spun polyester thread. Sewn into knitted wool fabric, after 50 000 cycles from 15 to 20% extension. (6) Mercerised cotton thread, sewn into knitted cotton fabric, after cycling from 10 to 15% extension and then extended to break.

24

YARN TESTING

When drawn continuous-filament nylon or polyester yarn is subject to a tensile test in the form in which it is supplied to the industry, with only slight twist or interlacing, each filament breaks independently, and the fractured ends are the usual ductile breaks, with a V-notch and a catastrophic region, **24A(1), (2)**.

If the yarn is twisted, even to a fairly low level, transverse forces develop when the yarn is put under tension: these change the stress in the fibres, but also cause the filaments to act together. As soon as one fibre breaks, this triggers the break of the whole yarn. Some fibres, probably the first to break, show ordinary tensile breaks, **24A(3)**; in others the ductile form is recognizable but very distorted, **24A(4)**. But many fibres break with the appearance of high-speed tensile breaks, **24A(5),(6)**, as a result of the rapid transfer of load in the final stages of break.

The appearance of break of a highly twisted yarn is shown in **24B(1)**. The individual filament breaks now include some which can be regarded as ductile tensile breaks, albeit highly distorted, **24B(2)**, or as simple high-speed breaks, **24B(3)**. But it is also apparent that the large forces, which occur in twisting, have caused surface damage to the fibres, **24B(4)**, and that this has complicated the forms of break, **24B(5),(6)**, which include both mangling due to transverse pressure and splitting.

The presence of high-speed breaks among the filaments of a broken yarn, even though the extension was carried out slowly, is an important warning that the appearance of broken fibre ends from a fibre assembly can be misleading in regard to the real cause of failure. In an actual product there may be hundreds, or thousands, or even millions of fibre ends across the break; and the majority of these will have broken as a result of picking up the load in the final stages of break, which is of little interest since the damage has already been done. What is important is what starts the failure.

Another study of nylon yarn breaks has been reported by Ogata, Dougasaki and Yoshida (1979).

Yarns have also been tested in fatigue. In an attempt to overcome the difficulties of carrying out biaxial rotation fatigue studies on single cotton fibres, mentioned in Chapter 18, a study was made of testing yarns by biaxial rotation over a pin. The yarn structure does complicate the response, but the method is a useful one. When tested in air at 65% r.h., 20°C, the broken yarn ends thin down, but there are significant differences between ring-spun yarn, **24C(1)**, and open-end (rotor) spun yarn, **24C(4)**, with wrapped fibres being prominent in the latter yarn. The cotton fibres remain separate and show the usual biaxial rotation fatigue by multiple splitting, **24C(2),(3),(5)**. When tested in water the fibre structure tends to be smeared out, with sheets of material pulling away and sticking the fibres together, **24C(6)**. A similar effect is found after prolonged washing of cotton fabrics, as shown by examples in Chapters 32 and 34.

Another yarn test, illustrated in Fig. 10.6, was introduced as a means of evaluating resistance to surface sliding in continuous-filament yarns used to make ropes. Two portions of yarn twisted together are pulled backwards and forwards, so that they suffer abrasion against one another. It was found that there were two failure modes depending on whether the test conditions were mild, leading to a long life in the test, or severe, leading to a very short life. In 'slow' abrasion, for example with failure after 35 000 cycles, fibres break with considerable splitting, **24D(2)**, and get pushed back along the yarn to expose more fibres to abrasion, **24D(1)**. In 'rapid' abrasion, for example with failure in two cycles, there is a more localized and immediate failure, **24D(3)**, with squashed fibre ends and considerable complication of snap-back effects following breakage, **24D(4)**.

In conditions which are not as extreme as in the previous examples, in terms of giving very high or very low numbers of cycles to failure, individual fibres in yarns which have not been taken to failure show extensive multiple splitting in mild conditions, **24E(1)**, but more squashing of the fibre under severe conditions, **24E(2)**.

The yarn-on-yarn abrasion test can be adapted so that the yarns are immersed in liquid, or preliminary treatments can be applied before testing. The most rapid breakage occurred with yarns which had been soaked in salt solution, or sea water, and then dried. The yarn damage is severe, **24E(3)**; and salt crystals, which are visible on the fibre surface, **24E(4),(5)**, presumably act as an abradant, gouging the fibres until they break, **24E(6)**. This is an important observation in regard to use of ropes and sails since although there is some difference, particularly in nylon, between fatigue resistance in wet and dry conditions, the really damaging situation is wetting and drying in a marine environment.

Detailed examination of fibre damage in yarn-on-yarn abrasion tests, except under very severe conditions, shows that the breakdown of nylon filaments mostly starts as peeling of strips from the fibre surface, **24F(1),(2)**, and develops into multiple splitting, **24F(3)**, as a result of the shear stresses. Similar effects are found in polyester filaments, **24F(4)–(6)**.

Under high tension the inter-fibre pressures can severely deform fibres, **24G(1)**, and this must be a factor in fibre breakage. However, even under relatively mild conditions, the fibres can be broken more by squashing than by splitting, **24G(2)**, and may be fused together, **24G(3)**. Possibly these effects occur in the later stages of the test, when the fewer filaments will be under higher tension. The occurrence of melting, **24G(4)**, and corrugating, **24G(5)**, is found in severe conditions, whether due to salt crystals or high tension. Finally, mushroom ends are found, **24G(6)**, but these probably form either when the load is rapidly taken up in the last cycle, which breaks the yarn, or are due to snagging of individual filaments. Some of the observed forms in **24G(3),(5),(6)** are probably a consequence of snap-back after break.

Many filaments are involved in yarn-on-yarn tests, and they are deformed, damaged and break in different ways at different stages of the test as the loading conditions on the fibres change. Then broken fibres tangle up, disturb the yarn structure, change the inter-yarn forces, and may themselves be further damaged. Thus the examination of broken yarns shows up a great complexity of forms, of which only a small sample has been illustrated here. The great distortions of the severe tests which lead to breakage in a very few cycles are perhaps of more academic than practical interest. But the surface peeling and multiple splitting, shown in **24F** and found in the milder conditions, is similar to that found in ropes, as shown in Chapter 39. The yarn-on-yarn abrasion test, carried out under the right conditions, is thus both a useful practical way of evaluating rope yarns, and a way of carrying out basic research on the surface shear and peeling mode of fibre failure, discussed in Chapter 14 and identified, rather loosely, as type 13 in Fig. 1.5.

Although yarns are an intermediate form, subsequently woven or knitted into fabric, there is some direct use of one-dimensional textile structures as braids, cords and ropes. A polyester braid was progressively load-cycled on an Instron strength tester from a base loading of 3% of its breaking load, with maximum loading increasing each cycle from 10%, 30%, 50% to 70% breaking load and then finally to break, which is shown in **24H(1)**. The break is complicated, with severe effects of snap-back, **24H(2)**. Most fibres have broken as high-speed breaks, **24H(3a)**, but some show other forms, **24H(3b)**, which are similar to tensile fatigue breaks.

If the braid is cycled for 1 hour between 3% and 70% of its breaking load prior to break, the broken filaments show evidence of melting, **24H(4)**, and there are interesting changes in fibre surfaces, **24H(5a),(5b)**. If the braid is cycled up to 90% of its breaking load, when it may fail after a short time, the surface damage is even more marked, **24H(6)**.

A nylon braid, fatigued to failure in water, is shown in **24I(1)**. There is evidence of surface peeling, **24H(2),(3)**, which would be due to surface rubbing, but other breaks, **24I(4),(5)** look more like tensile fatigue failure (see Chapter 11).

In **24D–G**, there are examples of yarn-on-yarn abrasion testing in continuous filament nylon and polyester yarns. Further tests have been carried out in which spun cotton yarns were subjected to yarn-on-yarn abrasion. When tested dry, **24J(1)**, the broken ends of fibres splay out from the yarn, and individual fibre ends show smearing wear with some fibrillation, **24J(2)**, or more extensive fibrillation, **24J(3)**. In the wet state, the intense smearing wear has caused the fibres to stick together in a mass, **24J(4)**. Individual fibres show some fibrillation, **24J(5)**, or complicated twisted forms, **24J(6)**.

The damage has similarities to that found in fatigue of cotton yarns by biaxial rotation over a pin, as shown in **24C**.

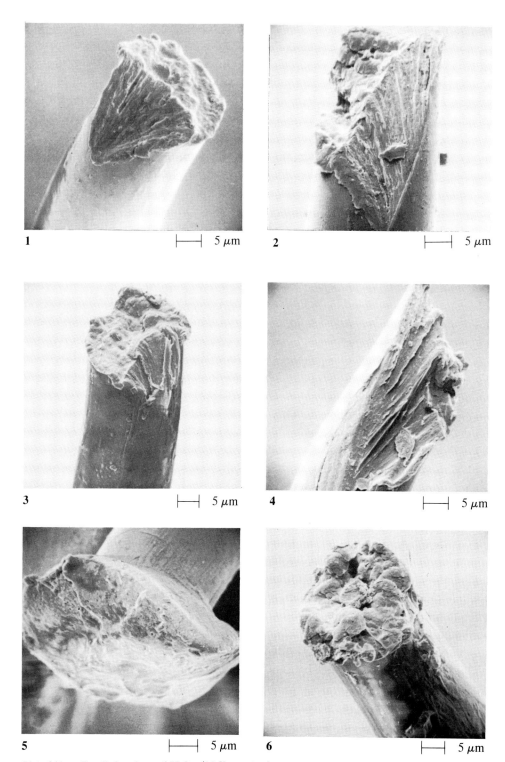

Plate 24A — Tensile breakage of 77 dtex/16-filament nylon.
(1), (2) Filaments from yarn tested as supplied with little twist. (3)–(6) Filaments from yarn tested after insertion of 10 turns/cm.

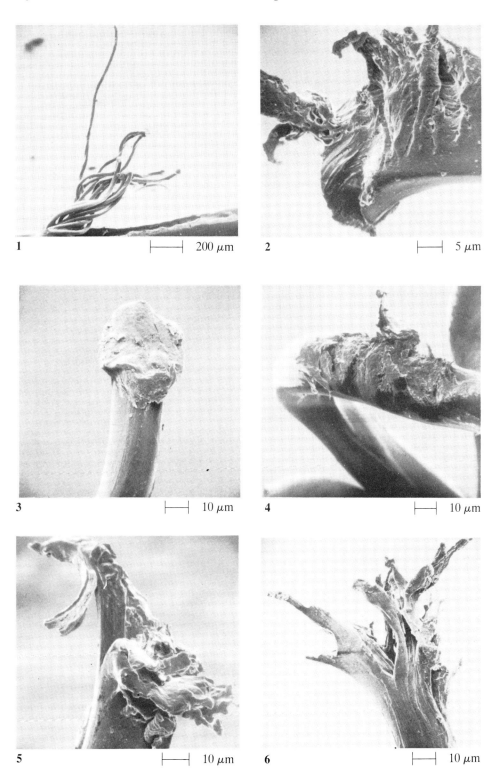

Plate24B — Tensile breakage of 77 dtex/16-filament nylon yarn, twisted to 100 turns/cm.
(1) Broken yarn. (2)–(6) Individual filaments in broken ends.

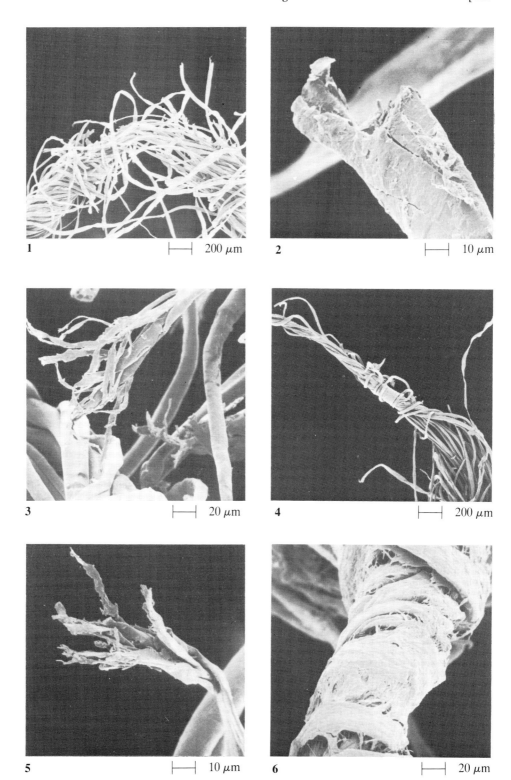

Plate 24C — Fatigue of cotton yarns by biaxial rotation over a pin.
(1) Ring-spun untreated yarn, tested in air, after 4628 cycles. (2) Broken fibre from the yarn (1). (3) Broken fibre in mercerized ring-spun yarn, tested in air, after 4188 cycles. (4) Open-end spun untreated cotton yarn, tested in air, after 5311 cycles. (5) Broken fibre from the yarn (4). (6) Open-end spun untreated cotton yarn, tested in water, after 885 cycles.

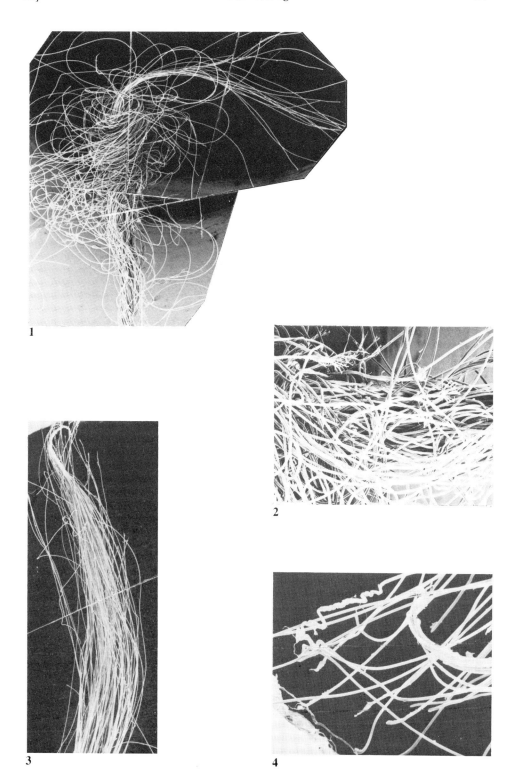

Plate 24D — Yarn-on-yarn abrasion testing, 1100 dtex industrial filament yarns, three wraps at wrap angle of 35°, 50 mm stroke, 52 cycles/min, in air at 65%, 20°C.
(1), (2) Nylon, with tension weight of 200 g, failed at 35 000 cycles. (3), (4) Polyester, with tension weight of 800 g, failed at 2 cycles.

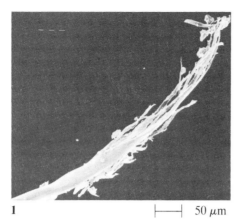

1   ⊢——⊣ 50 μm

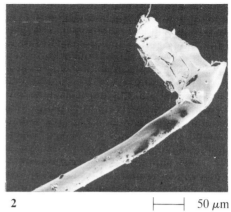

2   ⊢——⊣ 50 μm

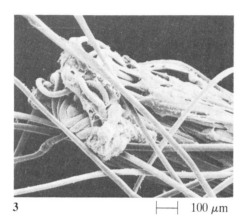

3   ⊢——⊣ 100 μm

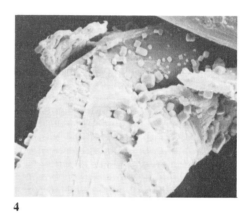

4

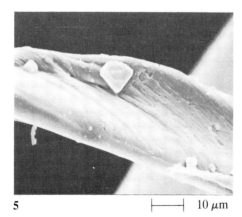

5   ⊢——⊣ 10 μm

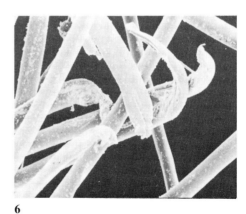

6

Plate 24E — Yarn-on-yarn abrasion testing: test conditions as in 24D.
(1) Nylon fibre, from mild test, 500 g weight, after 3000 cycles, before yarn failure. (2) Nylon fibre, from severe test, 800 g weight, after 50 cycles, before yarn failure.

Yarn-on-yarn abrasion of yarns soaked in sodium chloride solution and then dried: other test conditions as in 24D.
(3), (4) Polyester, with tension weight of 400 g, failed in 41 cycles. (5) Nylon, with tension weight of 55 g, failed in 169 cycles. (6) Nylon, with tension weight of 500 g, failed in 93 cycles.

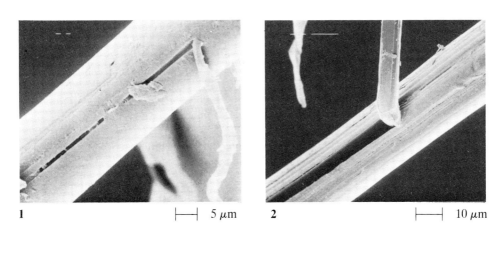

1   ⊢─┤ 5 μm   2   ⊢───┤ 10 μm

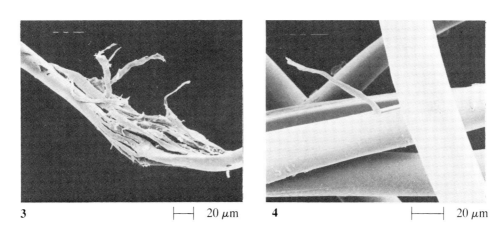

3   ⊢──┤ 20 μm   4   ⊢──┤ 20 μm

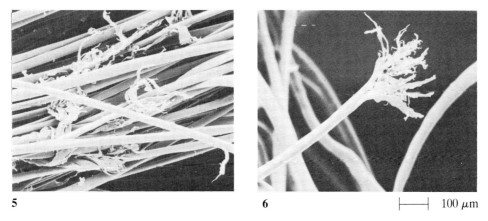

5   6   ⊢───┤ 100 μm

Plate 24F — Yarn-on-yarn abrasion testing in mild conditions: test details as in 24D (except as stated).
(1), (2) Nylon, with tension weight of 500 g, failed after 2600 cycles. (3) Nylon, with tension weight of 500 g, failed after 17000 cycles. (4) Polyester, with tension weight of 400 g, failed after 1600 cycles. (5) Polyester, with tension weight of 400 g, failed after 4900 cycles. (6) Polyester, with tension weight of 500 g, but lower wrap angle of 25°, failed after 13000 cycles.

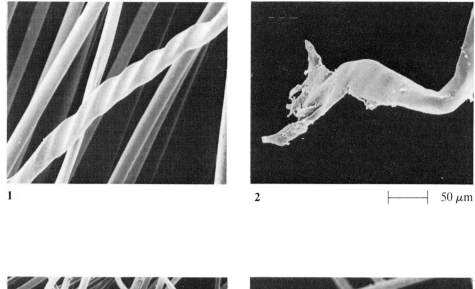

1

2   ⊢————⊣ 50 µm

3   ⊢——⊣ 100 µm   4   ⊢—⊣ 50 µm

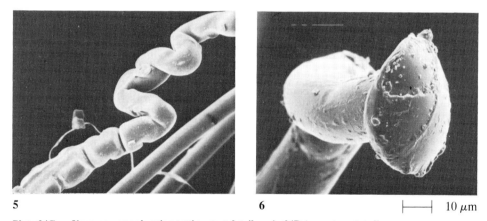

5   6   ⊢——⊣ 10 µm

Plate 24G — Yarn-on-yarn abrasion testing: test details as in 24D (except as stated).
(1) Polyester in water, with tension weight of 925 g, failed after 13 cycles. (2), (3) Polyester, with tension weight of 400 g, failed after 4800 cycles. (4) Polyester, dried from salt solution, with tension weight of 400 g, failed after 24 cycles. (5) Polyester, with tension weight of 800 g, failed after 2 cycles. (6) Polyester, dried from synthetic sea-water, with tension weight of 400 g, failed after 8 cycles.

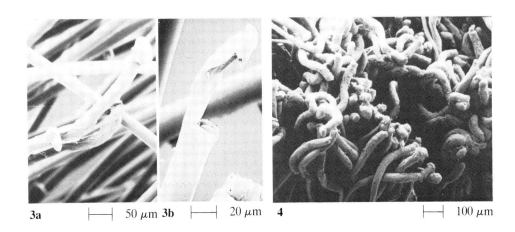

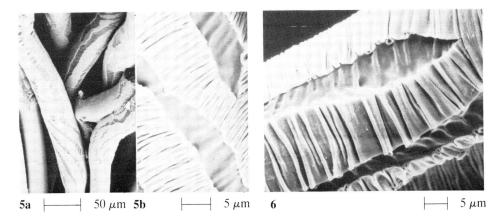

Plate 24H — Cyclic strength testing of polyester braid.
(1) Broken end of braid. (2) Coiled filament, snap-back. (3a, b) Filament breaks. (4) Evidence of melting in break. (5a, b) Core softening, melting, splits and wrinkles. (6) Severe core melting and skin contraction.

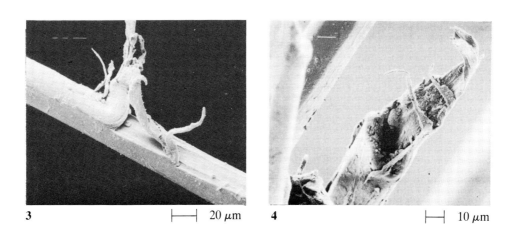

Plate 24(I) — Nylon braid fatigued tested in water, failed after 1374 load cycles.
(1) Broken braid. (2), (3) Surface peeling of filaments. (4) Broken filament with tail. (5) Groove in broken filament.

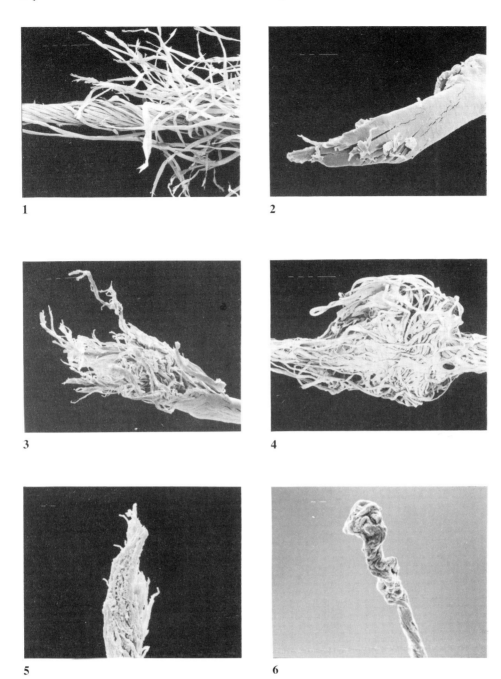

Plate 24J — Cotton yarn-on-yarn abrasion.
(1)–(3) Tested dry. (4)–(6) Tested wet.

25

FABRIC TESTING

Until the early years of this century textiles were designed and evaluated on the basis of experience in use. But, once a more technical approach was established, one of the first types of laboratory test to be tried was fabric abrasion as a guide to wear in use. Often, in order to speed up the test, very severe conditions were employed. The correlations were rarely good, and then only for very limited changes in fabric specification, and were often badly misleading. Over the last 75 years many different wear-test devices have been suggested and used, but it remains difficult to get reliable and instructive information.

Detailed examination of the way in which damage occurs in abrasion tests provides valuable additional information. If the mode of breakdown is quite different to that found in real wear in use, then the test is not likely to be of much significance. If the forms of breakdown are similar, there is a chance that the information is meaningful. Apart from providing evidence on whether the lifetimes found in laboratory tests are likely to be a good guide to lifetimes in use, the detailed studies can help to elucidate the mechanisms of failure, and so lead to design improvements.

In studies using a WIRA abrasion tester with a standard worsted fabric abradant, a cotton fabric wore away at the yarn crowns, leaving fibre ends sticking up at the interstices of the weave, 25A(1), in a way similar to that found in use. The fibre breakdown of the cotton fibres in fabrics treated in three different ways — desize and scour, bleach and mercerize, dye and resin-treat — all show failure by multiple splitting, 25A(2)–(4), as found in single-fibre and yarn biaxial rotation fatigue and flex fatigue tests (Chapters 12, 13 and 24) and commonly in use. The break is somewhat sharper in the embrittled resin-treated material. Paper, whether rough or smooth, is a more severe abradant: it has a smearing effect on the fibres in the yarn crowns, 25A(5), and results in more rounded fibre ends, 25A(6), produced from the multiple split ends.

One of the problems of abrasion testing is that the abradant, as well as the test sample, suffers damage. The rough paper, 25B(1), becomes considerably worn away and smoothed after comparatively few rubs, 25B(2). The worsted fabric, 25B(3), also begins to break down, with fibre breaks by multiple splitting, 25B(4). The scales on the surface of the wool fibres, 25B(5), are worn away, 25B(6).

The many different types of laboratory abrasion tester operate in different ways, and it is beyond the scope of this book to describe them. But examples of the forms of damage are appropriate. See 25C, which illustrates the effect of the Boss abrasion tester, 25C(1)–(3), the Martindale tester, 25C(4), the Stoll blade, 25C(5), and the Stoll bar test, 25C(6). All of these methods involve fairly high pressure on the fabric surface.

A recent addition to the KESF fabric testers, developed by Professor Kawabata at Kyoto University to evaluate fabric hand, is a shear fatigue tester, which involves no pressure on the surface. The fabric is repeatedly sheared in either direction, and thus breaks down as a result of internal action rather than an external abradant. A wool fibre in a fabric shows a multiple splitting in a portion from which the scales have broken away, 25D(1). However, comparatively little damage has been done, even after 100 000 cycles. The test can be speeded up by shearing the fabric wet with carborundum powder as an internal abradant: this leads to breakage of many fibres, 25D(2), but most of the damage is by direct abrasion on the fibre surface, 25D(3), although some fibres show multiple splitting fatigue, 25D(4). This form of severe damage would only be a realistic guide to the behaviour of a fabric if it was to be used in circumstances when it became contaminated with grit. Polyester fabrics show fibres with both multiple splitting, 25D(5), and surface peeling, 25D(6), presumably caused by surface shear.

There is an increasing use of rental companies to supply workwear on contract. These companies need to evaluate materials carefully and to monitor their use. This is a source of valuable, well-documented samples for examination. Observations on material in use are included in Chapters 32 and 34, but the laboratory testing of fabrics was the source of the samples now described.

Fabrics abraded in the Accelerotor abrasion tester are shown in **25E(1)–(3)**. The polyester/cotton fabric tested dry shows heavy abrasion damage, with peeling and shredding of the cotton fibres, **25E(1)**, but shows little change from a control fabric after a test wet. Similarly a polyester/modal (rayon) fabric shows severe abrasion in the dry testing, **25E(2)**, but only slight abrasion and a few breaks of the rayon in the wet test, **25E(3)**.

The effect of the Martindale tester is somewhat different, as seen in the wear of a polyester/cotton fabric intended for protective clothing, **25E(4)–(6)**. There are many long bushy-ended polyester fibres on the surface, and broken cotton and polyester fibres in the crevices of the weave. The projecting polyester fibres can become tangled into pills on the fabric surface, which are held in place by a few anchor fibres (see Chapter 30). The multiple splitting of the polyester fibres can extend a long way back along the fibre, **25E(6)**.

In most contracts, wear resistance is only needed so that the garment does not become too worn, or weak, or unsightly to be usable, and repair of holes may be acceptable. But in the supply of all-over protective clothing for use in clean rooms, it is essential not only that the fabric shall prevent contamination coming from the wearer's clothing or person, but also that the fabric itself should not shed any fragments. A very high degree of wear resistance is therefore needed, and the fabrics must be thoroughly tested to make the right choice.

The results of abrasion of continuous-filament fabrics intended for clean-room garments are shown in **25F,G**, tested against a standard wool fabric by the rental company, and in a more severe test, when fabrics were abraded against themselves on a Martindale tester at UMIST for long enough (177 000 cycles) to cause damage.

A polyester (Dacron) taffeta, which has been surface-modified by calendering, showed almost no damage against wool, **25F(1a)**; but the self-fabric test causes light abrasion of the surfaces of some warp and weft yarns, **25F(1b)**, with splitting and peeling of filaments. The surface coating is broken in some places, and debris is trapped in the fabric. Near the edge of the test sample there was a line of severe damage, **25F(2)**, which may be at a crease in the fabric. Many of the broken filaments show typical multiple-split ends, **25F(3)**, but there is also some peeling, **25F(4)**, and the rounding of ends, **25F(5)**, which is a result of further wear after breakage.

Another polyester (Dacron) fabric in a herringbone weave had suffered slightly more damage in self-abrasion. Fibre breakdown appears to have started with surface peeling, **25F(6)**, with the fragments then piling up in the interstices of the weave. A few long fibre ends with bushy tips are present on the fabric surface.

In a nylon 6 (Celon) fabric, a few filaments have started to split even in the less severe abrasion against wool, and there was more pronounced splitting in the self-abrasion, **25G(1)**. There was an area of severe localized damage, **25G(2)**, probably at the edge of a crease, with multiple split fibre breaks sticking out.

In another surface-modified polyester (Terylene) fabric, the wear was worse. Against wool, filaments in the warp had broken, leaving the ends projecting from the fabric crevices, usually short but sometimes long, **25G(3a),(3b)**. In self-abrasion there is more breaking in the warp yarns, **25G(4)**, and the material in the crevices has become compacted together, **25G(5)**.

There was also severe damage in an uncalendered polyester fabric, with many broken ends in the abrasion against wool, and in self-abrasion, **25G(6)**. The fibre breakage is by multiple splitting. In this material the crimp in some filaments gives a looser packing, which may be beneficial for comfort but makes abrasion damage easier.

A situation in which it is justifiable to use a severe abrasion test is in the evaluation of webbings used in harnesses, rucksacks and similar situations where the webbing may be abraded by metal guides and buckles. An extensive set of tests was carried out by RAE, and samples became available for examination. The webbings were of nylon in a twill weave, with one type woven on a conventional loom with two conventional selvedges, and one on a shuttleless loom with one conventional selvedge and one tucked-in selvedge. These different selvedges influenced the abrasion resistance of the edges of the webbing. After 10 000 cycles of edge-abrasion, on a Hexbar tester, the conventional webbing had lost 40% of its original strength, whereas the shuttleless variant had lost only 2%.

Wear on the surface of the webbings appears similar in the two webbings, and in shown for conventional webbing in **25H(1)** in a zone where the interaction of yarn and fabric geometry is such that the filaments lie parallel to the length of the webbing. At higher magnification it can be seen that there is considerable flattening and smearing on the yarn crowns, **25H(2)**, which is a result of material being peeled off the surface of the fibres, **25H(3)**. In other locations the interaction of yarn twist and weave is such that the filaments are at an angle to the length of the fabric. This alters the topography of the webbing surface and leads to some difference in the nature of the wear, **25H(4)**.

Part of the selvedge of the conventional webbing showed considerable abrasion, with filaments twisted into loops and abraded, **25H(5)**. Damage to the filaments involved surface peeling, **25H(6)**. The shuttleless selvedge remained intact, even though it had suffered wear, **25I(1),(2)**. Damage to filaments depended on their location in the structure. The locking yarns of the weft showed failure by multiple splitting, **25I(3)**, but on the warp crowns scraping and peeling of filaments was the dominant effect, **25I(4)**, and filaments were broken by flattening and shearing, **25I(5)**. In one region of very severe damage to the variant webbing, many fibres had been broken, **25I(6)**.

Another form of fabric testing is for flammability, or, more generally, the effects of heat on textile materials. SEM studies have been reported by Goynes and Trask (1985, 1987) for cotton, polyester and wool fabrics, including blends, with and without flame-retardant treatments, subjected to 45° edge ignition tests. A typical test specimen is shown in **25J(1)**, with an area burnt away at the bottom, and the residual piece consisting of a completely charred area, surrounded by an unburnt area, with an intermediate zone between these two, where heat will have had some effect on the fibres.

A comparison of unburnt and burnt untreated cotton twill is shown in **25J(2)**. There is shrinkage on burning, and the wispy, fragile charred material shows severe distortion of the cotton fibres, **25J(3)**. When the cotton fabric had been treated with a THPS finish, containing bis[tetrakis(hydroxymethyl)phosphonium] sulphate, urea and trimethylolmelamine, in a way which distributed the flame retardant throughout the fibres, the burnt region was black and brittle, but the fabric structure was little changed, **25J(4a)**. The cotton fibres retained their external shape, **25J(4b)**, although cross-sectional views showed that they had become thin-walled with enlarged lumens in the centres.

After burning an untreated 50/50 cotton/polyester fabric, there was little shrinkage, the char was less fragile than for 100% cotton, and, at low magnification, the appearance was similar to the treated cotton in **25J(4a)**. At higher magnification the mixture of fibres can be seen in the unburnt fabric, **25J(5a)**, but, in the burnt fabric, the cotton is coated with fused polyester, **25J(5b)**. Some distance away from the charred area the polyester fibres have started to melt; and there is a gradual progression from fused ends, similar to those shown in **20D(1)–(3)**, through larger regions of melting, seen in **25J(6a)**, to areas of complete embedding of the cotton fibres in the polyester melt, shown by the yarn cross-section in **25J(6b)**. The chars produced in burnt THPS-treated cotton/polyester fabric were not much different from those of untreated material.

In 100% polyester or wool fabrics the burnt material was a fused mass with no retention of fabric structure. However, untreated 60/40 cotton/wool fabric gave charred regions, similar to the cotton/polyester fabric, although not as dense. In the unburnt region there is loss of scale and ballooning of the wool fibres, **25K(1)**, and in the hotter regions there is melting of the wool and coating of the cotton fibres, **25K(2)**. The visual effects in the THPS-treated cotton/wool fabrics were similar.

Study of a tri-blend fabric of 60/25/15 cotton/polyester/wool enabled the sensitivity of the different fibres to heat to be shown up. In the untreated material the first indication of damage well above the charred area consisted of melting of polyester fibre ends, **25K(3a)**. Closer in, the molten polyester formed droplets and flowed over the other fibres, **25K(3b)**. Still nearer to the heat, the wool fibres began to swell, lost the scale structure and ruptured, **25K(4)**.

The heated, but not charred, region of THPS-treated tri-blend fabric was similar in appearance to the untreated fabric. Detail of the damage to the wool fibres is shown in **25K(5a),(5b)**. There are some differences in the charred remains, **25K(6a),(6b)**, but both untreated and treated fabrics have the charred cotton fibres embedded in the melted polyester/wool residue.

A number of general studies of fabric testing have been included in this chapter. Some more specific examples, which relate to particular products, such as rental textiles (Chapters 32 and 34), carpets (Chapter 33), seat belts (Chapter 37) and ropes (Chapter 39), are better brought into the accounts of case studies in use.

A series of fabric tensile tests were carried out by Seo *et al* (1993) in order to determine how failures differed according to the spinning technology used to make the yarns. **25L(1)** shows breaks starting in isolated places in a twill fabric before the rupture of the whole specimen. The fabrics had been piece dyed and the local failure exposed some relatively undyed material, so displaying the break as a light streak. Yarn breaks were of two types. In **25L(2)**, the break occurs sharply over a short yarn length. This reflects *extremely local load sharing of constituent fibres facilitated by high lateral pressures*, and the breaks usually occurred at bent configurations in crossovers. High lateral pressure is indicated by the deformed fibre ends in a yarn break, **25L(3)**. In some yarns, there were several bunches of fibre ends at intervals along the yarn, which are the sites of partial yarn breaks. The other type of break, **25L(4)**, has individual fibres breaking at many places over a considerable yarn length. The final separation occurs when the fibre lengths have become so short that they are no longer gripped and the fibre ends slip over one another.

Seo *et al*, who point out differences between ring, rotor and air-jet yarns, between plain and twill fabrics, and between warp and fill (weft) direction include among their conclusions:

From observations of yarn failure in uniaxially tensioned fabrics, we see that in tests of fabrics in a displacement controlled test, there are numerous isolated yarn failures, accompanied by significant tensile load drop at each failure. In many cases, the magnitude of the load drop exceeded the average single yarn breaking strength.

Most of the isolated failures occurring in ring spun and rotor spun fabrics subject to warpwise loading originated at bend locations. These yarns tended to break abruptly, with few protruding fiber ends. . . . similar to the failure ends of yarns tested out of fabric at near zero gauge length. . . . Most isolated failures in ring spun fabrics tested fillingwise showed large amounts of long protruding fibre ends, leading to a long failure zone . . . similar to the failure ends of long gauge length yarns tested out of fabric.

The form of tearing of fabrics depends on the tightness of the weave. This is shown in a study of the tongue tear test by Scelzo *et al* (1994). In loose weaves with high mobility, which is also accentuated by low friction and flexible yarns, the tear strength is higher and the tear is accompanied by major distortion of the fabric over an appreciable area, **25L(5)**. With tighter fabrics, allowing little yarn movement, the tear is sharper and the distortion is localised, **25L(6)**, with a lower tear strength due to the reduced fabric deformation energy.

The remaining pictures reinforce views elsewhere in the book. In ways similar to examples in Chapter 30, pilling, **25L(7)**, with breaks by multiple splitting, **25L(8)**, occurs in Martindale abrasion testing of a knit cotton fabric. The abrasion of a woven wool/mohair fabric is shown in **25M**. These pictures clearly display the concentration of wear at the interstices between yarns and the sequence from multiple splitting of fibres, due to bending, through a wearing away of the projecting ends while leaving the splits visible, to smooth rounded ends.

Plate 25A — Laboratory abrasion of plain-weave cotton shirting fabric, 20 ends/cm × 20 picks/cm, on a WIRA abrasion tester.

Abraded against standard worsted fabric for 2000 rubs.

(1) Bleached and mercerized fabric. (2) Fibre from desized and scoured fabric. (3) Fibre from bleached and mercerized fabric. (4) Fibre from dyed and resin-treated fabric.

Abraded against paper

(5) Dyed and resin-treated fabric against rough paper for 200 rubs: yarn crowns. (6) Fibres from dyed and resin-treated fabric against smooth paper for 300 rubs.

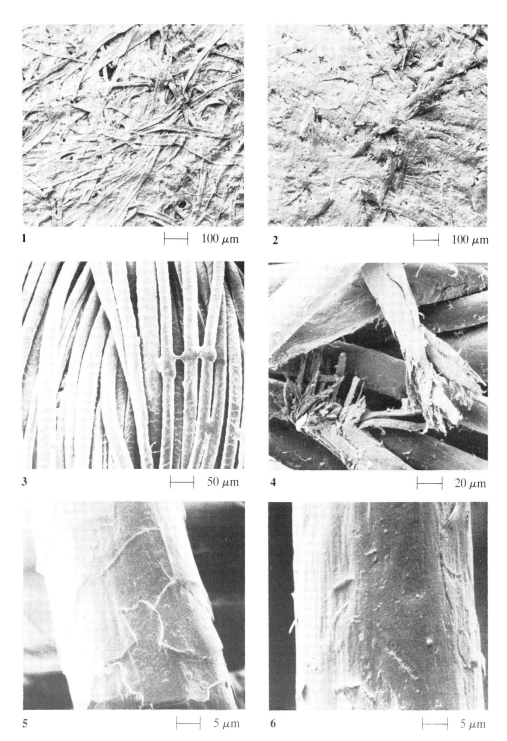

Plate 25B — Damage to abradant in testing of cotton fabrics: test details as in 25A.

(1) Rough paper before use as abradant. (2) Rough paper abraded against resin-treated cotton fabric for 200 rubs. (3) Worsted fabric before use as abradant. (4) Worsted fabric abraded against resin-treated cotton fabric for 2000 rubs. (5) Wool fibre from fabric before use as abradant. (6) Wool fibre from fabric after 2000 rubs.

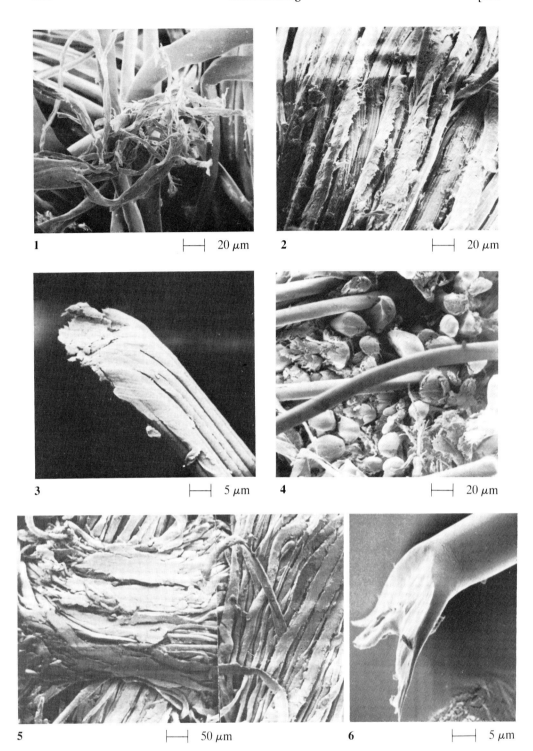

Plate 25C — Effect of different abrasion testers.
(1) Polyester/cotton sheet abraded on Boss abrasion tester using loomstate cotton canvas as abradant.
(2) 48% cotton/52% rayon sheet Boss abraded with mineral khaki cotton canvas abradant. (3) Sheared-
through viscose rayon fibre from Boss abraded cotton/viscose rayon sheet. Mineral khaki cotton canvas
abradant. (4) 80% cotton/20% nylon sheet abraded on Martindale tester, standard crossbred wool fabric
abradant. (5) Cotton sheet flex-tested against blade on Stoll tester. (6) Nylon fibre from cotton/nylon sheet
flex abrasion tested on Stoll tester.

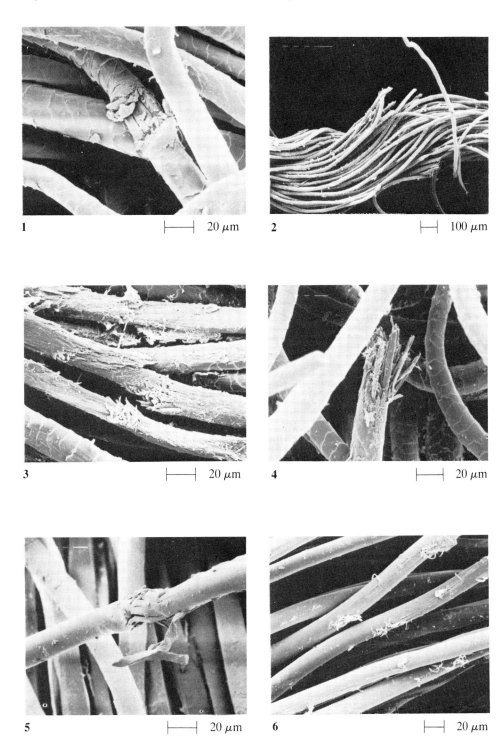

Plate 25D — Shear cycling of fabrics by Kawabata.
(1) Wool fabric, after 6×10^5 cycles. (2) Warp yarn from wool fabric tested with carborundum and water, after 10^5 cycles. (3), (4) Wool fibre damage after testing as in (2). (5), (6) Polyester fibre, after 10^5 cycles.

Plate 25E — Abrasion testing of overall fabrics using Accelerotor tester.
(1) Polyester/cotton fabric abraded dry; severe damage to cotton fibres in yarn crowns. (2) Polyester/
modal fabric abraded dry; more of a cutting action on fibres. (3) Same fabric as in (2) but wet abraded in
accelerotor. Break in modal fibre.

Using Martindale tester.

(4) Miraclean, polyester/cotton workwear after 75 000 rubs on Martindale. (5) Detail of fabric surface, as
in (4). (6) Long bushy end of polyester fibre, as in (4).

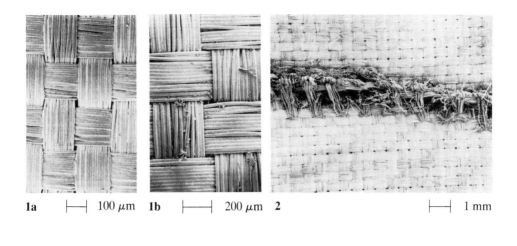

1a ⊢⟶⊣ 100 μm   1b ⊢⟶⊣ 200 μm   2 ⊢⟶⊣ 1 mm

3 ⊢⟶⊣ 200 μm

4 ⊢⟶⊣ 50 μm

5 ⊢⟶⊣ 50 μm

6 ⊢⟶⊣ 10 μm

Plate 25F — Abrasion testing of fabrics for clean-room garments: (a) less severe — against standard wool fabric by company; (b) more severe — against the test fabric itself for 177 000 cycles on Martindale at UMIST.
(1a) Polyester (Dacron) taffeta, against wool. (1b) Same fabric, against self. (2) Same fabric, against self, line of severe damage, possibly at crease. (3)–(5) Details of wear of fibres in region of (2). (6) Polyester (Dacron) herringbone, against self, start of fibre peeling.

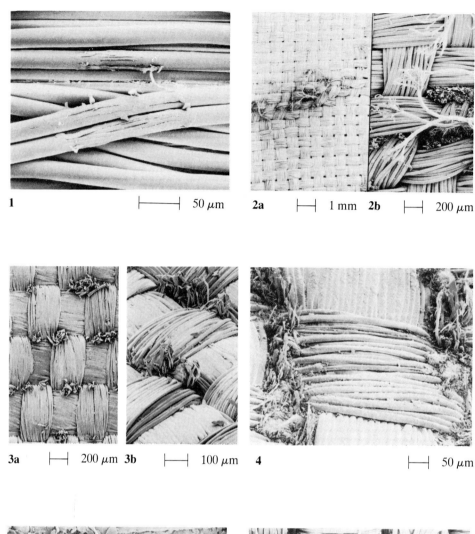

Plate 25G — Abrasion testing of fabrics for clean-room garments (continued).
(1) Nylon (Celon) fabric, against self, splitting of filaments at yarn crown. (2) Same fabric, against self, showing localized area of severe damage at edge. (3a,b) Polyester (Terylene) fabric, against wool, broken fibres at yarn interstices. (4), (5) Same fabric, against self, with compacting of broken fibres. (6) Partly textured yarn polyester fabric, against self.

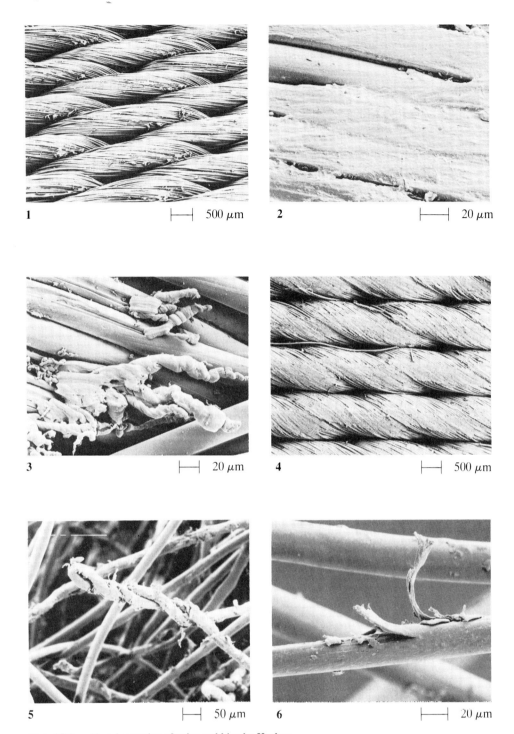

Plate 25H — Abrasion testing of nylon webbing by Hexbar.
(1) Surface wear of conventional webbing. (2) Scraped surface of yarn crown. (3) Rolling-up of surface peels at end of crown. (4) Surface smearing and scraping at variant selvedge. (5) Twisted-up, abraded filament near top of conventional selvedge. (6) Peeled filament at top of damaged conventional selvedge.

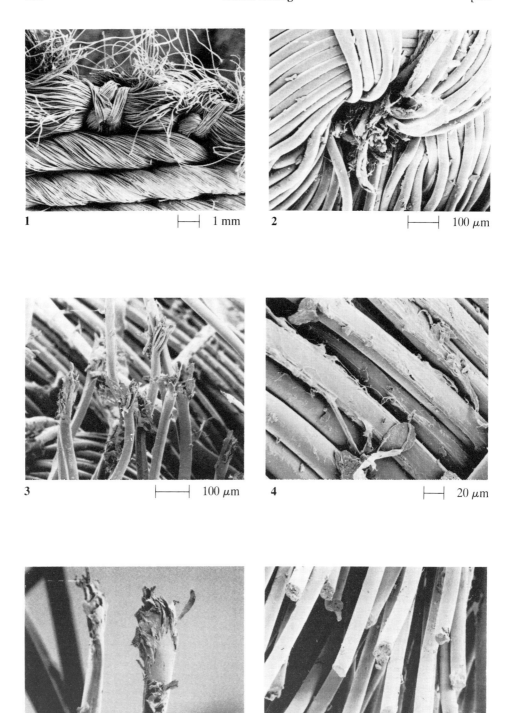

Plate 25I — Abrasion testing of nylon webbing by Hexbar (continued).
(1) Damage to variant selvedge (non-conventional) after 10^4 cycles against Hexbar. (2) Broken fibres in loop holding edge yarns of the selvedge. (3) Broken split ends in yarn lying along selvedge (edge of crown). (4) Surface peeling of yarn crossing over variant selvedge. (5) Sheared ends at variant selvedge. (6) Broken filaments at edge of variant selvedge.

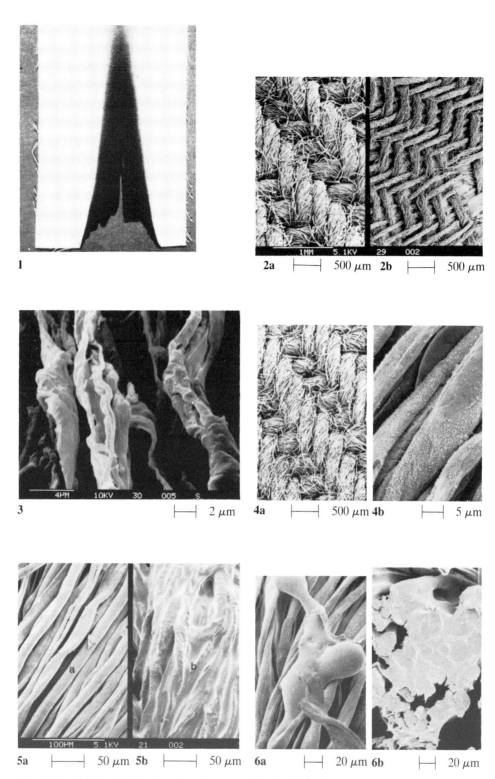

1

2a ⊢———⊣ 500 μm 2b ⊢———⊣ 500 μm

3 ⊢—⊣ 2 μm

4a ⊢———⊣ 500 μm 4b ⊢—⊣ 5 μm

5a ⊢———⊣ 50 μm 5b ⊢———⊣ 50 μm 6a ⊢——⊣ 20 μm 6b ⊢——⊣ 20 μm

Plate 25J — Fabrics after burning (from Goynes and Trask, 1985, 1987).
(1) General view of test sample. (2) Comparison of (a) unburnt and (b) burnt untreated cotton fabric.
(3) Charred untreated cotton fibres, from (2b). (4a,b) Charred THPS-treated cotton fabric, and enlarged
view of fibres. (5) Comparison of (a) unburnt and (b) burnt untreated cotton/polyester fabric. (6a) Fused
polyester fibres on fabric surface away from the charred region. (6b) Yarn cross-section with cotton in
polyester melt.

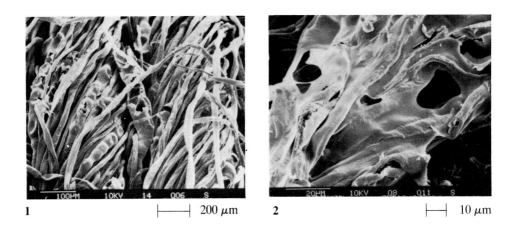

1 |——| 200 μm   2 |——| 10 μm

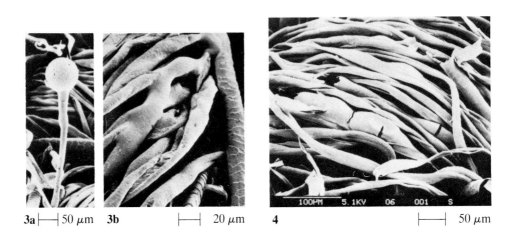

3a |——| 50 μm   3b |——| 20 μm   4 |——| 50 μm

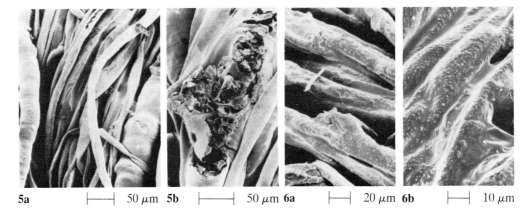

5a |——| 50 μm 5b |——| 50 μm 6a |——| 20 μm 6b |——| 10 μm

Plate 25K — Fabrics after burning (continued).
(1) Untreated cotton/wool fabric away from the charred region. (2) Melting of wool in hotter region.
(3a,b), (4) Moving progressively closer to charred region of untreated cotton/polyester/wool fabric.
(5a,b) Moderate and more severe damage to wool fibres in THPS-treated tri-blend fabric. (6) Charred
region of (a) untreated and (b) treated tri-blend fabric.

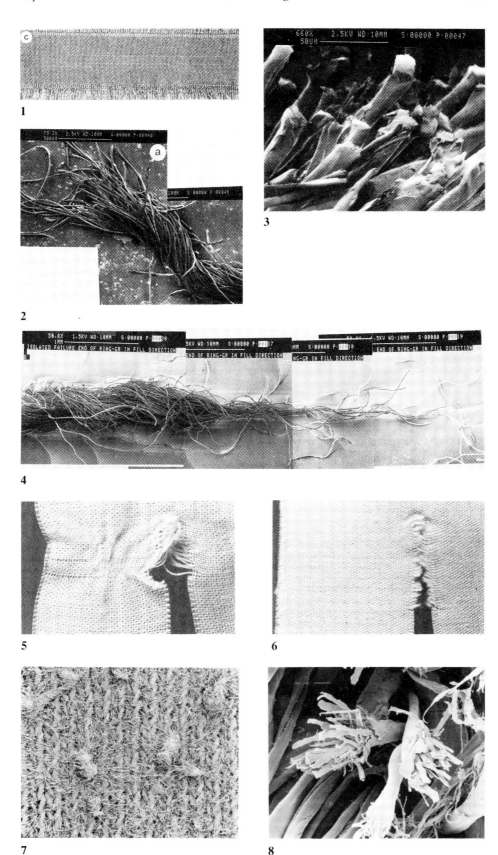

Plate 25L — Break of yarns in woven fabric, Seo *et al* (1993).
(1) Isolated failures in a ring-spun twill fabric. (2) Break of ring-spun cotton/polyester yarn in a twill fabric tensioned warpwise. (3) Detail of fibre breaks. (4) Break of similar yarn tensioned fillwise.
Tongue tear of woven fabric, Scelzo *et al* (1994).
(5) Loosely woven fabric. (6) Tighter fabric.
Martindale abrasion of knit cotton.
(7) Development of pills. (8) Break by multiple splitting.

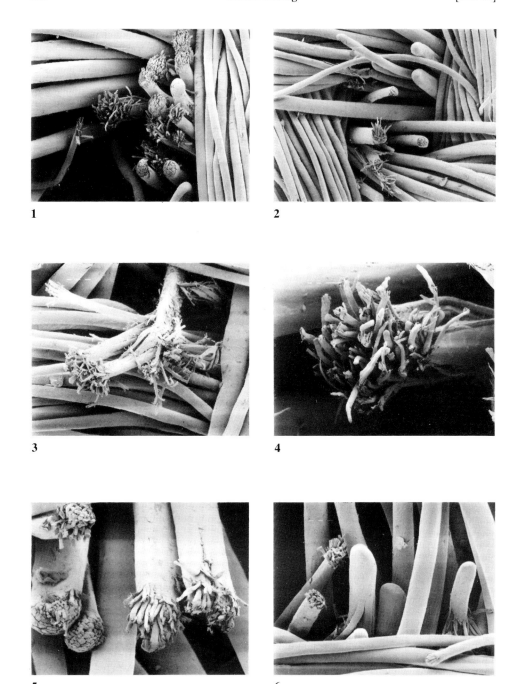

Plate 25M — Abrasion of woven wool/mohair fabric.
    (1),(2) Wear at interstices between yarns. (3)–(6) Detail of fibre splitting and wear.

26

COMPOSITE TESTING

Although most textile fabrics are used without any material added to the fibres, except for very thin layers of surface finish, there is an appreciable usage of fibres and fabrics in composite materials. This chapter gives some examples of failure testing in the composite form, both flexible and rigid. Some examples of failure in use of composites are included in Chapter 40.

As a means of evaluating the behaviour of flexible rubber/textile composites, as used for example in tyres, special laboratory specimens may be made up. For example, in order to test the effectiveness of adhesion between rubber and tyre-cord fabric, a thick sandwich is made with two fabric layers and three layers of rubber. The sandwich can then be torn apart, to give the form of surfaces indicated in Fig. 26.1, by the action shown in Fig. 26.2. A macrophotograph of the torn surfaces, **26A(1)**, shows that the failure is divided between different positions in the cross-section. In order to make identification easier, their locations are outlined in Fig. 26.1. Over the largest part of the surface. A, the separation is within the rubber, with occasional breaks through to the fabric on one side, B, or the other side, C. But there are clearly defined islands, where the separation is entirely between fabric and rubber, fabric side at D, rubber side at E. Naturally, the islands match on either side. The bare fabric surface appears lighter, and is lower than the surrounding area, A, while the mirror-image on the other side stands proud, and is identifiable by the regular lines of the replica of the yarns, which contrast with the more irregular tearing of the rubber in the surrounding region. Detail of failure at the fabric surface is shown in **28A(2)**. The fabric was a typical tyre-cord fabric with a strong warp of nylon, loosely held by a skeleton cotton weft. A typical cotton fibre tensile break is shown in **26A(3)**. The nylon fibres show either surface peeling, **26A(4)**, or a granular-type tensile break, **26A(5)**. The occurrence of a granular, rather than a ductile break is surprising: it might be due to some degradation or to the suppression of the usual crack propagation in the bonded composite.

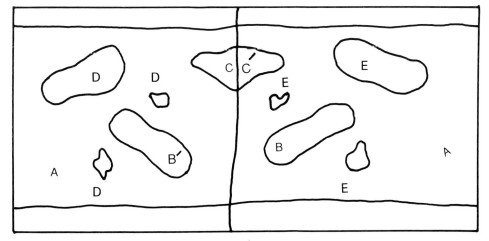

Fig. 26.1 — Identifications of regions in macrophotograph, **26A(1)**. Over most of the surface A failure is in the rubber, with occasional break-through to fabric. Failure is at the rubber–fabric interface in regions B and C, with the corresponding regions with fabric imprints at B′ and C′. Complete failure between rubber and fabric at positions D and E.

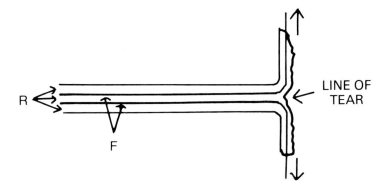

Fig. 26.2 — Tearing of test piece, with two layers of fabric F, sandwiched with three layers of rubber, R.

Another example is flex fatigue in a three-dimensional buckling mode of a thin rubber-coated fabric used in a metering device. It appears that failure starts with a loss of adhesion between rubber and fabric, and that the increased strain which this allows leads to rupture of the rubber layer, **26B(1)**. There may be some damage to fibres in the underlying fabric, **26B(2)**, but this is not a primary cause of failure.

Another test of a composite, PVC-coated polyester fabric used in flexible structures was designed to investigate chemical degradation. The investigation was carried out by Martin Ansell at the University of Bath and was briefly referred to in Chapter 16. In order to accelerate the degradation the material was boiled in water for several weeks and then broken in a tensile test. The control sample shows polyester fibre breaks with mushroom ends, **26B(3)**, typical of a high-speed break situation, probably resulting from transfer of load after break has started. After 3 weeks boiling, **26B(4)**, the breaks are mixed in form; but after 6 weeks, **26B(5)**, they are well-defined stake-and-socket breaks; after 8 weeks, **26B(6)**, embrittled rims become larger, and there are some changes in the appearance of the stake. These results were reported by Ansell (1983).

The study of fracture of rigid composites is a major subject in itself, and the forms of breakage depend on: (a) the type of fibre and matrix; (b) whether the composite structure has been made by dispersal of short fibres, by tape-laying of oriented pre-preg, by filament winding, by two-dimensional or three-dimensional textile structures, or by any other method; (c) the shape of the test specimen; (d) the form of loading. These larger-scale mechanical/geometrical aspects are beyond the scope of this book; and the account here is limited to some examples of failure at the level of fibre and matrix.

The main forms of failure have been categorized by Friedrich (1983) in a study of the fracture of a composite of short glass fibre in a polyethylene terephthalate matrix. A 'compact tension' specimen, illustrated in Fig. 26.3, was used. The fracture crack propagates from the tip of the preformed crack, AB, when the two arms are put under tension. This is termed mode I fracture, Fig. 26.4(a), because the crack opens under a tensile stress acting in a direction perpendicular to the plane of the crack.

Because the fibres are randomly arranged in all directions, the test shows examples of fracture both with the fibres lying across the crack and subject to a tensile stress along their length, and with the fibres lying in the crack plane so that the tensile stress is transverse to the

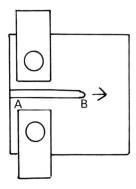

Fig. 26.3 — Compact tension test.

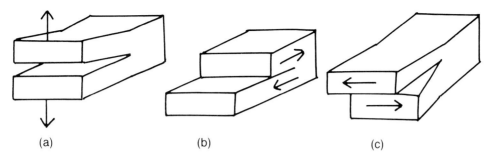

Fig. 26.4 — Forms of crack growth. (a) Mode I fracture under tensile stress perpendicular to the crack. (b) Mode II fracture under shear stress along the crack. (c) Mode III fracture under shear stress across the crack.

fibre axes and between fibres. In **26C(1)–(6)** and **26D(1),(2),** the left-hand set shows SEM pictures taken in profile of the polished surface of a cut perpendicular to the plane of the crack, and the right-hand set shows the corresponding fracture surfaces.

The effects observed, and located on Fig. 26.5, are as follows:

  (A) mechanical overload causing brittle fibre fracture, **26C(1),(2)**;
  (B) fibre pull-out, without rupture, **26C(3),(4)**;
  (C) delamination between fibre and matrix, **26C(5),(6)**;
  (D) plastic deformation and rupture of the matrix, **26D(1),(2)**.

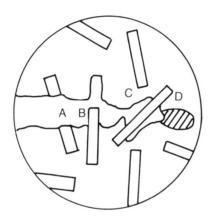

Fig. 26.5 — Mechanisms of failure at an advancing crack: (A) fibre fracture by overload; (B) fibre pull-out; (C) delamination; (D) matrix flow.

In addition, attack by a corrosive environment can lead to multiple fibre cracking, **26D(3),(4)**.

The effect of temperature on the fracture of the glass fibre/PET composite was also examined: the viscous flow of the matrix is much more pronounced at +60°C, **26D(5)**, than at −60°C, **26D(6)**.

Although the range of fibre orientations and local geometries causes different forms of failure to occur in the same short-fibre composite, larger-scale structural features do not appear in such a dispersed system. However, when yarns are regularly arranged in the composites, the structure shows up in the failure.

For example, both delamination and fibre breakage is found in the break of a cross-ply, [90,0,90,0]$_s$, laminate of carbon fibre in the form of continuous-filament yarn, with a matrix of nylon 12, applied as a powder and then thermally consolidated. Delamination in the plies containing filaments perpendicular to the tensile stress and fibre breakage in filaments lying parallel to the stress is shown in **26E(1)**.

If the reinforcement is a fabric the weave structure shows up at low magnification, as seen in **26E(2)** for a woven glass/polyamide composite. At higher magnification, **26E(3),** the fibre fracture and delamination is more clearly shown. Similar effects were found in a woven carbon/epoxy composite, **26E(4)**.

More complex effects are shown with three-dimensional woven structures, illustrated here in work by Guenon (1987) on carbon fibre in epoxy. The test method used is a double cantilever beam with a pre-formed crack, but with Z-direction reinforcement it was necessary

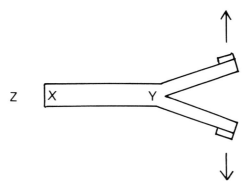

Fig. 26.6 — Double cantilever beam test (DCB).

to modify the usual test specimen, illustrated in Fig. 26.6, by sticking aluminium tabs on to the composite, in order to cause the crack to propagate along the specimen. There is then delamination for the two yarn layers parallel to the plane XY of the fabric, and rupture and pull-out of yarns in the Z-direction, **26E(5),(6)**.

The form of delamination is shown up in studies by Crick *et al.* (1987), of uniaxial APC-2 composites of carbon fibre in polyetheretherketone (PEEK) after crack propagation in a DCB test. At first there is a slow, stable crack growth, but this is followed by a fast, unstable fracture. Low-magnification views of fracture surfaces in the two regions are shown in **26F(1,2)**. More detail of the delamination is shown by cutting a cross-section through the crack in a fractured specimen embedded in an acrylic medium, polishing and then etching the surface, and finally dissolving away the acrylic in chloroform. In **26F(3),(4)**, the way in which the crack boundary follows a complicated profile within the matrix around the fibres can be seen. Particularly in the unstable region, there are large re-entrant cavities. There is a contrast between the considerable matrix flow in the stable region and the cleaner fracture in the unstable region, which is seen in more detail at the higher magnification, **26F(5),(6)**. The matrix surface shows evidence of spherulitic texture, and, especially in the stable region, of cracks or crazes, shown up by the etching, parallel to the fracture surface.

The glass and carbon fibres in the rigid composites illustrated in **26C-F** have roughly equal strength in all directions and break with sharp brittle fractures. A different situation exists with highly oriented linear polymer fibres, which easily split axially and fibrillate, as demonstrated by the work of Matsuda (1987) on composites of the aramid copolymer fibre Technora with an epoxy matrix. In a DCB test on a uniaxial laminate, there is substantial fibre and fibril bridging across the crack, **26G(1)**. Detail of the fibrillation, together with some delamination, is shown in **26G(2)**.

In addition to examining the mode I fracture, shown in Fig. 26.4(a), with the tensile stress opening the crack, Matsuda also used an end-notch flexure (ENF) specimen to study mode II fracture. In this test method, shown in Fig. 26.7, change in curvature generates the shear stress along the crack, Fig. 26.4(b), and causes the crack to grow. An overall low-magnification view of the fractured sample, **26G(3)**, shows the Teflon film separator, a mode I pre-crack, then the mode II fracture starting in stable growth, and beyond a critical crack depth becoming unstable. In the stable region there is extensive fibrillation, **26G(4)**, but in the unstable region there is matrix flow and delamination, **26G(5)**.

Matsuda also examined Technora/epoxy composites in compression, and the failed specimens show the large-scale kinking, which occurs in a uniaxial composite compressed in the orientation direction, **26G(6)**, and the delamination and slip in the 90° test, **26G(7)**.

Mode II fracture by the ENF test has also been studied by Trethewey (1986), for carbon fibre composites, and the SEM pictures show up the difference between the mode I pre-crack and the mode II fracture surface for AS4 carbon fibre in epoxy, **26H(1),(2)**, and APC-2 carbon fibre in the thermoplastic PEEK, **26H(4),(5)**. The appearance is somewhat different in mode II fatigue fractures, **26H(3),(6)**.

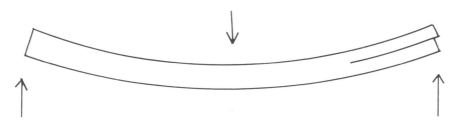

Fig. 26.7 — End-notch flexure test (ENF).

It has been found by Becht (1988) that the appearance of mode III fracture in carbon fibre/ epoxy laminates is similar to mode II.

In test situations where fibres break, the simple and characteristic forms of fracture within composites are seen as brittle fracture in glass fibres, **26H(7a,b),** from Valentin, Paray and Guetta (1987), and as granular fracture in carbon fibres, **26H(8)**, from Beaumont (private communication).

However, under some types of stress, there can be an axial splitting of carbon fibres. This implies that the transverse strength of the fibre is less than that of the fibre/matrix interface, and occurs with pitch-based carbon fibres. An example of crack propagation, partly across and partly around the carbon fibres, is shown in **26I(1),** which is a polished section which has been subject to severe thermal shock. The characteristic structure of pitch-based fibres, illustrated in Fig. 26.8, can be seen in **26I(2),** in which the fibres have broken over a transverse cross-section. Some delamination can also be seen. It is this layered form, which leads to the axial splitting shown in **26I(3),(4)**, as a result of transverse tension. If the layers are perpendicular to the tension, there is splitting, but, if they are in line with the tension, there is none. Axial splitting of fibres has also been seen in pitch-based fibre/epoxy composites.

There are differences between different types of pitch-based fibre, as seen in **26I(5)**, which shows a different texture of failure surface from **26I(2)**. In a PAN-based fibre composite, illustrated in **26I(6),** thermal shock causes cracking with delamination but no fibre splitting.

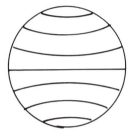

Fig. 26.8 — Characteristic structure of pitch-based carbon fibres.

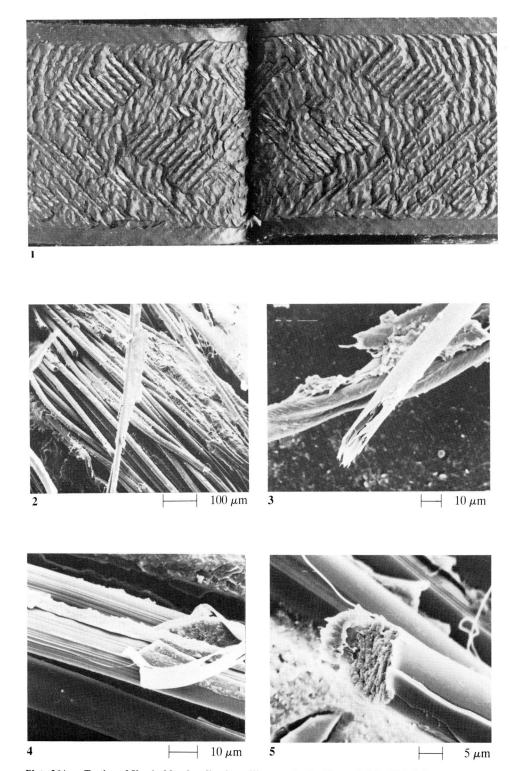

Plate 26A — Testing of fibre/rubber bonding by pulling apart a double sandwich of fabric in rubber.
(1) Macrophotograph of torn surfaces. See Fig. 26.1 for identification of regions of failure. (2) From
failure zone between rubber and fabric. (3)–(5) Detail of damage to fibres.

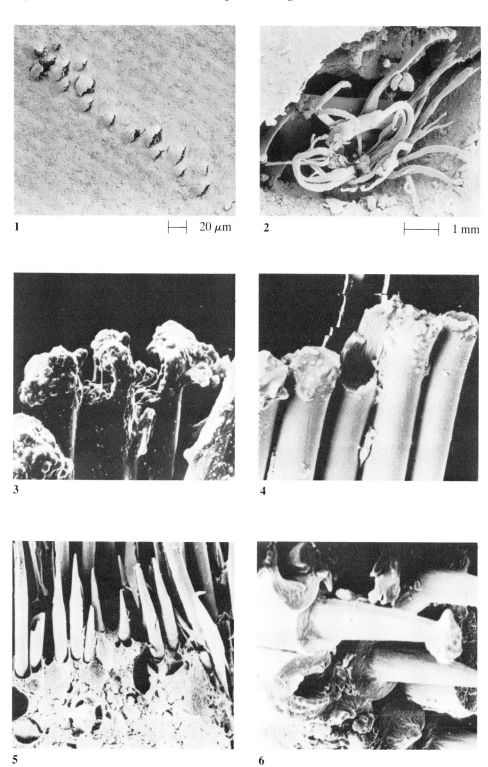

1 ⊢—⊣ 20 μm   2 ⊢———⊣ 1 mm

3   4

5   6

Plate 26B — Three-dimensional buckling fatigue of thin rubber-coated fabric.
(1) Initial appearance of damage. (2) Detail of fabric.
Accelerated degradation of polyester fabric coated with a 1mm layer of pigmented and stabilized plasticized PVC; fabric weight was 280 g/m² and coating was 600 g/m²; boiled in distilled water (100°C); then subjected to a tensile test. SEM pictures by courtesy of Martin Ansell, University of Bath.
(3) Control sample, not boiled. (4) After 3 weeks boiling. (5) After 6 weeks boiling. (6) After 8 weeks boiling.

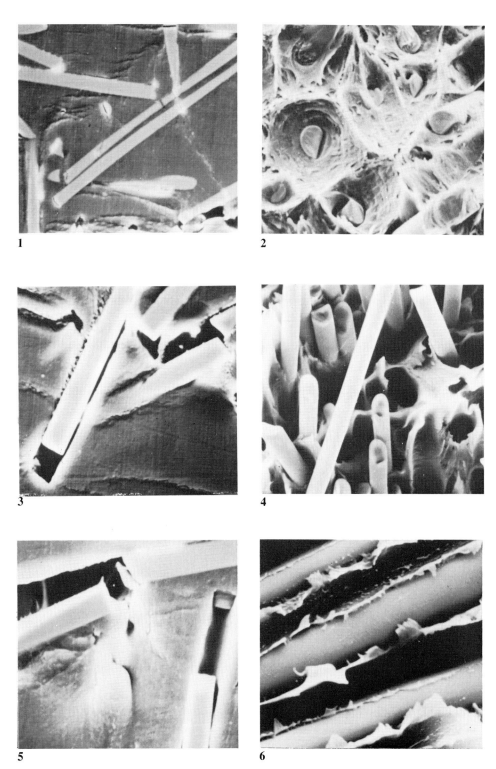

Plate 26C — Tensile fracture at an advancing crack tip in a compact tension specimen of a short glass fibre/ PET composite (from K. Friedrich, 1983).
Note: Left-hand set (odd numbers) are polished surface profiles and right-hand set (even numbers) are fracture surfaces. Locations, A to D, are shown in Fig. 26.5. (1),(2) Fibre fracture by mechanical overload, A. (3),(4) Fibre pull-out, B. (5),(6) Delamination, C.

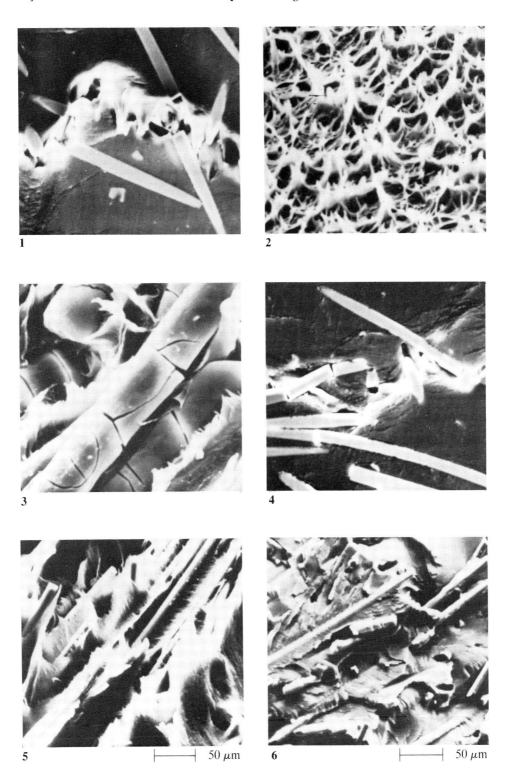

Plate 26D — Fracture at crack tip in glass fibre/PET composite (continued).
(1),(2) Plastic deformation of matrix, D. (3),(4) Multiple fibre cracking due to an additional corrosive
environment. (5) Fracture surface, tested at +60°C. (6) Fracture surface, tested at −60°C.

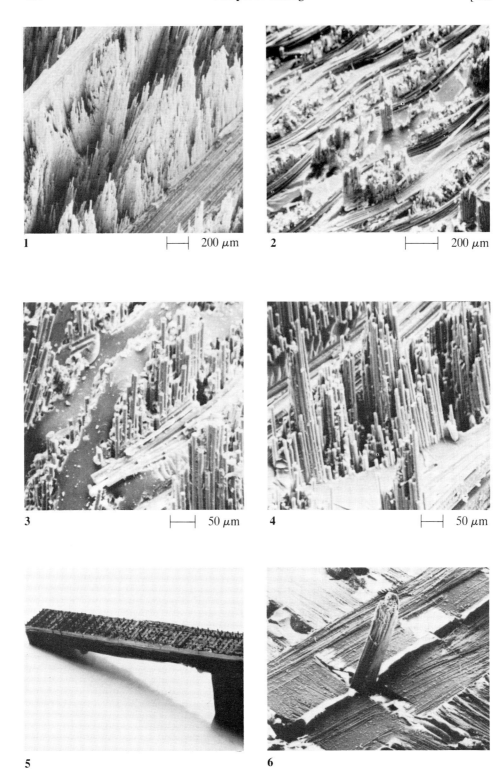

Plate 26E — Effects in structured composites (from K. Friedrich, 1983).
(1) Cross-ply carbon fibre/nylon 12 composite. (2),(3) Woven glass (continuous filament) fabric in polyamide. (4) Woven carbon (continuous filament) in epoxy.
Fracture of a three-dimensional woven carbon/epoxy composite with crack propagation in a DCB specimen (from V. A. F. Guenon, 1987).
(5) Photograph of failure. (6) SEM view of detail, with X and Y directions running along diagonals, and a Z direction yarn projecting.

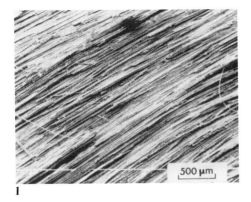
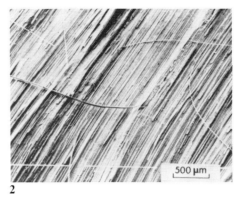

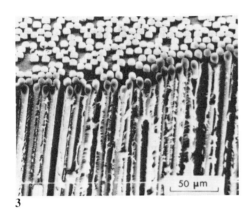
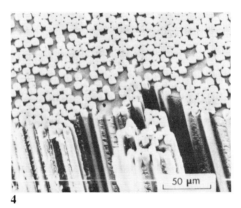

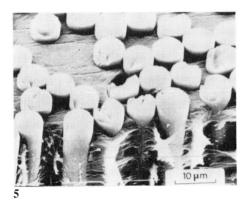
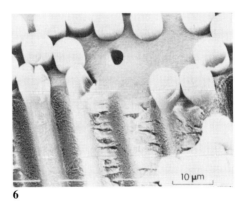

Plate 26F — Fracture of carbon fibre/PEEK composites in DCB test (from Crick *et al*, 1987).
(1) Stable crack growth region. (2) Unstable crack growth region.
Intersection of polished and etched cross-sectional surface with fracture surface.
(3),(5) Stable crack growth region. (4),(6) Unstable crack growth region.

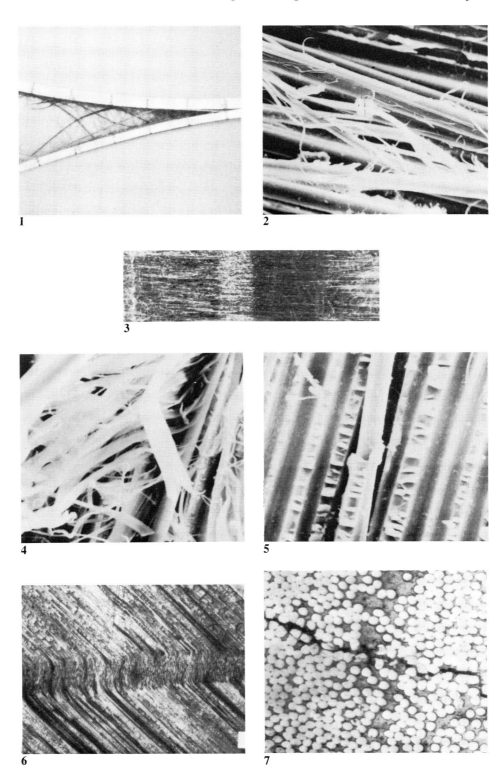

Plate 26G — Fracture of uniaxial Technora/epoxy composites (from T. Matsuda, 1987).
(1),(2) DCB test, mode I fracture. (3) ENF-fractured specimen showing, from left to right (a) Teflon film, (b) mode I precrack, (c) stable mode II crack growth, (d) unstable mode II failure. (4) Stable mode II fracture. (5) Unstable mode II fracture. (6) Compression failure in 0° test. (7) Compression failure in 90° test.

Plate 26H — Fracture of carbon fibre composites (from B. R. Trethewey, 1986).
(1) ENF test of AS4/epoxy, mode I pre-crack. (2) ENF test of AS4/epoxy, mode II. (3) Mode II fatigue test of AS4/epoxy. (4) ENF test of APC-2/PEEK, mode I pre-crack. (5) ENF test of APC-2/PEEK, mode II. (6) Mode II fatigue test of APC-2/PEEK.
Bending failure of glass fibre/nylon 66 composite (from Valentin, Paray and Guetta 1987).
(7a,b) Brittle fracture of glass fibres.
Torsion failure in carbon fibre/polyester resin composite with poor bonding (from P. W. R. Beaumont, private communication).
(8) Granular fracture of carbon fibres.

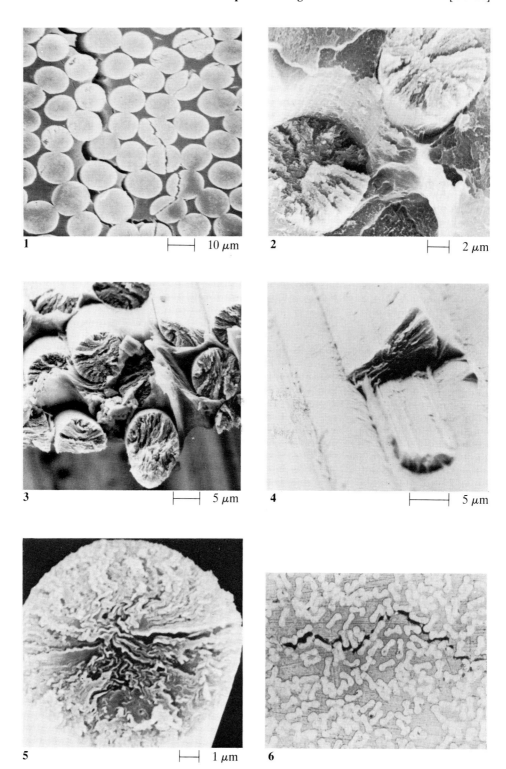

Plate 26I — Fractures in experimental Thornel P75 pitch-based carbon fibre/PEEK composites (from Barnes, private communication).
(1) Polished section of ±30° composite, cycled ten times in and out of liquid nitrogen. (2) Unidirectional composite, failed in longitudinal flexure. (3),(4) Unidirectional composite failed under transverse tension in a cantilever beam test.
Fractures in other experimental carbon fibre/PEEK composites (from Barnes, private communication).
(5) Nippon XN 50 pitch-based fibre in unidirectional composite, failed in longitudinal flexure. (6) Optical micrograph of polished section of Hoechst/Celanese GY70 PAN-based fibre in 0/90 composite, after cooling to room temperature.

Part VI
Case studies:
clothing and domestic uses

27

INTRODUCTION

We now turn to the study of materials worn in real-life situations, starting with case studies selected from the traditional uses of textiles in clothing and in household products, such as sheets, towels and carpets.

Although it is sometimes useful to make quick exploratory studies, any proper investigation, unless it is a particular example of an application which has been previously investigated in detail, requires a very thorough study by a variety of techniques, as described in Chapter 2, and takes a long time. Inevitably, therefore, the examples included here cannot be regarded as more than a somewhat arbitrary selection from the vast range of different types of textile fabric used in a vast variety of products and situations. The choice was dictated partly by the availability of interesting and adequately documented samples. No sexist bias was intended, but it turns out that more men's garments than women's were examined.

28

TROUSERS AND JACKETS

Our first example of a case study of textile failure arose from a customer complaint. A pair of men's trousers, made of a blend of wool, polyester and rayon, had worn into a hole, **28A(1)**, after only a few days of use. Examination at higher magnification shows fibres failing by multiple splitting, **28A(2),(3)**, and breaking with bushy ends, **28A(4)**. This is characteristic of failure by bending and twisting, as found in the laboratory biaxial rotation fatigue test. Further wear may cause fibre ends to become rounded off, **28A(5)**. There is also some evidence of central splitting, **28A(6)**, which may be a result of pure flex fatigue by the shear mechanism described in Chapter 12. This example of textile failure was not the subject of major study, and details of fabric construction and conditions of use are not known, but it does demonstrate the typical pattern of wear and the characteristic form of fibre failure.

More detail was available of four pairs of men's trousers subject to a wear trial by IWS. All the trousers, which differed in fibre composition or weave, had worn through in the seat, close to the back seam and in or near the crotch, although the precise pattern of wear did vary. One pair of trousers was made of a grey pin-stripe material blended from a mixture of black and white wool fibres, with the pin-stripe formed by two continuous-filament polyester yarns. There was a hole in the crotch region on one side of the centre seam, with the fabric worn thin on the other side of the seam. The polyester yarns had not broken but continued across the hole, undamaged except for slight peeling on the filament surfaces, even though warp and weft wool yarns had worn completely away. The wool fibres start to break down at several places along their length, **28B(1)**, with the development of multiple splitting, **28B(2)**, leading to break with a bushy end, **28B(3)**, and then to rounding off with more wear, **28B(4)**.

Another pair of trousers was a charcoal-grey blend of wool and polyester staple. This pair had pilled badly on both sides of the centre back seam near the crotch, and this had caused the fabric to wear thin on one side of the seam and into a hole on the other. The fibre damage is the typical multiple splitting at many places. This takes place in both wool and polyester fibres, **28B(5)**, and is followed by rounding off, **28B(6)**.

The other two pairs of trousers from the IWS trial were looser tweeds, made from all wool. There were some differences in the exact location of wear, and another effect observed was the loss of nap from the surface, which made some parts of the fabric appear bald or threadbare, **28C1(b)**, compared with the relatively unworn trouser legs where the surface hairiness is evident, **28C(1a)**. A characteristic feature in these wool tweeds was the loss of yarn in one direction in the crotch region but not in the other direction, **28C(2)**. Nevertheless, the sequence of fibre damage — splitting, bushy ends and rounding off — was similar, but not as severe or widespread as in the wool/polyester trousers.

In all four pairs of trousers similar fibre damage was found, not only extensively in badly worn regions, but also in a few fibres in the trouser legs, where the fabric did not show obvious signs of wear. The fibre breakdown is occurring throughout the material, but is more severe in particular regions.

The remaining examples in this chapter are garments subject to ordinary wear, and not from special wear trials.

The mechanism of fabric wear by pilling is discussed in Chapter 30, but is also illustrated by a woven wool/polyester fabric used in ladies' trousers. The pills, which appear on the fabric surface, are tangled balls of fibres, **28C(3)**. They are surrounded by material which is deficient in fibres. The fibres in the pills show evidence of much multiple splitting, **28C(4)–(6)**. In all wool fabrics the pills easily break off, but the stronger polyester fibres in the blend hold them on the fabric more strongly, so that higher pill densities develop.

The next case study is of a pair of men's trousers from a suit made from an extremely durable worsted material, containing hard twisted wool yarns, tightly woven in the fabric. It proved impossible to wear holes or thin places in the material, but the suit was eventually discarded because the seat of the trousers had become excessively shiny. It is thus a complete contrast to the first case study, in which a hole formed after very little use.

A typical view of fibres in the outer surface of the seat is shown in **28D(1)**. Fibres in the plane of the fabric at the outermost (upper) level have had their surfaces worn away; but this is not so for fibres deeper in the fabric. Fibres which project out at an angle to the plane of the fabric are broken with multiple splits. A clear example of a fibre with a surface worn flat, so that it reflects light and contributes to the shiny appearance, is shown in **28D(2)**. Detail of a broken fibre from this region is shown in **28D(3)**: it has broken by multiple splitting, but is beginning to be rounded off. On the inner seat surface, **28D(4)**, there are broken fibres, some of which have developed rounded ends. On the knee both fibre breakage, **28D(5)**, and surface wear, **28D(6)** can be seen.

The material used in the pockets of this pair of trousers was also examined. This was a durable continuous-filament polyester fabric, but it had formed a hole near the bottom seam, **28E(1)**. Even away from the hole there was damage to the fibres, with many zones of multiple splitting, **28E(2)**. Detail of fibre breakage is shown in **28E(3),(4a,b)**.

The label on the waistband of this pair of trousers was made of viscose rayon, and also showed evidence of wear, with broken fibres projecting from the interstices of the weave, **28E(5)**. These fibre ends had become rounded off by prolonged wear, but the actual breaks of rayon were sharp, **28E(6)**, and resembled tensile breaks. Rayon is exceptional in not showing failure by multiple splitting.

The last example of trousers is of cotton cord jeans. There is a clear contrast between relatively undamaged fabric, **28F(1)**, and fabric from a worn area on the right knee, **28F(2)**. Fibre ends in the little-worn fabric, **28F(3)**, are typical of fibres cut when the cord fabric was made; but in the worn material there is entanglement and fibre splitting, **28F(4)**.

Finally, we give one example of failure in a jacket. This was a wool/polyester school blazer which had become shabby after use. The wool fibres fail by multiple splitting, **28F(5)**, with the ends gradually being rounded off as they suffer further wear, **28F(6)**.

Yokura and Niwa (1990) report changes in mechanical properties, as measured on the KESF system, of 20 men's summer suiting fabrics after laboratory fatigue tests. They also calculated *total hand values* (THV) and *total appearance values* (TAV) from predictive equations proposed by Kawabata and Niwa (1980) and Niwa and Kawabata (1988). Suits were made from two of the fabrics: N36, a 50/50 wool/mohair blend, and N37, a 35/65 wool/polyester blend. These suits were worn at work in a laboratory for 800 hours over six seasons with dry-cleaning after each season. The changes in mechanical properties were measured in samples from different parts of the garments, and they also show photographs of the jackets, before and after wear. The comment on the appearance of the jackets was:

The jacket tailored from fabric N36 showed a beautiful seam line coming from natural overfeeding and a smooth curve at the shoulder before and after the wear test. On the other hand, jacket N37 showed fabric distortion at the shoulder before wear. The puckering and distortion at the shoulder and sleeve head of jacket N37 became severe after 800 hours of wear.

The overall conclusion was:

Shape retention during wear of a suit jacket tailored from fabric with a total appearance value (TAV) of 4.39 was superior to that of a jacket tailored from a fabric with a TAV of 2.15. The decrease in the TAV of fabrics with a high TAV was less than that of fabrics with rather low TAV after the simulation test. These results suggest that the TAV of the fabrics can be used to characterize the appearance of a suit jacket after wear.

The hysteresis properties of fabrics such as bending (2HB) and shearing (2HG, 2HG5) increased markedly with wear, suggesting that fabric fatigue phenomena can be quantified by the increase in mechanical hysteresis properties. For the wool and wool/mohair blend fabrics, there was a linear relationship between the decrease in the total hand value (THV) after the simulation test and the increase in the 2HG measured after the deformation test of 10^4 cycles along the weft direction. Clearly fabrics with good handle durability show a small increase in the 2HG in the deformation test of 10^4 cycles.

In a second paper, Yokura and Niwa (1991) describe the results of wear for 1500 hours in men's summer trousers made from fabrics N36 and N37 and also from NZ100, an all-wool fabric. Changes in properties and structure of the fabrics, yarns and fibres as a result of wear were measured and explanations of the changes in fabric properties were suggested. SEM pictures of the fibres were taken before and after wear. **28G(1),(2)** show kid mohair fibres from the weft of fabric N36. As a result of wear, the fibres have lost their scales and become roughened on the surface. Fabric NZ100 contained a mixture of merino wool (22 micron) and coopworth (35 micron). **28G(3),(4)** displays a clear wearing away of the scales of the fine merino fibres, but there was little obvious change in the coarser fibres. The slightly coarser merino fibres in fabric N37 also show wear of the scales, **28G(5),(6)**, while the polyester fibres are roughened and debris has accumulated between the fibres. Some multiple splitting of wool fibres was found in the warp of N37, but otherwise there were few broken fibres.

The conclusions from this work include the statements;

Crimp in both yarns and fibers decreased due to considerable extension of the fabrics by the mechanical action of wear and abrasion. As crimp decreased, the yarns became flattened and increased the lateral pressure and contact region at yarn crossover points in the fabrics. These increases are considered to produce increased interyarn friction. The shear torque parameter C_J, which represents the effect of interfiber friction, increased with wear. Increasing interyarn and interfiber friction is considered to govern the increased hysteresis properties of fabrics. The removal of fiber scales and the ragged fiber surfaces were apparent in the SEM observations, and we believe this degradation was influenced by the increased interfiber friction.

The yield stress and strain of fibers for tensile and torsional properties decreased with wear. Fibers were fatigued more markedly in the transverse direction along the fiber axis during wear, so torsional properties of fibers are more affected by the degree of fatigue than tensile properties.

. . . results suggest that the performance of coopworth fibers is superior to kid mohair

. . . in fabric N37 . . . we believe that the wool fibers were weakened by the abrasive action of neighbouring PE [polyester] fibers.

The experimental work reported in this study provides a basis for engineering improved fabric quality and for developing new kinds of wool fabrics.

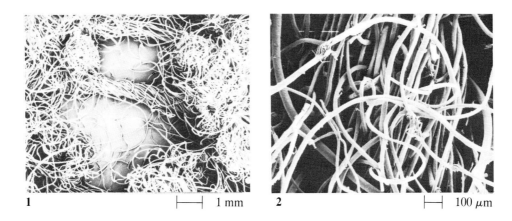

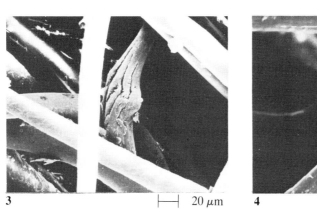

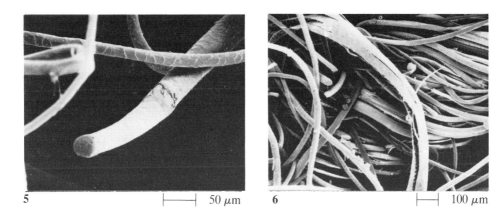

Plate 28A — Unsatisfactory performance of a pair of men's trousers.
(1) Hole formed after very little wear. (2) From region of wear close to the hole. (3)–(6) Details of fibre breakdown.

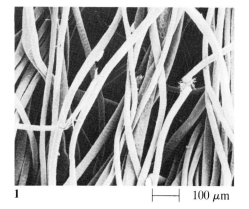

1   ⊢——⊣ 100 μm

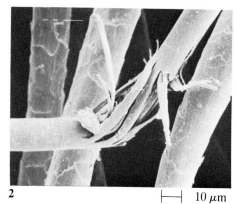

2   ⊢—⊣ 10 μm

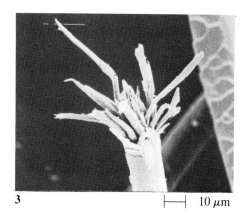

3   ⊢—⊣ 10 μm

4   ⊢——⊣ 50 μm

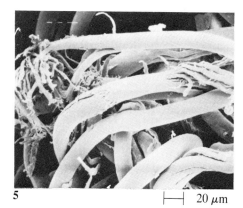

5   ⊢—⊣ 20 μm

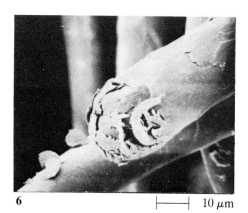

6   ⊢—⊣ 10 μm

Plate 28B — Study of men's trousers subject to wear trial by IWS.
(a) Grey wool with polyester pin-stripe.
(1) Multiple damage sites. (2) Axial splitting of wool fibres. (3) Bushy end of break. (4) Rounding off with further wear, with another fibre in multiple splitting state.

(b) Charcoal-grey wool/polyester blend.
(5) Multiple splitting in wool and polyester fibres. (6) Rounding off after more wear.

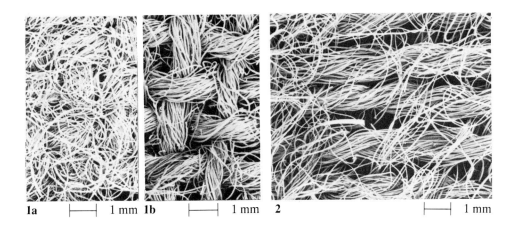

1a ├──┤ 1 mm 1b ├──┤ 1 mm 2 ├──┤ 1 mm

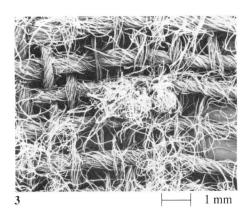

3 ├──┤ 1 mm

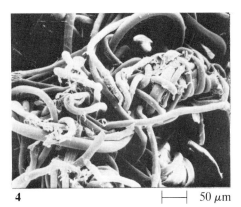

4 ├──┤ 50 μm

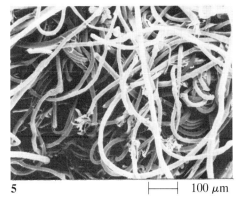

5 ├──┤ 100 μm

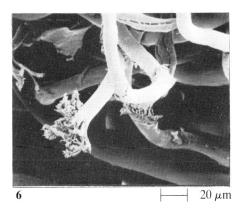

6 ├──┤ 20 μm

Plate 28C — Study of trousers subject to wear trial by IWS (continued).
(c) Fine wool tweed.
(1a) Appearance of relatively unworn fabric. (1b) Surface fibres worn away in seat region.

(d) Coarser wool tweed.
(2) Warp yarns worn away in seat region, leaving weft yarns crossing the gap.

Wear of ladies' wool/polyester trousers.
(3) Pill on fabric surface. (4) Detail of pill. (5), (6) Fibre damage by multiple splitting.

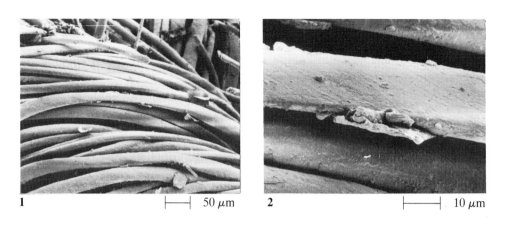

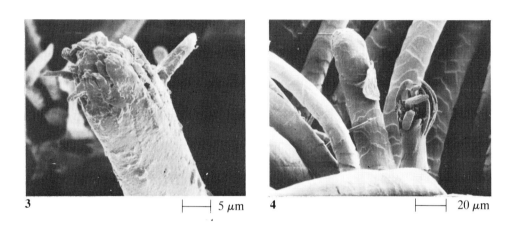

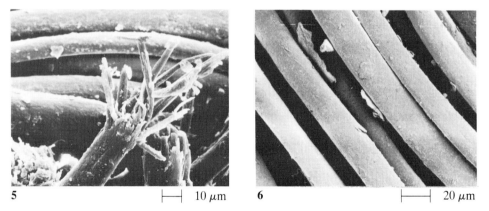

Plate 28D — Examination of worn trousers, made from a durable worsted fabric, after many years of wear.
(1)Fibres on outer surface of worn seat. (2) Detail of surface wear of fibres in worn seat. (3) Detail of fibre breakage in worn seat. (4) Fibres on inner surface of worn seat. (5), (6) Fibres in fabric at right knee.

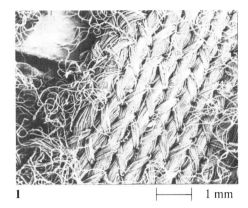

1 ⊢———⊣ 1 mm

2 ⊢———⊣ 50 μm

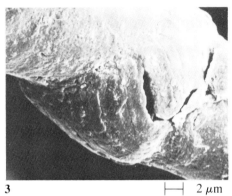

3 ⊢———⊣ 2 μm

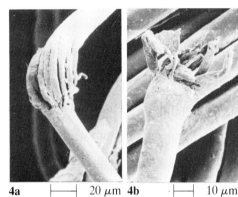

4a ⊢———⊣ 20 μm 4b ⊢———⊣ 10 μm

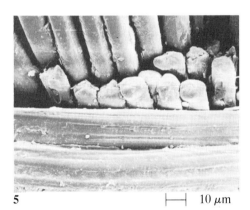

5 ⊢———⊣ 10 μm

6 ⊢———⊣ 50 μm

Plate 28E — Examination of pocket from trousers from 28D: continuous-filament polyester fabric.
(1) Edge of hole, near bottom seam. (2) Inner surface of pocket, away from hole. (3) Fibre at broken seam. (4a) Fibre at broken seam. (4b) Fibre at inner surface of pocket.

Label at waistband of the same trousers.
(5) Broken rayon fibres at fabric interstices. (6) Form of breakage of rayon fibres.

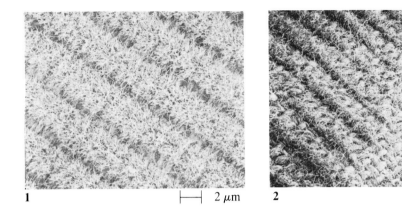

1 ⊢——⊣ 2 μm   2 ⊢——⊣ 2 μm

3 ⊢——⊣ 50 μm

4 ⊢——⊣ 50 μm

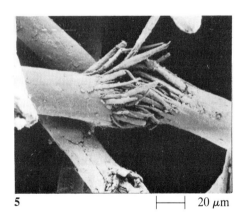

5 ⊢——⊣ 20 μm

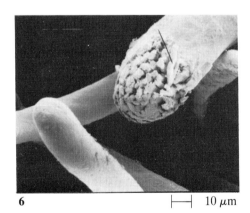

6 ⊢——⊣ 10 μm

Plate 28F — Worn cotton cord jeans.
(1) Relatively unworn fabric from pocket. (2) Fabric in well-worn area of right knee. (3) Cut fibres in little-worn fabric. (4) Fibres in worn area of right knee.

Worn wool/polyester school blazer.
(5) Wool fibre broken by multiple splitting. (6) Partial and complete rounding off of broken fibre ends.

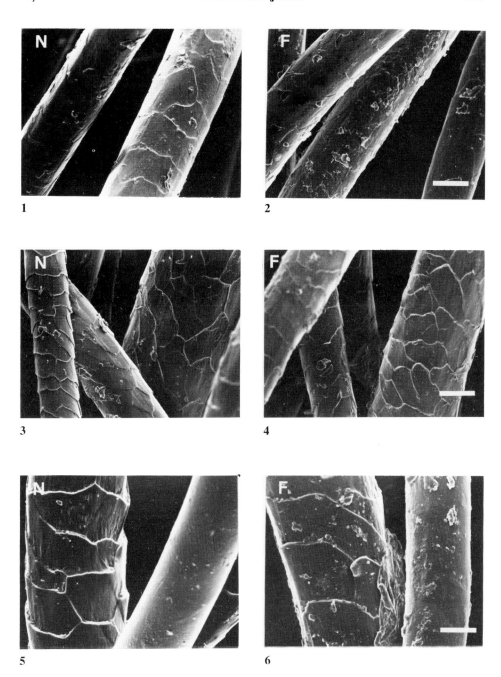

Plate 28G — Wear in men's summer trousers, from Yokura and Niwa (1990).
(1) Mohair fibres in fabric N36 before wear. (2) After wear. (3) Coopworth and merino fibres in fabric NZ100 before wear. (4) After wear. (5) Polyester and merino fibres in fabric N37 before wear. (6) After wear. [Bars are 10 μm.]

29

SHIRTS

Wear in shirts is influenced both by the material, design and make-up of the garment, and by the habits of the wearer. The worst wear commonly occurs along the collar fold, and in long-sleeved shirts, at the edges of the cuffs. In both these places it is associated with sharp folds, supported by interlinings, which are usually bonded to the fabric, and also in collars by point stiffeners. Wear of collars and cuffs spoils the appearance of shirts, so that they are often rejected, at least for fashionable wear, long before the rest of the garment has sustained much damage. Sometimes, presumably after a lot of desk-work, there is severe wear at the elbows. Wear can also occur at shirt front-openings, and there may be accidental tearing, either during wear or laundering.

A popular modern shirting fabric is woven from 65% polyester/ 35% cotton yarns, as illustrated in **29A(1)**. Note the bulbous ends of polyester fibres, which indicate that the fabric has been singed. The front of the shirt, as shown in **29A(1)**, has suffered little wear and hardly any fibre damage is apparent; but there is severe wear at collars and cuffs, both elbows have worn right through in places and there has been wear elsewhere.

The loss of yarn at the collar fold is shown in **29A(2)**, and in cuff edges in **29A(3)**, where the interlining adhesive and ground-in debris can be seen between the remaining weft yarns. The rigidity given to the fabric, as a result of being stuck to the interlining, is apparent from a section through a cuff, **29A(4)**. This rigidity intensifies the forces on the fabric, since they cannot be relieved by buckling of the fabric; and the adhesion concentrates the forces on individual fibres, which are not free to move out of the way of external pressure. These constructional features of the garment are an important factor in wear at collars and cuffs.

Elsewhere in the shirt the degree of damage depends more on the level and frequency of forces applied, either during use or in laundering. In the front the forces are small, and the damage, as seen in **29A(1)** is negligible. At the inside shoulder there is a little more damage. The polyester fibre breakdown is by multiple splitting, **29A(5a),(5b)**. This may be associated with pilling, as discussed in Chapter 30, and occurs at anchor points in the yarn, **29A(6)**.

The severe wear at the elbow is shown in **29B(1)**, and is a result of the rubbing of the fabric on solid surfaces during use. The warp yarns are wearing away, and piles of fibre lie on the surface. The most common form of fibre damage to be seen is localized multiple splitting, **29B(2),(3)**, similar to that found in laboratory biaxial rotation fatigue (Chapter 13), although there is some more extensive splitting and fibrillation, **29B(4)**, which may be due to peeling from the surface. Because most of the cotton has worn away in this region, the observed effects are mostly in polyester fibres.

The effects of wear at the elbow can also be illustrated by another polyester/cotton shirt. The relatively undamaged fabric from the front facing is shown in **29B(5)**, and contrasts with the worn fabric from the elbow, **29B(6)**, where some of the cotton has been lost and the material has stretched, leaving a more widely spaced scaffolding of threads.

Another example of severe wear on the edge of a cuff is illustrated in a 100% cotton shirt in **29C(1)**. The warp threads running along the edges have worn completely away, and some weft threads have broken. The fibre damage consists of fibrillation, **29C(2)**, and splitting, **29C(3)**, which is typical of cotton. Severe wear is also seen on the collar, **29C(4)**, particularly near the point. Wear has continued through to the interlining surface, where the yarns have partly worn through, **29C(5)**, leaving tufts of fibre ends at the interstices of the weave. Typical fibre damage in this region is shown in **29C(6)**.

Examination of one set of experimental cotton shirts with an easy-care finish from a wear trial directed by the late Mrs J. Lord, at the Shirley Institute, frequently showed not only some wear on the fabric surface but also failure by tears parallel to the warp and weft on either side of the front opening just below the collar, **29D(1)**. The general nature of the fibre damage on the surface is fibrillation and splitting, **29D(2).** The edge of the tear is shown in **29D(3)**, where the cotton fibres may break by splitting along the lines of the spiral structure, **29D(4)**, but often show much sharper angular breaks and cracks, **29D(5),(6)**. The latter are characteristics of resin-treated cotton, which can be unduly weakened and embrittled by the chemical cross-linking.

An example of more severe wear of the collar of a cotton/polyester shirt is shown in **29E(1)**. In addition to the wearing away of yarns, the appearance of the material is spoilt by the development of pills, which are tangles of polyester and cotton attached to the fabric by a few anchor fibres. In the severely worn region the cotton has disapppeared, so that the visible fibre damage is in the polyester fibres, **29E(2),(3)**. These pictures show the complete sequence: first in **29E(3)** from undamaged fibre, through multiple splitting, to break with a bush end; and then in **29E(2)** to a rounding of the broken end with further wear.

Fibre damage in a warp-knitted, continuous-filament nylon shirt shows a similar pattern, **29E(4)–(6)**.

A relatively undamaged region of another warp-knit nylon shirt is shown in **29F(1)**. However, even in this region, fibres are beginning to break, and show the typical bushy and rounded ends, **29F(2),(3)**. There is also evidence of peeling, **29F(4)**, which may be a consequence of shear on the surface of the fibres at the yarn crowns. More severe wear, with many broken fibres in the yarn, is shown in **29F(5)**, but the fibre breakdown is similar, **29F(6)**.

It is stated at the beginning of this chapter that wear is in part influenced by the material used. Most of the shirts examined have been made from a poplin weave fabric, a weave which has been most frequently used in shirtings for many years. In this weave, warp and weft yarns interlace alternately in both yarn directions as shown in **29A(1),(3)** and **29D(3)**. There are approximately twice as many warp as weft yarns and in consequence the fabric can be said to be warp faced. Weft yarns are more deeply seated within the fabric and only when the warp has been seriously eroded are weft yarns susceptible to wear.

The design of a shirt demands that collars and cuffs are cut along the warp direction; for example in a striped shirt the stripes run the length of the collar and around cuffs. Thus warp yarns at the parallel to a collar fold and cuff edge receive the brunt of the wear and only when they are virtually worn away do weft yarns begin to suffer serious damage and fail, **29C(1)**.

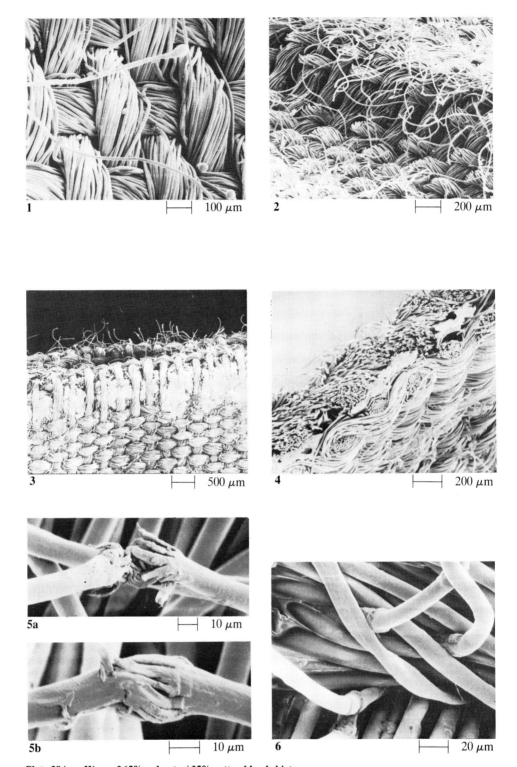

Plate 29A — Wear of 65% polyester/ 35% cotton blend shirt.
(1) Fabric from front of shirt, with little wear. (2) Worn collar fold. (3) Worn cuff edge. (4) Section through cuff, showing bonded interlining. (5a,b) Fibre damage from inside shoulder. (6) Anchor fibres in a pill on inside shoulder.

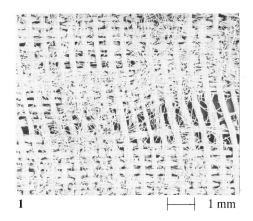
1   1 mm

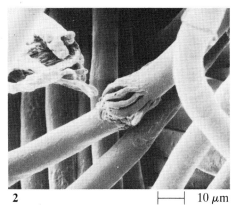
2   10 μm

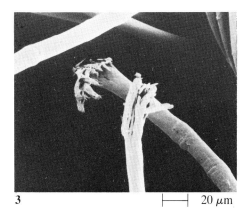
3   20 μm

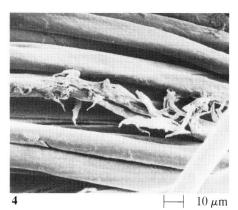
4   10 μm

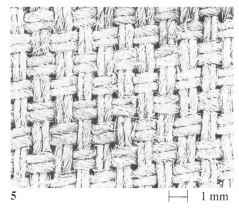
5   1 mm

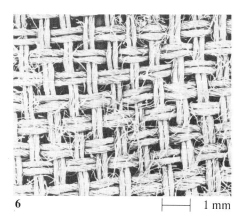
6   1 mm

Plate 29B — Wear of 65% polyester/ 35% cotton blend shirt (continued).
(1) Badly worn material at the elbow. (2)–(4) Details of fibre damage at elbow.
Wear in another polyester/cotton shirt.
(5) Front facing, with little sign of wear. (6) From elbow.

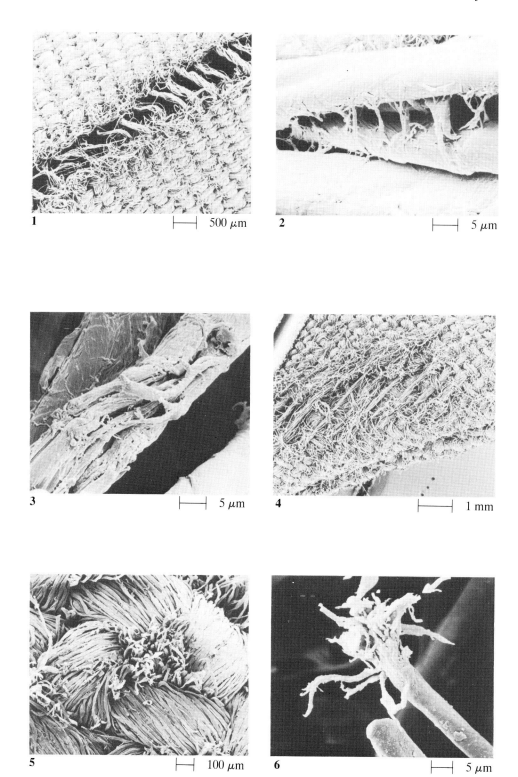

Plate 29C — Wear of cotton shirt.
(1) Edge of cuff. (2), (3) Fibre damage in cuff. (4) Damage at point of collar. (5) Partial breakage of yarn in collar interlining. (6) Fibre break in collar.

Plate 29D — Experimental resin-treated cotton shirt from a wear trial.
(1) Tears in fabric. (2) Fibre damage on fabric surface. (3) Edge of tear. (4)–(6) Detail of fibre breakage.

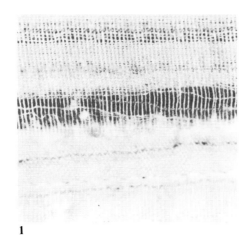

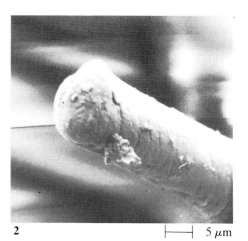

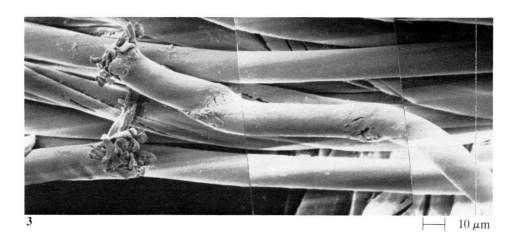

Plate 29E — Worn collar of 65% polyester/ 35% cotton blend shirt.
(1) Collar fold region, showing loss of yarn and pills. (2), (3) Polyester fibre breakdown.
Worn collar of warp-knitted nylon shirt.
(4)–(6) Fibre breakdown.

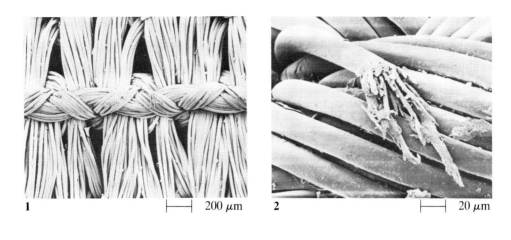

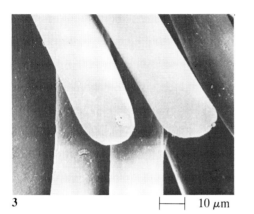

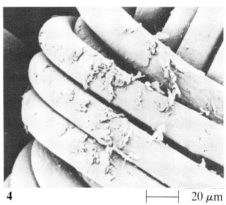

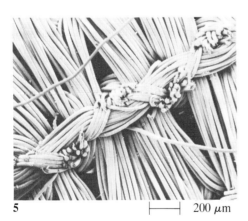

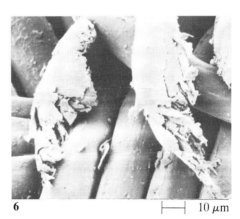

Plate 29F — Worn collar of another warp-knitted nylon shirt.
(1) Relatively undamaged fabric. (2) Fibre break by multiple splitting. (3) Rounding of end after further wear. (4) Peeling on fibre surface. (5) Region of greater wear. (6) Fibre breakdown.

30

WEAR AND PILLING IN KNITTED AND WOVEN FABRICS

Pilling is a form of change of fabric appearance, in which small tangled balls of fibre are distributed over the surface of the material. It is most prevalent in loosely constructed knitted or woven fabrics made of staple fibres, but can occur in other circumstances. The density of pills results from a balance between the rate of pill formation and the rate of loss of pills by their breaking off; and it can thus be minimized either by preventing pills forming, as a result of holding fibres more firmly in the fabric, or by allowing them to break off more easily, as a result of weakening the fibre.

Wear is associated with pilling in several ways. Firstly, the pilling itself is a form of deterioration of appearance brought on by use, and thus is 'wear' in the broad sense of the term. Secondly, as shown by the work of W. D. Cooke (1981, 1982, 1983, 1984, 1985), pilling can be a major source of fabric attrition, with the material becoming thinner, as pills form and break off, and may eventually lead to the formation of a hole. Thirdly, again from the work of Cooke, the mechanism of pill formation and loss is closely linked with the development of multiple splitting, as found in biaxial rotation fatigue (Chapter 13).

A general view of extensive pilling on a merino wool worsted-spun ladies' jumper is shown in **30A(1)**; and a microscopical view of the characteristic fatigue of wool fibres in pills is shown in **30A(2)**. These pictures are of samples from a large wear trial by IWS, but unfortunately a detailed SEM study was not made. Light wear on the back of a discarded wool jersey is shown in **30A(3)**, together with detail of the multiple splitting, **30A(4)**, which leads to breakage, **30A(5)**, and later rounding of the fibre end, **30A(6)**.

A much more detailed and carefully monitored study was made of the development of pilling in men's knitted underwear, made of cotton and various man-made fibres. A typical pill is shown in **30B(1)**, and the type of entanglement which occurs is seen in **30B(2)**. If a length of fibre is removed from a pill, it is found to contain a sequence of zones of multiple splitting, and a bushy broken end, **30B(3)**. This led Cooke to describe the mechanism of pilling as intermittent pull-out and roll-up. A schematic view of a pill is shown in Fig. 30.1, with the tangled ball connected to the fabric by anchor fibres. Bending and twisting at the anchor point cause multiple splitting fatigue, and the increased flexibility of the fibre then allows tension to be transmitted more effectively, so that another length of fibre is pulled out of the fabric. The excess length then gets rolled up into the pill. As the pill gets larger, the greater forces developed will cause zones with multiple splits to break; and when all the anchors have

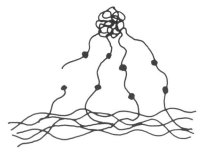

Fig. 30.1 — Pill near the end of its life, based on a schematic sequence drawn by W. D. Cooke.

ruptured, the pill will fall off. Other examples of the multiple splitting fatigue are shown in **30B(4)(5)**.

The last set of illustrations comes from a laboratory pilling test of woven overall fabrics. An early stage of pill development in a polyester/cotton fabric is shown in **30C(1)**, and a fully developed pill in **30C(2)**. Although some cotton fibres can be observed in the pills, the main component is polyester, in which multiple splitting fatigue can be seen, **30C(3)**.

A badly pilled polyester/modal (rayon) fabric showed considerable multiple splitting of polyester, **30C(4)**, together with broken fragments of the modal, **30C(5)**. The initial twisting and splitting of the polyester, which leads to pills, can also be seen in **30C(5)**. Little modal debris is found in the pills, but fragments are easily transferred to other fabric samples, **30C(6)**.

Wear in a wool cardigan shows characteristic wear and lifting of scales on the surface of the fibres, **30D(1),(2)**. Fibre breaks have been worn into the typical rounded end, **30D(3)**.

A hole in a cotton knit garment, **30D(4)**, illustrates the way in which fibres break up by multiple splitting into coarse fibrils. In contrast to this, a cotton T-shirt, in which wear will be due mainly to washing, shows the peeling away of fine fibrillar sheets, **30D(5),(6)**. This is similar to the wear in bed-sheets and towels, shown in **32A–D**. In some of the fibres, **30E(1),(2)**, larger chunks split off.

There are a few examples of the wear of silk in Chapter 31, but the break-up by peeling away of fibrillar layers is more clearly shown in **30E(3)–(6)**. The damage is most severe on the crowns of the yarns in the woven fabric.

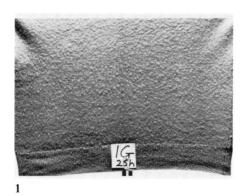

1

2                    $\longmapsto$ 10 μm

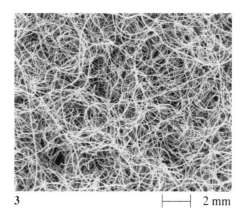

3                    $\longmapsto$ 2 mm

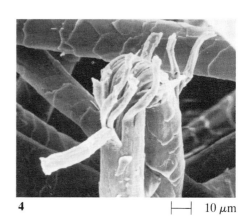

4                    $\longmapsto$ 10 μm

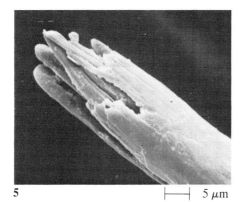

5                    $\longmapsto$ 5 μm

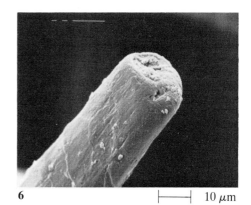

6                    $\longmapsto$ 10 μm

Plate 30A — Wear in wool sweaters.
(1) Ladies' merino worsted jumper, after being worn for 50 hours, with a wash at 25 hours. (2) Wool fibre in such a worn jumper. (3) Least-worn area on the back of a discarded man's jersey. (4)–(6) Detail of fibre failure.

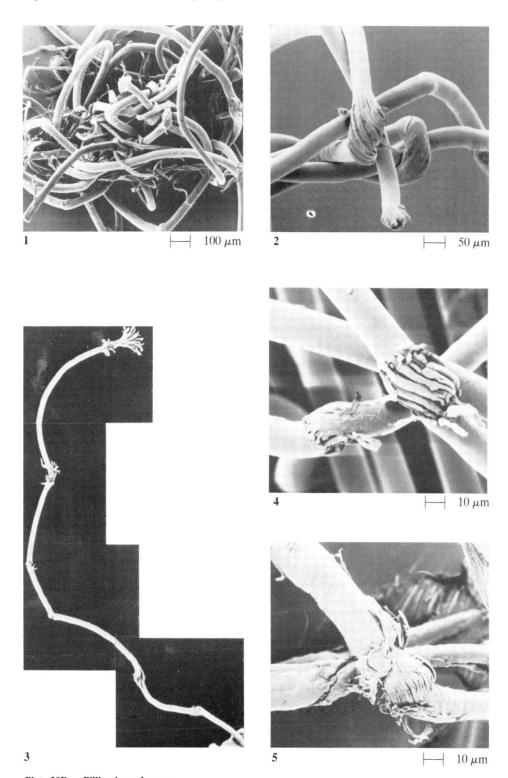

Plate 30B — Pilling in underwear.
(1) A typical pill. (2) Fibre entanglement. (3) A fibre from a pill. (4) Multiple splitting in polyester, (5) Multiple splitting in cotton.

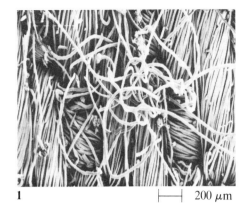

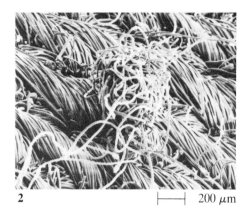

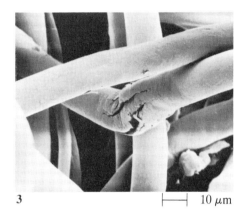

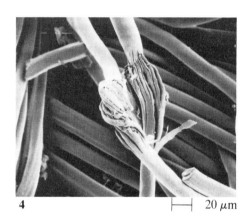

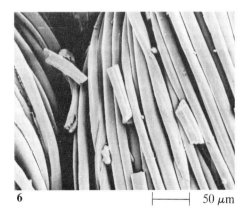

Plate 30C — Pill-testing of woven overall fabrics.
(1) Early stage of pill development in polyester/cotton fabric. (2) Well-developed pill in same fabric. (3) Detail of fibre splitting in another polyester/cotton fabric. (4) Polyester splitting in polyester/modal fabric. (5) Broken modal fragments, and initial twisting and splitting of polyester. (6) Modal fragment present on the surface of a polyester/cotton fabric.

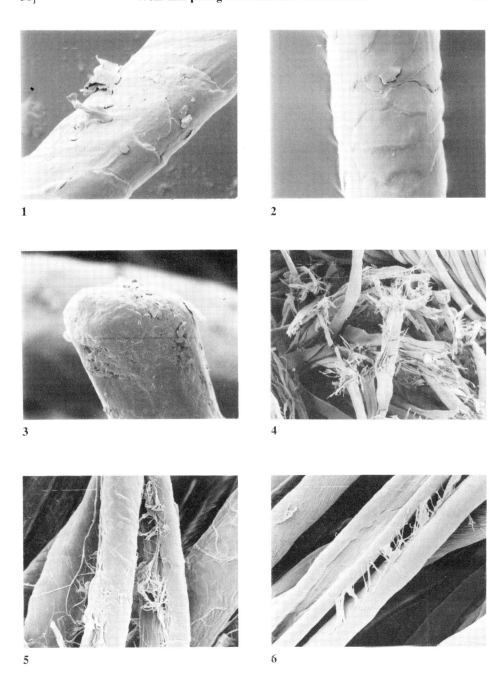

Plate 30D — Wear in a wool cardigan.
       (1),(2) Surface wear and lifting of scales. (3) Rounded end of broken fibre.
Hole in a cotton knit.
       (4) Fibre failure by multiple splitting.
Worn cotton T-shirt.
       (5),(6) Peeling of fibrillar layers.

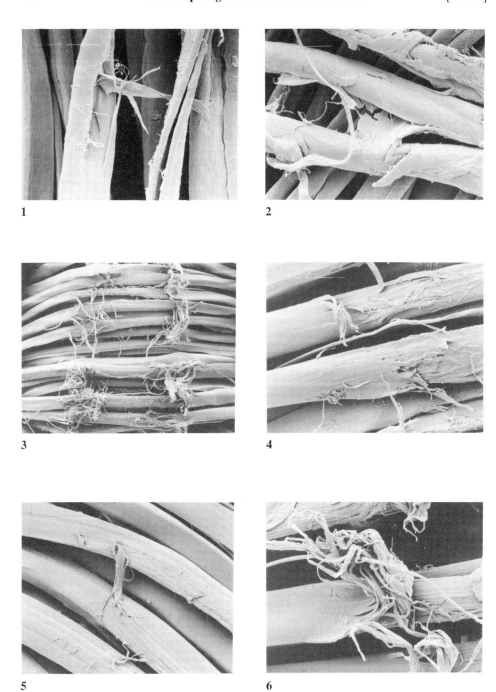

1

2

3

4

5

6

Plate 30E — Worn cotton T-shirt (continued).

(1),(2) Splitting away of larger pieces.

Worn silk fabric.

(3) Damage concentrated on yarn crowns. (4)–(6) Detail of fibre peeling.

31

SOCKS, UNDERWEAR AND OTHER ITEMS

Although knitted nylon socks are extremely durable, in contrast to wool socks, which once made darning a weekly household chore, they do eventually wear into holes, **31A(1)**, through extensive fibre breakdown by multiple splitting, **31A(2)**. The nylon fibres break by developing long axial splits, **31A(3)**, which may be due either to repeated bending (Chapter 12) or bending and twisting (Chapter 13). The individual portions of the split then break to give an end with many splits, **31A(4),(5)**. In a little-worn area on top of the foot of another nylon sock, there is evidence of an axial split, **31A(6)**, which might be due either to tensile fatigue (Chapter 11) or to peeling by surface shear (Chapter 14).

Thicker, absorbent sports socks are differently constructed. Traditionally, they would have been knitted in thick cotton or wool yarns, but a common modern method is to use a combination of continuous-filament nylon yarns, to give durability, and acrylic yarn, to give bulk, in a knitted pile construction. An example of a black nylon/red Orlon sock, which has worn away at the ball of the foot, is shown in **31B(1)** and illustrates the various stage of wear. Away from the hole, the material is only moderately worn and the loops of the pile yarn can be seen. Coming in towards the centre of wear, there is increasing wear of the pile and then a region where only the nylon yarns remain. There is an actual hole just outside the worn region, probably due to snagging on a projection. There was another hole in the heel of this sock. The breakdown of the acrylic fibres is by multiple splitting, **31B(2)**.

A moderately worn region in a white nylon/Acrilan sock is shown in **31B(3)**, and demonstrates that the loss of the acrylic fibres is associated with the formation of pills, which is the attrition mechanism discussed in Chapter 30. Examination of a region where the Acrilan has been completely lost shows no damage to the nylon fibres, **31B(4)**. The fibre breakage is frequently by the multiple splitting, **31B(5),(6)**, which is characteristic of combined bending and twisting. However, it can also occur with a central split, **31B(7),(8)**, which was found in simple flexing, as discussed in Chapter 8. Sometimes there appears to be a combination of both effects, **31B(9)**.

Knitted nylon tights can suffer damage by snagging. Any broken fibres may slip back into the knitted structure, or become tangled up in pills, **31C(1)**. The individual fibre breaks are a rare example in actual use in clothing of failure in the high-speed tensile break mode with mushroom ends, **31C(2),(3)**. The reason is that this is a situation in which a filament is caused to break rapidly by a sudden pull, when it catches and snags.

A silk vest which had been used for many years finally deteriorated to the point where the knitted structure became thin, **31C(4)**, from loss of fibre. The silk fibres show evidence of surface peeling, **31C(5)**, and some fibres had broken from tensile loading of the weakened structure, **31C(6)**.

In woven cotton/polyester underpants, the fabric in relatively undamaged regions is compact, **31D(1a)**; but in regions of high wear, the cotton has mostly been lost from the yarns, leaving a thin open weave of residual polyester, **31D(1b)**, with some tangled cotton fibres on the surface. In regions of intermediate damage, the cotton fibre is seen to break down by multiple splitting, **31D(2)**. Eventually, even the polyester goes from one yarn direction, **31D(3)**. The early stages of damage are shown in **31D(4)**, with fibre details in **31D(5)(6)**.

A similar loss of cotton, which is present in the little-worn regions, **31E(1a,b)**, is found in knitted cotton underpants, where the fabric is strengthened by knitting a fine continuous-filament nylon yarn in together with the cotton yarn. In worn regions, **31E(2a,b)**, only the filament yarns remain in the knit structure, with some tangles of cotton fibre attached. The cotton fibre breaks down, probably to a considerable extent in washing, by peeling off of fibrillar layers, **31E(3)**. However, it is also interesting to see that the nylon suffers damage by surface peeling, **31E(4)–(6)**.

As a conclusion to the chapters on ordinary clothing, and before going on to household products and industrial workwear, we give two examples of wear in accessories.

A tie made from the expensive high-fashion continuous-filament yarn, Qiana, which is now no longer made, developed excessive pilling in use. Little-worn fabric, from under the label, is shown in **31F(1)**. On the front of the tie, the appearance had deteriorated, **31F(2)**, due to the pills, **31F(3)**. Detailed examination shows some fibre breaks, **31F(4)**, with bulbous ends similar to high-speed tensile breaks, perhaps due to pulling of the pills, but none of the multiple splitting of anchor fibres, which is reported for other types of fibre in Chapter 30. The strength of Qiana prevents the pills breaking off, and so leads to the large unsightly accumulation of pills on the surface.

A worn silk tie is damaged by multiple splitting of the fibres, **31F(5)**, followed by break to give a bushy end, **31F(6)**, in the common pattern of wear.

Cotton handkerchiefs usually become thin and tear with continual use and laundering. This can be accentuated along creases, which are pressed during ironing, as illustrated in **31G(1)**. Detail of the way in which the yarn has worn thin is shown in **31G(2)**. The breakdown of the cotton fibres starts by splitting along the helical lines between fibrils, **31G(3)**, which will be in both Z and S senses in different parts of the fibres, **31G(4)**. A pronounced split is shown in **31E(5)**, and an almost broken fibre in **31E(6)**. Although handkerchiefs are frequently used surface attrition of fibres is slight, the main causes of damage being laundering and ironing.

Plate 31A — Worn knitted nylon socks.
(1) Hole in heel. (2) Fibre splitting near hole. (3)–(5) Detail of fibre break. (6) From region of little wear on top of foot of another sock.

1

2 ├─┤ 100 μm

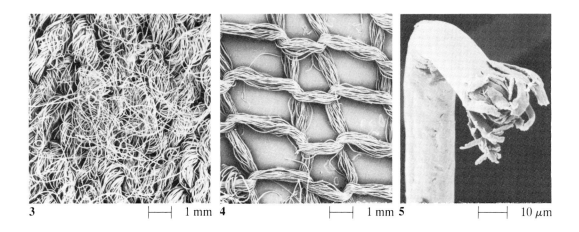

3 ├─┤ 1 mm **4** ├─┤ 1 mm **5** ├─┤ 10 μm

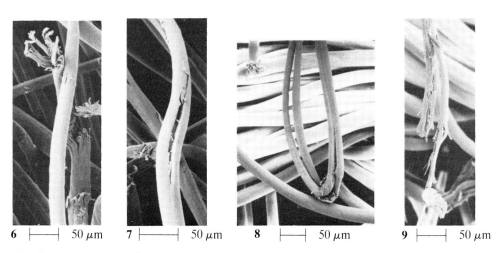

6 ├─┤ 50 μm **7** ├─┤ 50 μm **8** ├─┤ 50 μm **9** ├─┤ 50 μm

Plate 31B — Wear in nylon/Orlon sports sock.
(1) Macrophotograph showing worn area at ball of foot. (2) Orlon fibre breakage.
Wear in nylon/Acrilan sports sock.
(3) Pill in region of moderate wear. (4) Nylon yarns in severely worn region. (5)–(9) Details of Acrilan
fibre breakage.

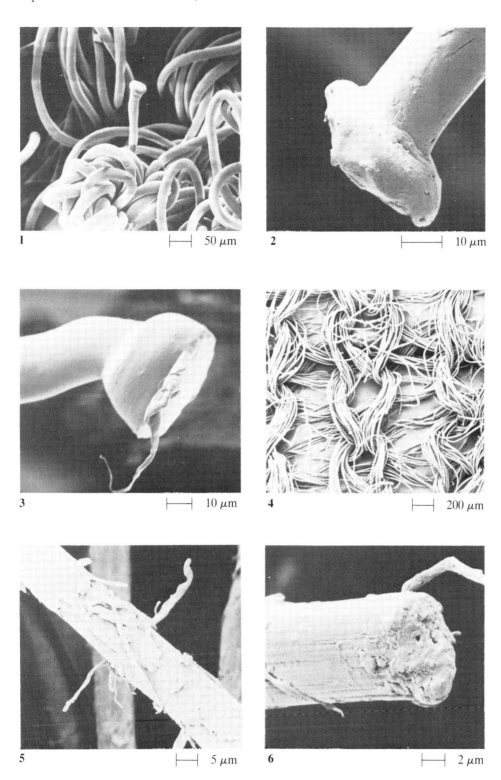

Plate 31C — Ladies' nylon tights.

(1) Pill formed after snagging. (2), (3) Details of fibre break.

Worn silk vest.

(4) General appearance. (5) Surface peeling. (6) Fibre break.

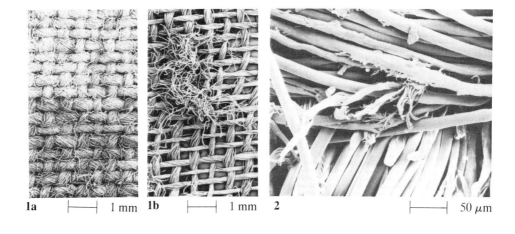

1a ├───┤ 1 mm   1b ├───┤ 1 mm   2 ├───┤ 50 μm

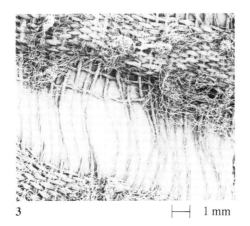
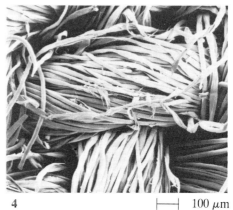

3 ├───┤ 1 mm   4 ├───┤ 100 μm

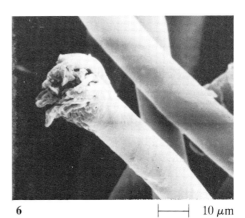

5 ├───┤ 20 μm   6 ├───┤ 10 μm

Plate 31D — Worn woven cotton/polyester underpants.
(1a) Relatively undamaged region. (1b) Region of severe wear. (2) Form of fibre breakdown. (3) Region of very severe wear. (4) Region of intermediate wear. (5) Multiple splitting of fibres. (6) After breakage, partial rounding of bushy end.

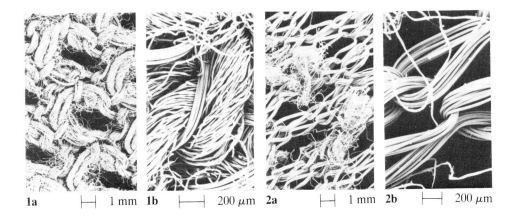

1a ⊢—⊣ 1 mm **1b** ⊢—⊣ 200 μm **2a** ⊢—⊣ 1 mm **2b** ⊢—⊣ 200 μm

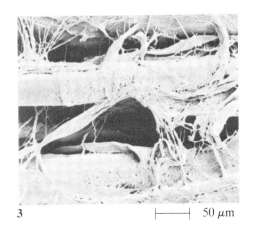

3   ⊢——⊣ 50 μm

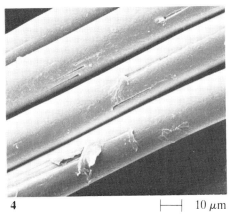

4   ⊢——⊣ 10 μm

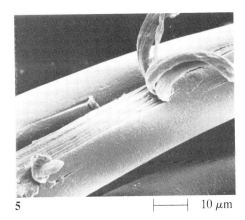

5   ⊢——⊣ 10 μm

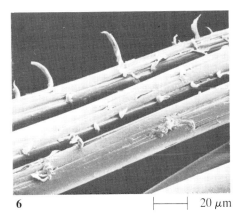

6   ⊢——⊣ 20 μm

Plate 31E — Worn knitted cotton/nylon underpants.
(1a,b) Little-damaged fibre. Note nylon reinforcing yarn. (2a,b) Badly worn fabric. (3) Cotton fibre
breakdown. (4)–(6) Surface peeling of nylon.

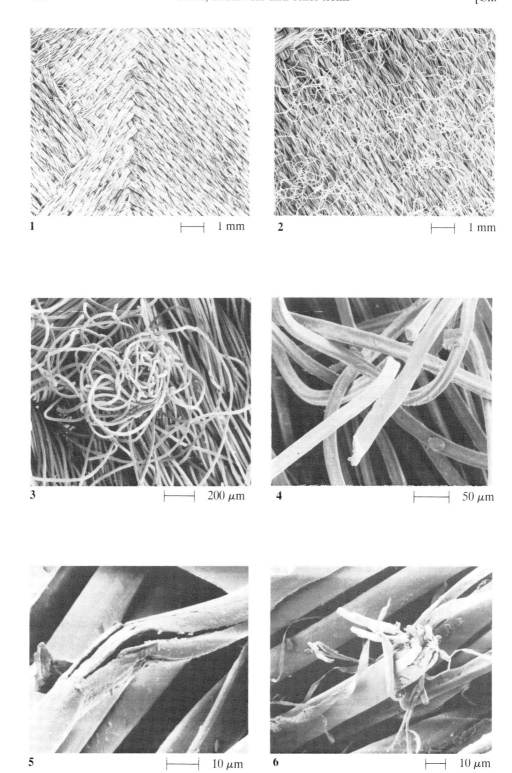

Plate 31F — Worn Qiana tie.
(1) Little-worn material, under the label on the back of the tie. (2) Material on front of tie. (3) Detailed view of a pill. (4) Detail of fibres.
Worn silk tie.
(5) Multiple splitting of fibre. (6) Fibre break.

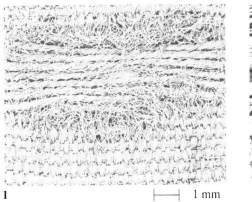

1   ⊢━━┥ 1 mm

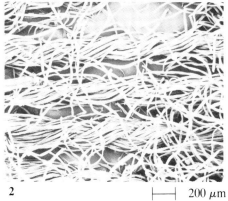

2   ⊢━━┥ 200 μm

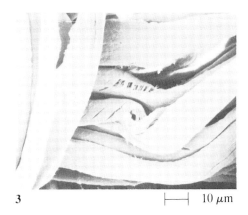

3   ⊢━┥ 10 μm

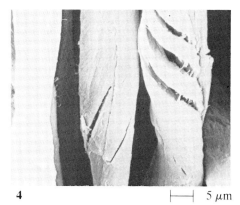

4   ⊢━┥ 5 μm

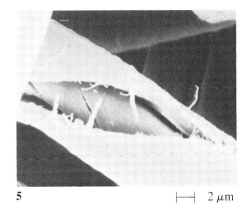

5   ⊢━┥ 2 μm

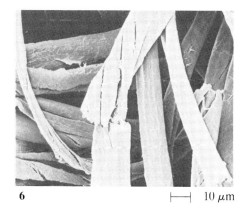

6   ⊢━┥ 10 μm

Plate 31G — Worn cotton handkerchief.
(1) Worn area near crease put in by ironing. (2) Detail of fabric wear. (3)–(6) Breakdown of cotton fibres.

32

HOUSEHOLD TEXTILES

Textiles are used in many places in the home. Articles collected after severe wear, which is common before they are discarded, can be examined to show forms of breakdown; but the detailed investigation of comparative wear requires well-designed trials.

An extensive and carefully monitored study of five types of bed-sheets was carried out at the Shirley Institute, and some samples were made available for SEM examination. Four different organizations and types of laundries were used in the trial, which lasted for six years. One feature of the study was to see how much loss of strength could be attributed to laundering alone, compared with use plus laundering, and tests were also made to separate the incidence of chemical and mechanical damage. Reference should be made to the paper by Lord (1971), for the details of this study; but Table 32.1 shows that large differences result from fabric type and laundering conditions.

A laundered-only cotton/nylon blend sheet after 30 cycles of laundering is shown in **32A(1)**: there is some peeling away of layers of fibrils in the cotton, but not the sort of damage which would cause much loss in strength. In contrast to this, the use and laundering of an all-cotton sheet after the same number of cycles does show appreciable splitting of cotton fibres, **32A(2)**. More extensive separation of lamellar layers in cotton fibres is shown after more use and laundering in **32A(3)**. Splitting of cotton can be seen in the nylon/cotton sheet after use, **32A(4)**, but the nylon is undamaged. Rayon fibres in a blend show peeling of strips of material, **32A(5)**, and this is marked even after rather little use, **32A(6)**.

There is an interesting difference between the apparent extent of damage as revealed by different types of examination of the sheets. More damage is visible in SEM pictures of the sheets after use at the men's hostel, and this is caused by severe mechanical wear, although the splitting does not necessarily weaken the fibres unduly. But the sheets after use at the school are actually weaker, as shown by their lower tear strength: in these sheets the damage is chemical, and the material gives higher values in a cotton fluidity test, which is an inverse measure of degree of polymerization. The greater chemical damage was due to the use of more bleach in order to get the schoolboys' sheets clean. The internal chemical damage is not shown up in the SEM pictures.

Wear in ordinary use of a patterned domestic towel, made of cotton in the usual loop-pile woven terry fabric, is illustrated in **32B**. There was an expected variation of wear in different parts of the towel, but, in addition, the severity of wear in any location varied with the colour of the yarn. In a region of least wear the loops are clearly visible on the fabric surface, **32B(1)**, but some fibrillation of the cotton fibres is apparent, **32B(2)**, as would be expected from repeated washing, even without the effects of use. In regions of severe wear, top left of **32B(3)**, the loops have almost disappeared, leaving the base weave clearly visible. An interesting feature of **32B(3)** is that it shows the boundary between two differently coloured areas of the pattern: the bottom right still shows the relatively undamaged loops. In the severely worn region there is considerable damage to the cotton fibres, **32B(4)**, with pronounced splitting, **32B(5)**, and fibrillation, **32B(6)**.

The monitored use of towels was only available for cabinet towels supplied by a rental company. One sample was a smooth woven cotton fabric, shown as a control after one washing, in **32C(1)**. The initial processing in preparation for use has already led to the peeling away of some layers from the cotton fibres, **32C(2)**. After being washed 50 times (without use), the peeling is more extensive, **32C(3)**. A cotton/Vincel (rayon) blend after 100 washings shows considerable smearing of layers from the cotton, **32C(4)**. The effect of 200 washing cycles on all-cotton fabric is shown in **32C(5),(6)**.

After 100 washes the cotton/rayon blend shows some peeling of cotton and the first signs of damage in the rayon, **32D(1)**, but after 200 washes the Vincel (rayon) is showing massive

Table 32.1 — Comparative life expectancy of sheets
Number of wash cycles giving a reduction of strength to 100 gf/thread, which is about 25% of
original value. UL = used and laundered; LO = laundered only.
The lowest and highest values, UL and LO, are shown in bold type.

Organization		Sheet type				
		100% cotton	65% cotton/ 35% rayon	48% cotton/ 52% rayon	89% cotton/ 11% nylon	80% cotton/ 20% nylon
School/ private laundry	UL	**53**	56	62	55	60
	LO	236	194	**182**	214	220
College/ local authority laundry	UL	105	100	119	108	**132**
	LO	216	196	198	264	310
Hostel/ commercial laundry	UL	101	105	114	104	126
	LO	295	200	207	332	**1021**
Nurses' home/ hospital laundry	UL	88	77	91	80	105
	LO	280	217	197	296	450

splitting, **32D(2)**. After the same number of washes polyester fibres in a blend with cotton have no signs of damage, **32D(3)**.

The final illustrations in this chapter are an example of uncontrolled wear: a cotton curtain, used for about six years in a men's cloakroom, in conditions of considerable exposure to light. A general view of a tear is shown in **32D(4)**, with details of fibre breakage in **32D(5)(6)**. The chemical changes, associated with tendering by light, have led to sharp breaks at an angle to the fibre axis.

Plate 32A — Wear in bed-sheets.
(1) 80% cotton/20% nylon sheet, 30 cycles commercial laundering only. (2) 100% cotton sheet, 30 cycles, used in hostel, commercial laundering. (3) as (2), 50 cycles. (4) 80% cotton/20% nylon sheet, 30 cycles, used in hostel, commercial laundering. (5) 48% cotton/52% rayon sheet, 50 cycles, used in school, private laundry. (6) as (5), 10 cycles, used in hostel, commercial laundering.

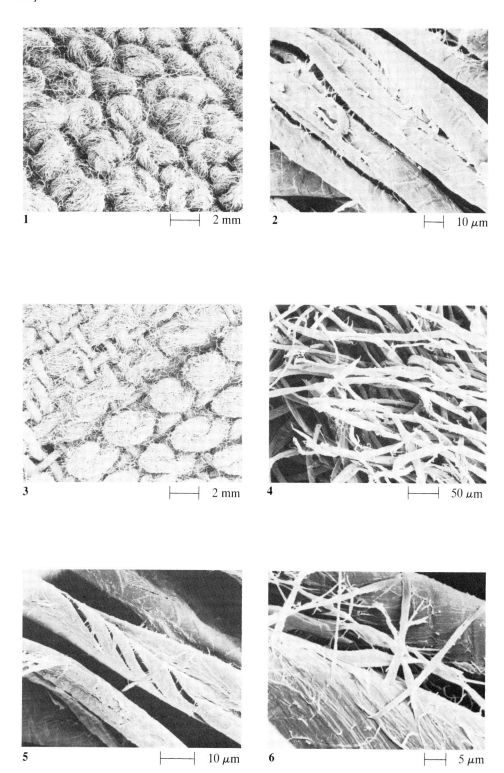

Plate 32B — Uncontrolled wear of a domestic bath towel.
(1) Area of least wear (pink). (2) Detail of peeling of cotton fibres in this area. (3) Boundary of colour difference in towel, with severely worn area in top left, and less worn in bottom right. (4)–(6) Detail of damage in badly worn area.

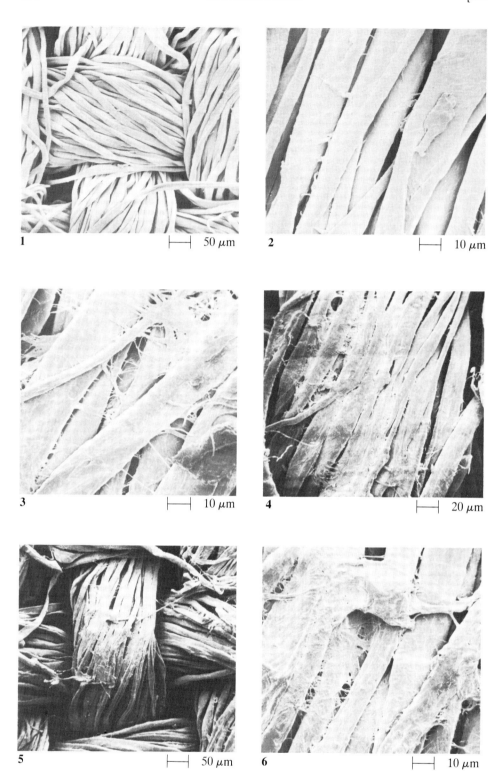

Plate 32C — Washing-only trials of cabinet towels.
(1),(2) 100% cotton, after one wash. (3) 100% cotton, after 50 washes. (4) 50% cotton/50% Vincel
(rayon), after 100 washes. (5),(6) 100% cotton, after 200 washes.

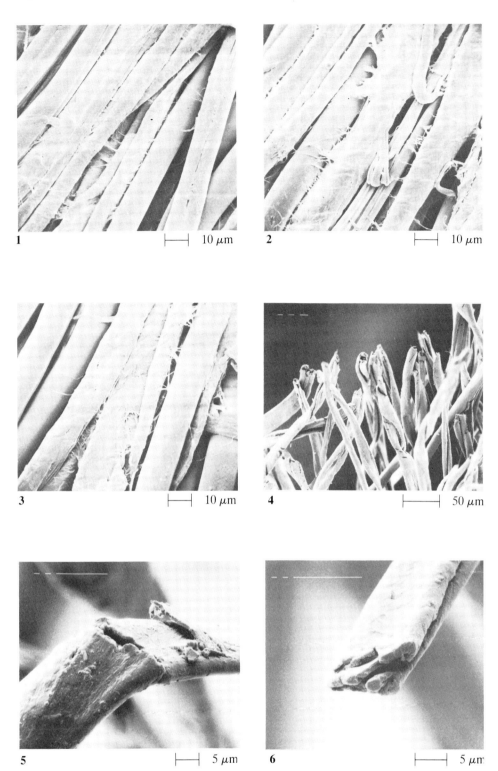

Plate 32D — Trials of cabinet towels (continued).
(1) 50% cotton/50% Vincel (rayon), after 100 washes. (2) As (1), after 200 washes. (3) 50% cotton/50% polyester, after 200 washes.
Light-tendered cotton curtain after six years' use.
(4) Edge of tear. (5),(6) Detail of fibre breakage.

33

CARPETS

Like most modern textiles, carpets are now made from a wide range of fibres in a variety of constructions. They are used in a great diversity of circumstances, ranging from the protected environment of a bedroom to the demanding requirements of an airport departure hall. There will thus be diverse patterns of wear; and the studies of failure mechanisms, which have been made since the SEM became available, have only touched on part of the subject.

Because of the interactions of art and function, visual and tactile aesthetics, cost and useful life, together with differences in fibre properties and structural arrangements, carpets illustrate very clearly the complexity of the appreciation of wear, which is found in some degree in all consumer textiles. Wear may be manifested in the simplest way as a loss of fibre from the pile, which eventually reveals bare patches of the carpet backing. But the carpet can also change appreciably owing to permanent deformation of fibres without any fibre breakage. The simplest example of this is a flattening of the pile, which may or may not recover in time, but other examples are shading, where pile angle changes direction in large areas of the carpet, often beginning when tuft tips bend backwards and change of appearance due to uncrimping of fibres. Other changes, such as loss of sharp tuft definition, may be due to shifts in the packing of fibres and the formation of entanglements, which can appear as pills on the surface. Even more remote from the theme of this book, deterioration due to such effects as soiling by dirt or spillage and fading of areas exposed to sunlight. Many of these changes, whether caused mechanically or chemically, may be very small in themselves and would hardly be detectable if they occurred over the whole carpet, but become dramatically apparent when they occur on selected patches or tracks.

The impact of changes on the observer is most serious in modern plain carpets in clear, bright colours with tightly defined tufts, and is much reduced by patterning, muted colours and shaggy pile. Finally, it must be noted that the life of a carpet may be ended — or pass into a poorer situation — for reasons unconnected with even the broadest definition of wear, but rather because of a mere desire for change, a new fashion or the termination of a contract.

In former times, it was difficult to take an engineering approach to carpet wear, because the evolution of technology was very slow and carpets lasted for a lifetime, but now that there are rapid changes in technology, and the planned life of a carpet may be as short as five years, it has become necessary to develop effective ways of testing and monitoring carpet wear. In this chapter we shall concentrate on inputs to these studies which are particularly associated with fibre failure. All the observations are on carpet samples provided by the International Wool Secretariat (IWS) or the Wool Research Organization of New Zealand (WRONZ), but they include man-made fibre carpets for comparative purposes.

The main study covered a variety of carpets which had been subject to exposure to wear as follows:

(a) unworn;
(b) a floor trial in a straight-walk situation;
(c) a floor trial at a location of turning-walk;
(d) on WRONZ 6S (W6S) laboratory tester in turning mode.

It must be emphasized that these were exploratory studies of 2-inch square samples from much larger test pieces, and were intended only to show the forms of damage occurring. A proper comparison of quantitative effects would require more controlled sampling.

Severe wear, with exposure of carpet backing, is shown in 100% wool cut-pile carpet from the turning trial, **33A(1)**; and the dominant cause, namely breakage of fibres following multiple splitting, is shown by the detail in **33A(2)**. An overall comparison of the effect of the wear test is given by the appearance of complete tufts, **33A(3)–(6)**. The W6S test, **33A(4)**,

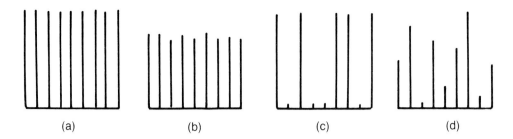

Fig. 33.1 — Schematic indications of possible patterns of loss of pile. (a) Original tuft.
(b) Hypothetical: wear from tip of tuft, almost uniformly on all fibres. (c) Hypothetical: break
from base of tuft. (d) Closer to reality: breaks randomly distributed through pile.

shows only a minor degree of wear, in comparison with the unworn tuft, **33A(3)**; and the
straight walk, **33A(5)**, gives a similar low level of wear, although there appears to be more
compacting of the tuft, and an interesting change of twist level, tighter on one side and looser
on the other.

The turning trial, **33A(6)**, gives much more severe wear and provides an important
indication of the manner of loss of fibre from the pile. A few fibres remain at, or very close to,
their original full length, and the fibre density then increases down the tuft. This suggests that
individual fibres break at different depths within the pile, either owing to points of weakness or
to random mechanical stressing within the fibre bundle, and varied lengths are lost, so that
there is a gradual thinning from base to tip of the tuft. The pattern is illustrated schematically
in Fig. 33.1. The mass of pile, unworn in Fig. 33.1(a), is not reduced either by wearing away of
fibre ends, Fig. 33.1(b), or by breaking of whole fibres from the base, Fig. 33.1(c),but the wear
is closer to a random distribution of breaks, Fig. 33.1(d), although there may be some bias with
position in the pile, and repeated breaks in the residual fibre will give an evolving pattern of
loss of fibre.

Examples of various stages of the breakdown of wool fibres in carpets are shown in **33B.**
Even though the general wear is not severe, the SEM pictures show that there is fibre damage
in the W6S test, **33B(1)**, and the straight walk, **33B(2)–(5)**. The breakage follows multiple
splitting, and is similar to that caused by the laboratory biaxial rotation test, **19D(6)**, which
involves bending and twisting, but might also be caused by flexing without twisting. Both
intermediate stages and final rupture are seen in **33B(1)**.

Possible causes of failure are suggested by **33B(2)**, where the splitting has developed at a
location where one fibre is bent round another. If the bent portion rolls along the fibre, it will
suffer the same type of deformation as in biaxial rotation over a pin (Chapter 13), with
opposite sides of the fibre going alternately into tension and compression, and torque
developing owing to friction and hysteresis. Alternatively, if the fibre is pulled backwards and
forwards, the effect will be similar to simple flex fatigue without twisting (Chapter 12), which
can also cause splitting.

The sequence of damage is illustrated in **33B(3)–(5)**. First there is multiple splitting, **33B(3)**,
then some splits break, **33B(4)**, until finally they have all broken to give the characteristic
bushy ends, **33B(5)**. The scales on the surface of the wool fibre do not split in the same way, but
do get broken off in the damaged zone and worn away beyond. The final appearance
resembles the damage that occurs in a rope made from yarns in a plastic sheath.

Although SEM examination is needed in order to show the detailed form of fibre damage,
the regions of splitting or rupture can be identified in the light microscope, **33B(6)**, and this is
an easier way of studying the extent of damage in a larger amount of material. These pictures
also show that splitting can occur at several places along the same fibre, usually at bends or
points of inter-fibre contact, with the most severe causing breakage.

The relative durability of nylon and wool is shown up by examination of an 80% wool/20%
nylon cut pile carpet. Compared with an unworn tuft, **33C(1)**, a tuft from the severe turning-
walk trial, **33C(2)**, shows the nylon fibres projecting as a fringe at their original length, but the
wool fibres broken off at various depths within the tuft. This is even more clearly seen in a view
of a more badly worn tuft, **33C(3)**. In the light microscope, **33C(4)**, it is possible to distinguish
the bushy ends of the shortened wool fibres from the smeared or bulbous ends of the nylon
fibres, as formed by the cutting of the pile. The only visible damage to the nylon is some
formation of kinkbands, **33C(5)**, whereas the wool fibres have the usual multiple split break,
33C(6).

Similar differential wear is found in an acrylic/nylon blend, **33D(1)**. In the turning trial the
acrylic fibre has been lost almost to the base of the tuft, but the nylon fibres appear to be intact.
Detail of the fibre damage is shown in **33D(2)**, and is similar to failure mechanisms in wool.
This form of fibre damage is also seen in the much less severely worn straight-walk and W6S
samples, **33D(3)–(5)**. After the turning trial the wear of some fibres has passed to the third
stage, in which the bushy end becomes rounded, **33D(6)**.

In 100% nylon carpets there is almost no loss of material through fibre breakage, although there are changes of appearance due to fibre deformation, rearrangement, and entanglement, associated with considerable flattening of the pile, distortion of tufts, and loss of tuft and pattern definition on the surface. Examples of damage in trilobal nylon fibres after the severe turning trial are some mangling of fibre ends, **33E(1)**, acute bending, **33E(2),** and the formation of pits and cracks in the fibre surface, **33E(3)**.

The observations on an 80% polyester/20% nylon carpet with round fibres showed a similar response, but there was some cracking, **33E(4),** and wrinkling of skin in the fibres, probably through snagging at pilling on the surface of the carpet.

In a polypropylene carpet the fibre damage has a different form, and is predominantly peeling from the fibre surface, **33E(5),(6)**. This is similar to that found in surface shear (Chapter 14), and must reflect great sensitivity to this mode of breakdown in polypropylene.

Rayon shows yet another form of fibre damage. The carpet examined was of Evlan carpet rayon, with a small amount of trilobal nylon, which remained unbroken. Cut fibres in the unworn carpet have the form shown in **33F(1)**, although some have not been completely severed, **33F(2)**. Fibre breaks are shown in **33F(3)**, and come from transverse cracks, **33F(4),(5)**, leading to fractured ends perpendicular to the fibre axis, **33F(6)**.

One situation leading to localized wear of carpets is rubbing under chair castors, and a particularly severe example occurs with office chairs. We have examined samples provided by WRONZ from both actual wear in a newspaper office after 13 months trial and a test-method (not developed by WRONZ) called the 'Bamburg castor-chair' test.

In the office test of 100% wool cut-pile carpet, there was a moderate, but not an excessive, amount of fibre damage by the usual mode of multiple splitting, **33G(1),(2)**. In 100% nylon loop pile there is some flattening of pile, but no damage to the trilobal fibres, except for an occasional squashed or split filament, **33G(3)**. In contrast to this the Bamburg test after 25,000 cycles causes much more damage in wool carpets than is seen in use, **33G(4)**, and appreciable damage, by squashing and flattening plus some splitting, in nylon, **33G(5)**. Another feature of the test, which is not found in use, is the accumulation of a considerable amount of debris, packed deep into the pile of the wool carpets. Debris can be removed from the surface of the sample by dabbing with an adhesive surface: an example from a blend carpet is shown in **33G(6)**, where the long lengths are nylon and the short ones are wool. These observations indicate that the Bamburg test is not well related to wear in use, and misrepresents the wear of wool and wool-rich carpets in comparison with nylon carpets.

At the time of our first studies of carpet wear, reported earlier in this Chapter, we were impressed by the multiple splitting failures, which were similar to those found in our concurrent studies of biaxial rotation fatigue, which combine bending and twisting. In particular, **33B(2)** seemed to be explained by a similar mechanism. However, it was recognised that bending alone also leads to multiple splitting where there is variable curvature; **33B(1)** could be interpreted in this way. However, a closer examination of all the data, including more recent studies, suggests that, in most cases, the splitting is a secondary feature and not the initial form of damage.

Carnaby (1981,1984,1985) and Tandon *et al* (1990) have proposed a model in which carpet wear is attributed to the breakage of fibres at random heights in the pile, with a subsequent loss of broken pieces. Quantitative studies have been carried out at UMIST by A. Sengonul and P. Noone in order to test this hypothesis. Optical microscopy combined with image analysis enabled a large number of fibres to be studied. Tufts were taken from more and less worn parts of the carpet sample, and then, in order to avoid bias, a procedure for the random choice of fibre from the tufts was used.

The first study was of a piece of all-wool tufted carpet from a straight-walk trial. As shown in **33H(1)**, the wear is not severe. The purpose of the investigation was to determine the type and location of damage in each of the selected fibres. Preliminary studies enabled the damage to be divided into eight types. Type A, **33H(2),(3)**, was the mildest form and appears as a slight bending with some flattening. Type B, **33H(4),(5)**, consists of a bulbous distortion, which is accompanied by kink bands and cracks; a variant called type I has just the bulbous form. Type C, **33H(6),(7)**, has a deep gash in the side of the fibre, often half the width of the fibre, with little fibrillation. Type D, **33I(1),(2)**, has a high degree of splitting. Type E, **33I(3),(4)**, starts as a deep kink-band crack on one side of the fibre, from which an axial slit may develop. Type F, **33I(5),(6)**, is a less severe form, which may also be found in unworn carpet fibres, and consists of very clear kink bands. Finally there are a variety of forms of broken ends, listed as type G and shown in **33J**. These range from simple forms, **33J(1)–(3)**, to varying degrees of axial splitting, **33J(4)–(7)**. The optical micrographs do not show the details of damage as clearly as the SEM pictures, but they are adequate to identify the type of damage. Having noted the type, its position is recorded by an image analysis system. The results can then be displayed and used to check the theory.

For the second study, the carpet had been subject to a turning-walk trial and, as shown in **33K(1)**, the damage was more severe and was comparable to that in **33A(1)**. A detailed quantitative investigation was carried out as before. In addition, a few tufts, such as the one in **33K(2)**, were examined in the SEM. The major source of damage appeared to be a sharp kinking, which yields effects similar to those found in the buckling test of wool yarns as shown in **12J,K**. An example of such a sharp bend is shown in **33K(3)**, with a crack developing on the inside of the bend. A small crack is seen in **33K(4)**. A larger crack in the 45° direction expected for a kink band is shown in **33K(5)**, which also has indications of another crack in the same direction and one in the other 45° direction. Substantial cracks on the inside of bends are apparent in **33K(6)** and **33L(1)**. In **33L(2)**, the crack is on the outside of a bend, but this has probably been bent back in the reverse direction. A clean fibre end is shown in **33L(3)**; this might be break along a kink band, but is more likely to be a fibre that has been cut on the face of the pile. Break usually proceeds from the kink-band crack to develop axial splitting, as seen in **33L(4)–(6)**. Broken ends with multiple splitting are shown in **33M(1)–(4)**. Further wear leads to rounding of the end, starting in **33M(5)** and well developed in **33L(6)**.

The overall conclusion from the quantitative studies is that the location of damage and the change in length of the fibres, which in the turning trial comes down from a modal length of 12 mm to 4 mm in a highly worn tuft, is compatible with fatigue and loss of fibre occurring at random locations. The common sequence of damage is formation of kink-bands, which turn into cracks and then develop axial splitting before rupture occurs.

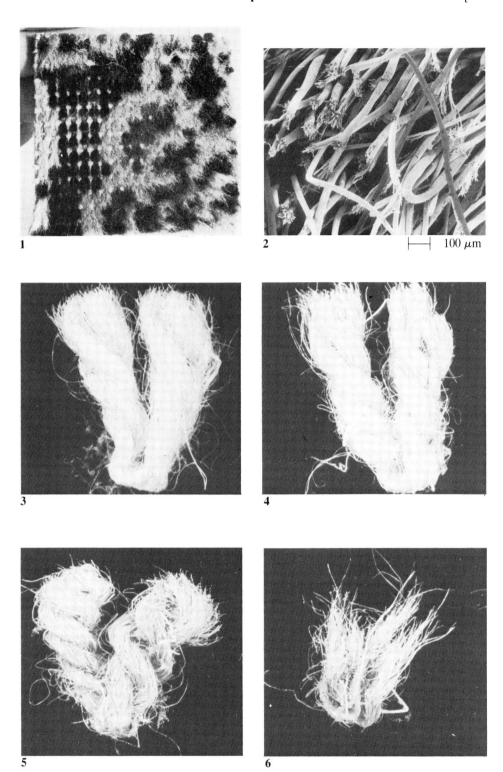

Plate 33A — 100% wool cut-pile carpet.
(1) Appearance of 2-inch square piece after turning walk trial, at edge of the region of severe wear, which
exposes carpet backing (photograph). (2) Detail of fibre damage after turning trial.
Tufts removed from carpet sample (macrophotography).
(3) Unworn. (4) After WRONZ 6S laboratory test. (5) After straight-walk trial. (6) After turning-walk
trial.

Plate 33B — Wool fibres in worn carpets.
(1) 100% wool, cut pile, WRONZ 6S test. (2) 80% wool/20% nylon, cut pile, straight walk. (3)–(5) 100% wool, loop pile, straight walk.
Optical micrograph.
(6) 80% wool/20% nylon, cut pile, turning trial.

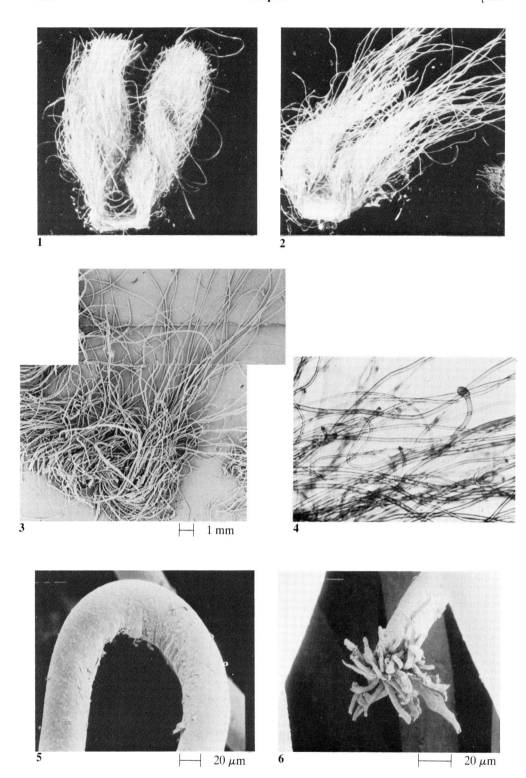

Plate 33C — 80% wool/20% nylon, cut pile carpet.
(1) Tuft from unworn carpet (macrophotograph). (2) Tuft after turning trial (macrophotograph). (3) Another tuft after turning trial. (4) Fibres after turning trial (optical micrograph). (5) Nylon fibre, after turning trial. (6) Wool fibre, after turning trial.

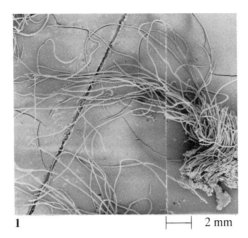

1 |— — — —| 2 mm

2 |— — — —| 200 µm

3 |— — — —| 100 µm

4 |— — — —| 100 µm

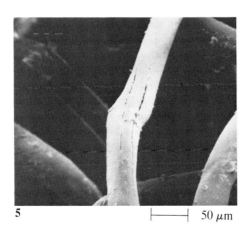

5 |— — —| 50 µm

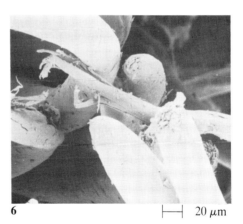

6 |— —| 20 µm

Plate 33D — 80% acrylic/20% nylon, cut-pile carpet.
(1) Tuft after turning trial. (2) Fibres after turning trial. (3) Fibres after straight walk. (4) Fibres after W6S test. (5) Early stage of splitting of acrylic fibre, after straight walk. (6) Fibre ends subject to further wear after rupture, from turning trial.

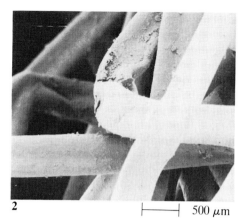

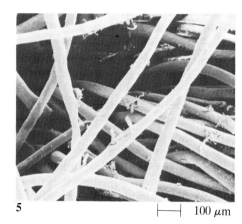

Plate 33E — 100% trilobal nylon, cut pile.
(1) Fibre ends, after turning trial. (2) Sharp bend, after turning trial. (3) Fibre surface, after turning trial.
80% polyester/20% nylon, cut-pile carpet.
							(4) Fibre cracking, probably polyester, after turning trial.
100% polypropylene, loop pile.
							(5),(6) Surface peeling, after straight walk.

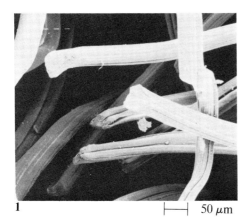

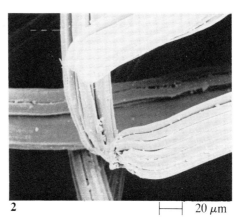

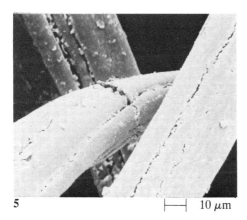

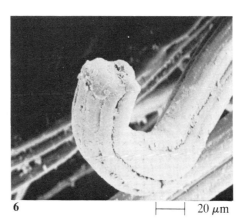

Plate 33F — Evlan (rayon) with some nylon, cut-pile carpet.
(1) Fibre ends in unworn carpet. (2) Incompletely severed end in unworn carpet. (3),(4) After turning trial. (5),(6) After straight walk.

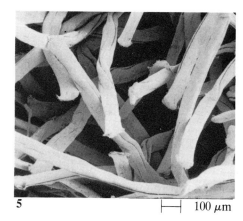

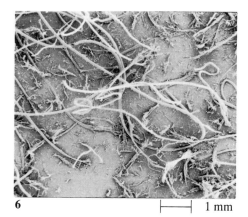

Plate 33G — Office trial at a desk with castor chairs.
(1),(2) 100% wool, cut pile. (3) 100% nylon, loop pile.
Bamburg test.
(4) 100% wool, cut pile, 25 000 cycles. (5) 100% nylon, cut pile, 25 000 cycles. (6) Debris from carpet surface, 80% wool/20% nylon, cut pile, 25 000 cycles.

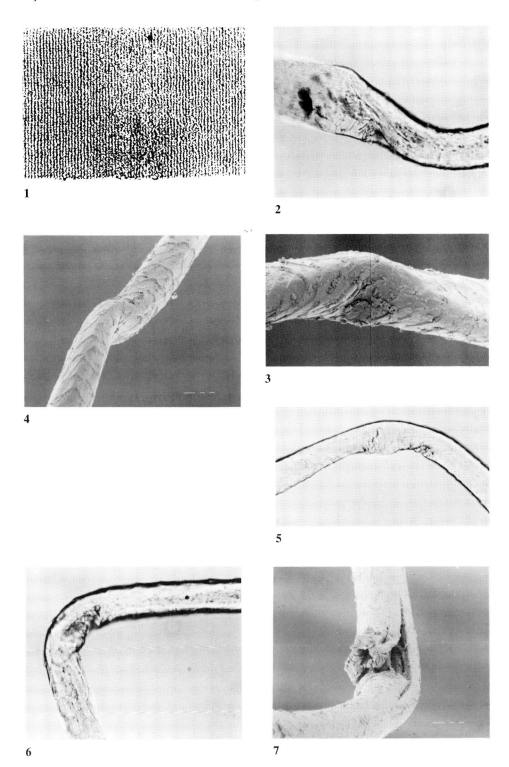

Plate 33H — Straight-walk wool carpet trial.
(1) Central worn part of carpet. (2),(3) Type A damage. (4),(5) Type B. (6),(7) Type C.

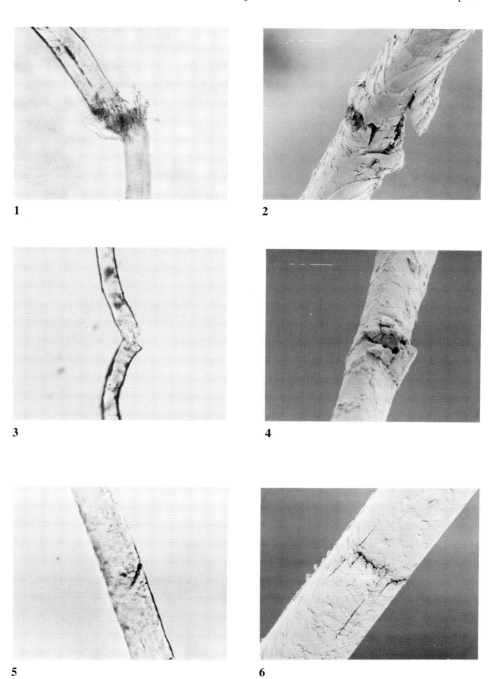

Plate 33I — Straight-walk wool carpet trial (continued).
(1),(2) Type D damage. (3),(4) Type E. (5),(6) Type F.

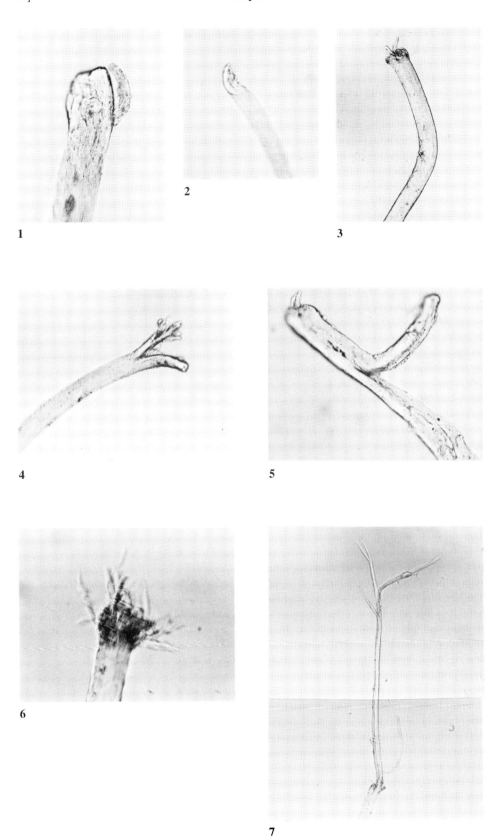

Plate 33J — Straight-walk wool carpet trial (continued).
        (1)–(3) Simple breaks. (4)–(7) Breaks with axial splitting.

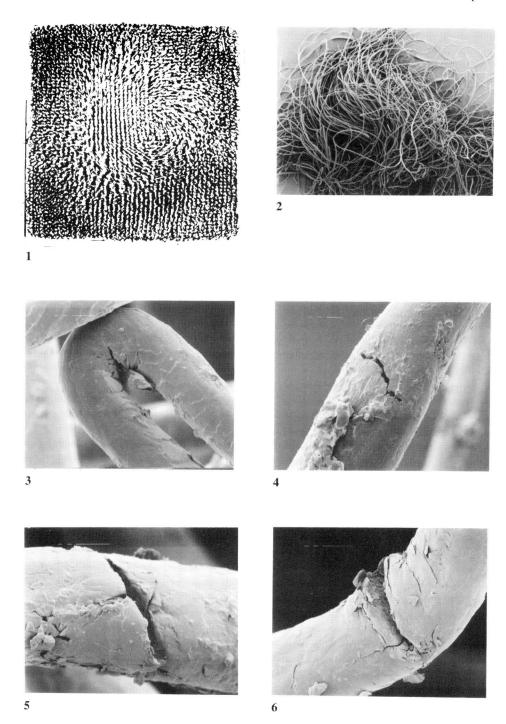

Plate 33K — Turning-walk wool carpet trial (continued).
(1) Worn sample of carpet. (2) Tuft from which selected fibres were observed in detail. (3) Sharp bend.
(4)–(6) Cracks in fibres.

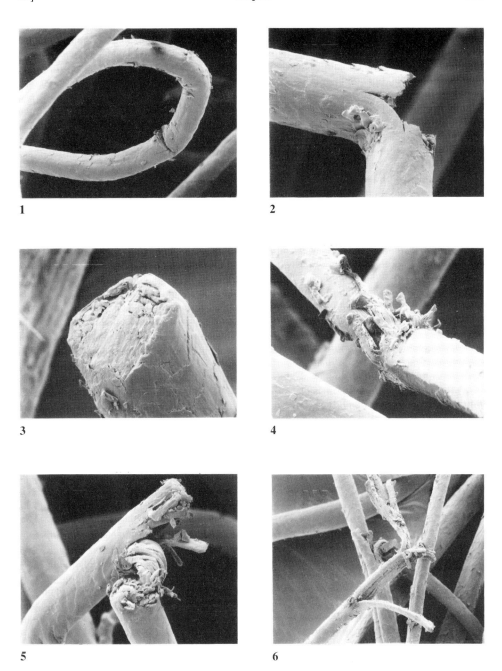

Plate 33L — Turning-walk wool carpet trial (continued).
(1) Crack on inside of bend. (2) Crack appearing on outside of bend. (3) Probably a cut end. (4)–(6)
Development of axial splitting.

Plate 33M — Turning-walk wool carpet trial (continued).
    (1)–(4) Breaks with axial splitting. (5) Beginning of rounding of end. (6) Rounded end.

34

INDUSTRIAL WORKWEAR

Industrial clothing provides a productive means of studying wear in textile materials for two particular reasons. Firstly, the forms of product and of wear are varied, with uses ranging from the delicate environment of a clean room to the rigours of a coal-mine, but are usually well defined. Secondly, when the employer provides and launders the garments, their life-usage is known: the monitoring and recording is particularly detailed when the garments are supplied under contract by a rental company. Furthermore, rental companies need to build up a body of knowledge and test procedures, so that they can select fabrics and garment designs which will last for the period of a contract, but no longer, except for a certain safety margin. If the garments, or at least any appreciable fraction of them, become unacceptably worn before the contract finishes, typically after two years, then replacement is a high cost. If the garments last too long, then they are over-designed and thus would normally be more expensive.

Where the working environment is not severe, laundering may be the major cause of wear. A washing machine is a good laboratory wear tester; and repeated laundering of garments is a rapid means of evaluation, as illustrated by the overalls from which the fabric specimens shown in **34A** were taken. After 250 launderings of a 100% cotton overall under standard conditions, there is appreciable disturbance of the fabric, **34A(1)**, and sheets of fibrils are peeling away from the cotton fibres, **34A(2)**, which, in some places, are badly disintegrated, **34A(3)**. In a comparable 50% cotton/50% modal rayon (Vincel) overall, similar fabric disturbance and damage to the cotton fibres is apparent, **34A(4),(6)**, but the rayon fibres have suffered little in comparison with the cotton, **34A(5)**. This does not imply that 100% rayon would be more durable, since the cotton fibres in the blend may well be taking most of the load, and protecting the rayon. In addition, some fibrillation, as seen in **34A(5)**, still leaves much of the cotton fibre intact, whereas any breakdown of rayon is usually more severe. In the language of fracture mechanics, the peeling apart of the cotton fibres is a way of absorbing a lot of energy before ultimate failure.

Actual wear in use is illustrated in **34B,** which shows the appearance of material in a navy-blue twill weave 100% cotton coverall after more than six years use in a dirty industrial environment. Damage and loss of colour was most severe in regions below the waist, particularly in the seat, thigh, knee and shin areas, and was markedly absent on the chest, back and collar. In any part of the garment there was preferentially severe abrasion on the raised surfaces of the seams, accentuated on the front seam by the rubbing of metal buttons.

The material had been repaired with a patch below the pocket on the right thigh, and by darning at the corresponding place on the left thigh. There was also mending by darning of splits at the crotch seams, and in a few other places. The fabric is severely abraded at the bottom of the legs, and has been patched on the left side. Some holes in the trouser seat can be attributed to general wear, but other small holes in the back, with no associated loss of colour, are probably due to some specific localized cause. The sleeve cuffs had been folded back, so that wear occurred on the inner fabric surface, especially at folds and seams. There was little wear at the elbows, but rather badly abraded seams under the arms, presumably due to fabric surfaces rubbing together.

The moderately severe wear on the seat of the coverall shows fibre breakage on the crowns of the warp yarns causing the development of a fuzzy line of fibre ends along the crevices of the weave, **34B(1)**, in a manner typical of cotton fabrics. The breakage of the cotton fibres is by multiple splitting, **34B(2)**, similar to that shown by many types of fibre in bending and twisting fatigue (Chapter 13). The splits start along the helical lines of the cotton fibre structure, and continue to have this bias, but are less influenced by the fibrillar structure of cotton than is found with the milder laundering treatments. There is no peeling away of sheets of fibrils, as seen in **34A(3),(5)**.

The wear on the underarm seam is different, and shows abrasive flattening and compacting of the yarns, **34B(3)**. At a higher magnification, **34B(4)**, it can be seen that the individual fibre identity has been completely lost by smearing out of material on the yarn crowns, and is partly lost in the crevices between yarns, although fibre ends with multiple splits can still be discerned. This form of damage results from shear between two fabric surfaces, rubbing against one another, probably aided by the effects of perspiration, and without any opening up of the yarn and fabric structure by bending.

The seam at the centre back is less severely worn, and does not show either broken fibres in the crevices or the surface smearing. However, a few places on the cotton fibres show the start of multiple splitting breakdown, **34B(5)**, leading to break with a bushy end, **34B(6)**.

Localized damage is shown by a hole in the back of the coverall. There is severe disturbance over a small region, but the surrounding fabric is intact, **34C(1)**. There are many rounded fibre ends, **34C(2)**, and some split ends, but little evidence of fibre fatigue. Most probably, the hole was made by catching on a pointed projection: this would break fibres, and the ends would become rounded by subsequent wear and laundering.

The badly worn area at the knee shows the usual features of hard wear, **34C(3)**, with the many broken fibres between the yarns having bushy ends following multiple splitting, and with some smearing on the yarn surfaces. However, there is also much more extraneous matter adhering to the yarns, **34C(4)**: this is probably a mixture of grease, dirt and fibre particles.

The fraying of the hem fold at the bottom of the trouser leg is shown in **34C(5)**, with detail of fibre damage in **34C(6)**. The damage is similar to that found elsewhere, but very severe, with the splitting leading to fibrillation.

The information from examination of this old coverall helps to answer the two important questions: how does damage develop? what can be done to improve the product? Wear in cotton fabrics in severe environments is clearly a result of rubbing and abrasion, intensified by the loosening effect of water on the internal structure of the fibre. Where the fabric remains flat smearing predominates, but where it is repeatedly bent the fibres remain separate as they split and break. The twill weave puts durable hard-twisted warp yarns on the fabric surface, with the softer weft yarns inside. The route to greater durability is to have even harder-wearing yarns on the surface, such as blends with nylon or polyester, and to minimize freedom for fibre bending.

Blue cotton/polyester coverall, supplied under contract for an industrial location by a rental company, had the fabric appearance shown in **34D(1)** in a region of negligible wear. The singed ends of the round polyester fibres and the convoluted cotton fibres can be seen. In contrast to this view, fabric from moderately worn material in the middle of the back is more hairy and shows broken fibre ends at the yarn interlacing points, **34D(2)**. A higher-magnification picture, **34D(3)**, shows the usual multiple-splitting damage in the polyester fibres and the rounded ends which result from wear after the breakage. The splitting of the cotton fibres is more influenced by their characteristic structure. Much more severe damage, but of the same general type, is shown along the edge of the trouser hem, **34D(4)**.

A small hole in the right trouser knee, **34D(5)**, was evidently accidental damage which resulted in fusing of the polyester fibres. The exposed cotton fibre surfaces show no influence of the heat, although the fibres have been trapped by the molten polyester. The fusing was limited to an area close to the edge of the hole, indicating that the cause of damage was very localized. There were also two small holes in the seat, near which the polyester fibres had formed bulbous ends, **34D(6)**, which suggests damage by heat or perhaps a chemical agent.

The pictures shown in **34E** and **34F** are taken from a study of a coal-miner's coverall, which had been withdrawn from service because the rental company considered it to be no longer of the standard required. The fabric was a polyester/modal (rayon) twill weave. The material had been given an easy-care finish, which cross-linked the cellulose and changed its mechanical properties. The material had been dyed orange, but had faded considerably on exposed surfaces and was stained in heavily worn areas. There were many holes, some of which had been repaired. The damage was located where expected from the nature of the work, but details, as given for the first blue coverall, will not be repeated here. The largest hole on the knee appeared to be due to a cut.

Fabric from the reverse face of the collar, **34E(1)**, is almost indistinguishable from samples of original fabric, which had been examined previously. This illustrates how control material can be found within a used garment, where it is protected from damage. The fabric has been singed to give a smooth surface, which is not hairy, and the bulbous ends of the polyester fibres can be seen. In contrast to this, the rayon fibres have square ends, although most of these will be due to breakage and only a few from cutting in manufacture.

Examination of the back of the fabric at low magnification shows little appearance of wear, but a closer look indicates that there are some broken fibre ends between the yarns, **34E(2),(3)**. The two polyester fibres in **34E(3)** have failed in the usual way by multiple splitting to give a ragged end; but the modal rayon fibres have broken cleanly, as if they had been snapped by a brittle break. Similar effects can be seen in **34E(4a)**, where the polyester fibres have suffered splitting, but the modal fibres appear undamaged except where they are broken, or where a crack has gone part way across the fibre, **34E(4b)**. This form of breakage of modal, rather than the peeling which often occurs in rayon, probably results from the application of an easy-care finish, which embrittles the cellulose and makes it easy to break. This was confirmed by the presence of much orange-coloured dust, consisting of small fragments of broken modal fibres, in the laundry.

Severe damage at the edge of a hem is shown in **34E(5),(6)**. Close investigation shows that the modal rayon fibres have almost completely disappeared from the yarn crowns in such badly worn areas, as pieces have broken off and left the fibre ends projecting from between the yarns, together with the bushy broken polyester fibres. At the heel, wear at the edge of the hem had destroyed the fabric integrity.

Another region of severe wear is on the knee, where the fabric has become very hairy, especially around holes, **34F(1)**. The polyester fibres break down owing to bending and twisting, with many bushy ends, **34F(2)**, which develop from multiple splits, **34F(3)**, very similar to a laboratory failure in a biaxial rotation fatigue test (Chapter 13).

Apart from general wear, there may be other forms of accidental damage. Two small holes, darkened round the edges, were found in the front of the left trouser leg. The damage is very localized, **34F(4)**, and shows much fibre fusion, **34F(5)**. The fibres in one yarn end, which would have passed through the middle of the hole, are completely melted together. The melting is not confined to the holes, but is also found to a lesser extent in nearby regions, **34F(6)**. This damage is clearly caused by very localized heat, and a likely cause is the impact of sparks generated in welding.

Workwear may also be worn in situations where they are subject to chemical attack. This can lead to unusual forms of damage, as we found in a blue polyester overall used in a virology laboratory. The fabric will have experienced chemical and biological contamination, but the most serious factor was that the garments were sterilized in an autoclave both before and after each period of use, with cleaning between the two autoclave treatments. The presence of wet steam seems to have been particularly damaging. The back of the garment retained only 40% of its original strength; and the worst affected part of the front had lost over 90% of its strength! There were splits at creases and folds, and also considerable discoloration. The sleeves were easily torn.

Fabric from the overall back seems undamaged at low magnification, **34G(1a)**, but higher magnification reveals transverse cracking of filaments at yarn interstices, **34G(1b)**. Surface cracks are seen in **34G(2)** as is a filament broken by axial splitting: this suggests that the interior of the fibre was less damaged, and could break in the usual way. However, along a warp-way split in the fabric over the zip fastener, **34G(3)**, the breaks were of a different form, **34G(4)–(6)**. They were stake-and-socket failures, as found by Ansell, and shown in **16D(1),(2)** and **26B(5),(6)**, and result once again from exposure to hot, wet conditions with chemical present. It was commercially valuable to the supplier of the garments to establish that the cause of failure was the form of use, and not any fault in the product. Changes in the autoclave procedure led to less damage.

Very strict requirements are put on fabrics for garments used in clean room, whether in the micro-electronics industry or in medicine. The fabric is intended more to protect the environment from the wearer, than vice versa. The user, with some lower layers of regular clothing, is enclosed so that no contaminating particles can escape. This means that dense weaves of closely packed continuous-filament yarns, as in **34H(1)**, must be used. However, it is also vital that the fabric material itself should not shed particles: once again, the important feature is whether fibre damage harms the environment, and not whether it causes any significant deterioration in the fabric, which is the usual concern in wear studies.

As with the first case dealt with in this chapter, the actual wear in use is not severe, and more damage comes from servicing the garments, in this case by dry-cleaning. Such tightly woven fabrics are uncomfortable to wear and cause perspiration, and so must be changed and cleaned frequently. Samples of fabric from garments made from five different continuous-filament polyester fabric constructions were examined after the garments had been subjected to 50 and 300 dry-cleaning cycles, without use. Detailed analysis of the results enabled the fabrics to be ranked in order of merit, from slight peeling in fabric D after 300 cycles to severe peeling on every yarn crown, with long ragged pieces being stripped off, in fabric A.

General damage to the fabric was limited to surface peeling of filaments on the top of yarn crowns, as seen in a fairly severe form on fabric A after 300 cycles in **34H(1),(2)**, and at an early stage after 50 cycles in **34H(3)**. In contrast to this, the damage is slight in fabric D after 300 cycles, **34H(4)**. There is more severe damage at hems and seams, which is at its worst where a hem meets a seam, as in **34H(5)**. The large concentration of multiple splitting failure of fibres can cause complete breakage of yarns. Material may also be damaged by sewing, and **34H(6)** shows a position where a sewing needle has exited: fibres have been broken by the sewing action. The breakage which occurs in these regions of severe wear is an unwanted source of fibrous contamination of the atmosphere.

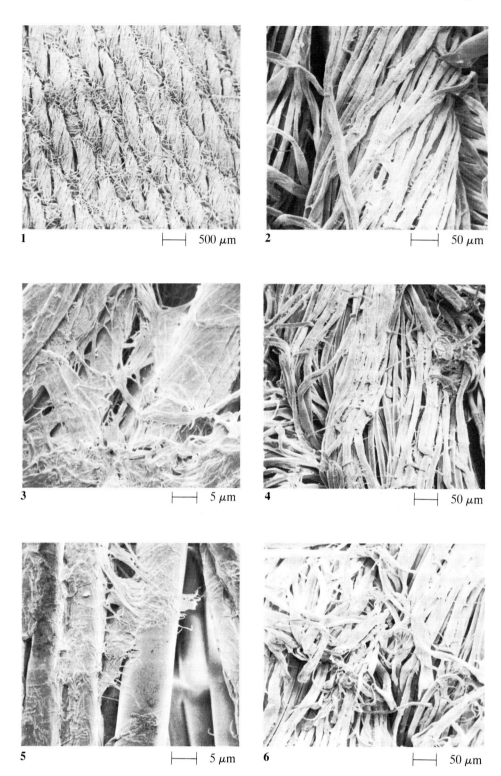

Plate 34A — Overalls after 250 launderings, without being worn.
(1)–(3) 100% cotton fabric, from face of material in front of overall, below pocket. (4),(5) 50% cotton/
50% modal rayon (Vincel fabric), from face of material in front of overall, below pocket. (6) ditto, but
from reverse side of material.

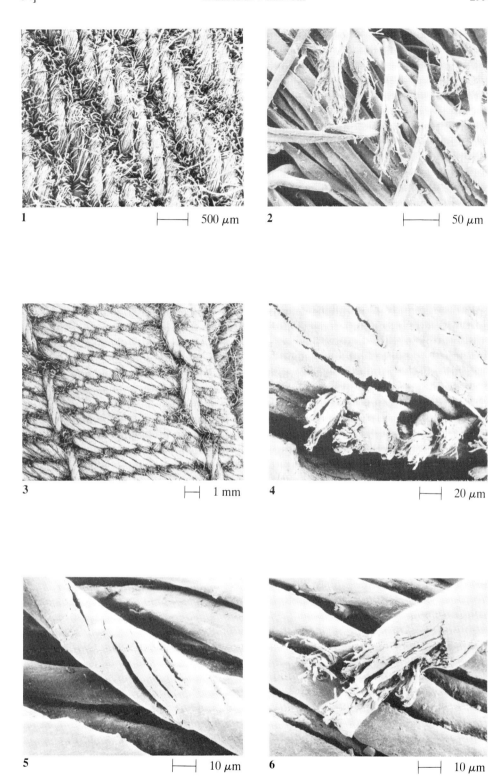

Plate 34B — Blue cotton twill coverall after industrial use for six years.
  (1),(2) From worn area in seat. (3),(4) From underarm area. (5),(6) From seam at centre back.

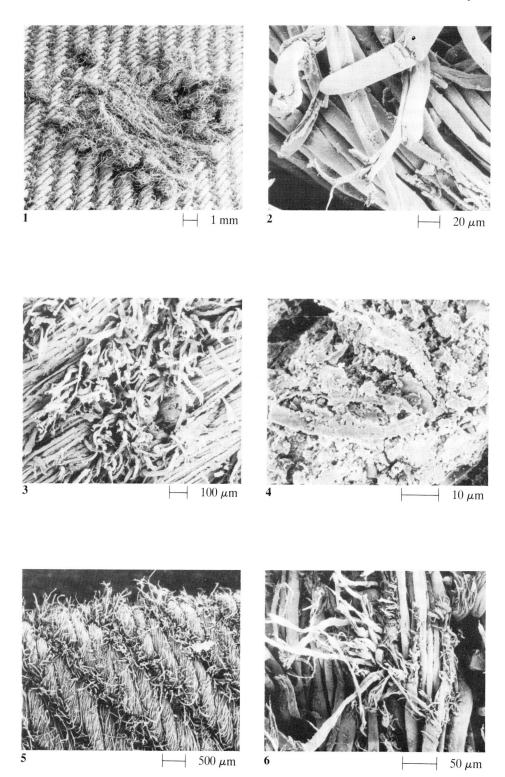

Plate 34C — Cotton coverall (continued).
(1),(2) Hole in back of garment. (3),(4) From badly worn area on knee. (5),(6) Frayed edge at bottom of trouser leg.

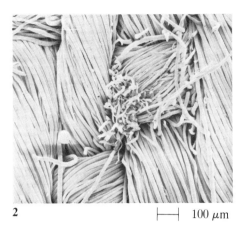

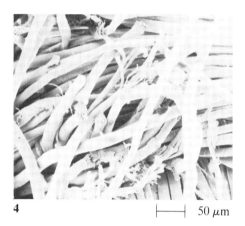

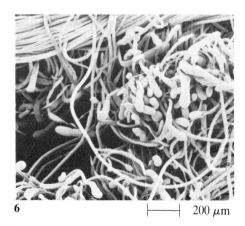

Plate 34D — A cotton/polyester coverall after industrial use.
(1) Fabric from inside breast pocket showing negligible wear. (2),(3) Fabric from centre of the back. (4) Edge of trouser hem. (5) Edge of hole in knee. (6) Small hole in seat.

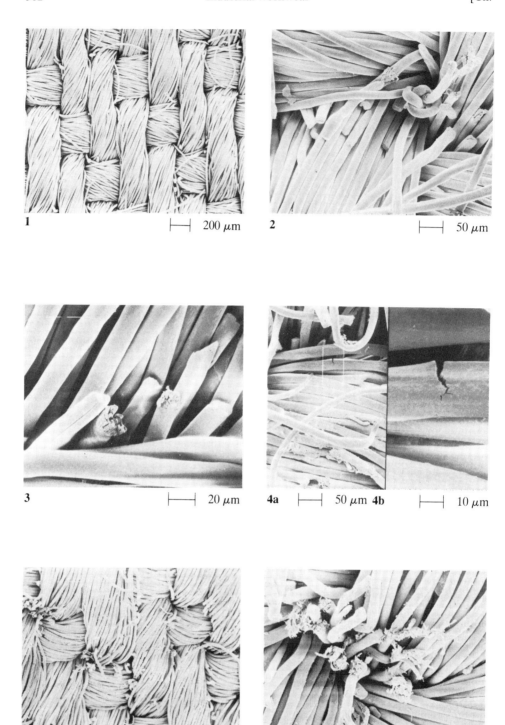

Plate 34E — Coal-miner's polyester/modal (rayon) coverall withdrawn from use because of poor appearance.
(1) Almost undamaged fabric from reverse side of collar. (2),(3) Slightly damaged fabric from centre back region. (4a) Fibre damage from centre back region. (4b) Enlarged view of part of (4a). (5), (6) Badly damaged fabric at edge of hem.

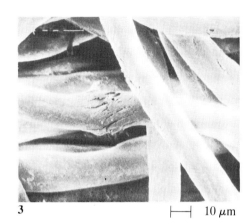

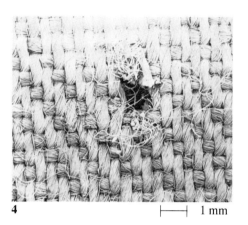

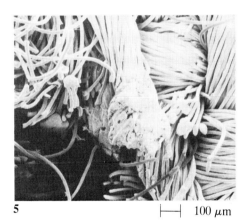

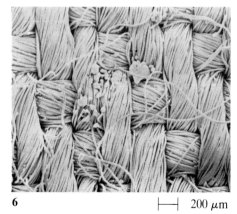

Plate 34F — Coal miner's polyester/modal coverall (continued).
(1) Hole in knee. (2),(3) Detail of fibre damage in the badly worn knee region. (4) One of two small holes in trouser leg. (5) Fibre fusion around the hole. (6) Fibre fusion in fabric near the small holes.

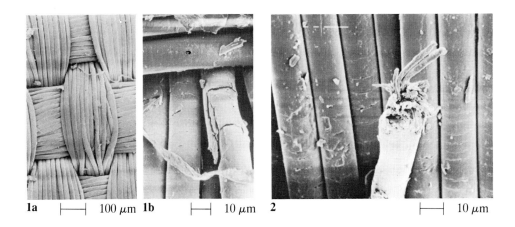

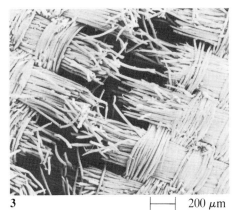
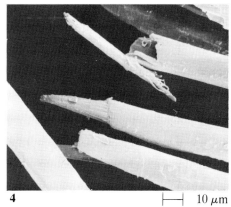

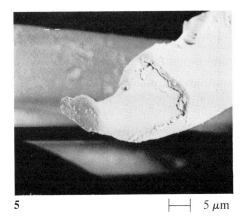

Plate 34G — Polyester overall worn in a virology laboratory, and autoclaved between each period of use.
(1a) Fabric from back of overall. (1b) Enlarged view of part of (1a) showing transverse cracking. (2) Fibre broken with multiple splitting. (3) Tear in fabric over zip fastener. (4) Stake-and-socket breaks from the tear. (5) Detail of stake. (6) Detail of socket.

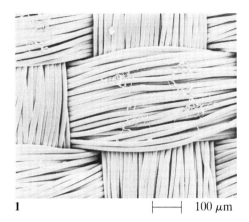

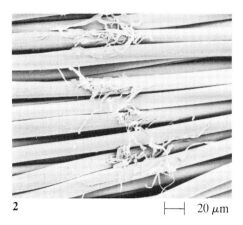

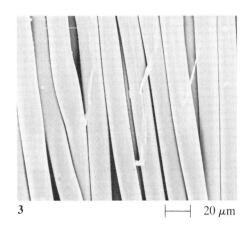

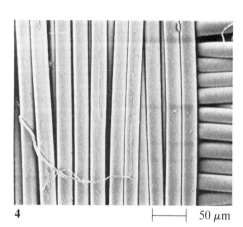

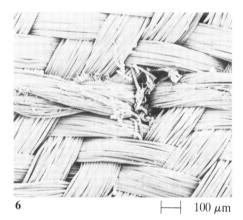

Plate 34H — Clean-room garments of continuous-filament polyester fabric after being dry-cleaned for 50 or 300 times.
(1),(2) Fabric A after 300 cycles. (3) Detail of fabric A after 50 cycles. (4) Fabric D after 300 cycles. (5) Fabric B after 300 cycles, near junction of hem and seam. (6) Fabric C after 300 cycles, at point of exit of sewing needle.

35

ARMY COVERALLS

Military clothing provides interesting case studies of garments subject to very severe wear, and also, because of a frequent need for thermal protection, includes the use of special treatments and fibres.

The first group of coveralls examined had been subject to normal army usage for a time in which major damage had been incurred. The pictures in **35A** and **35B** are of 100% cotton woven coveralls. In **35A(1)** the lower part of one garment is shown, which has worn into holes at each knee; there is another hole midway down the right shin, which has been patched. Parts of the garment which have suffered general wear, such as the shoulders, below the waist, seat and thighs have a rubbed and faded appearance. Throughout the garments, seams and raised edges are abraded and have lost colour. There is another hole near the front opening at the waist, and there are slit-like holes along the hem at the bottom of the trousers. The inner surface of the garment also shows rubbing along the seams and some pilling on the seat. Another coverall worn by the same corporal shows wear in the same places, but this is even more severe, with holes on both shins, at the ankles and on the trouser leg pocket. Some of these holes have been patched, and even the patches have been patched.

In **35A(2)** the break-up of the cotton fibres in a region of severe damage is shown. At lower magnification, **35A(3)**, it can be seen that there are many broken fibre ends, concentrated at the junctions of yarn cross-overs, while from higher magnification, **35A(4)**, it is clear that extensive splitting is the mode of breakdown of individual fibres.

Even on the inside surface of the fabric, **35A(5)**, there is some wear which shows up as peeling away of layers of the fibre surface, **35A(6)**. This type of damage may result from laundering, as illustrated in Chapter 34.

Successive stages of development of the splitting of the cotton fibres in regions of wear at creases in the fabric are shown in **35B(1),(2)**.

Two coveralls used by another corporal show less severe damage. The only holes are on the left knee of one garment, and towards the inside of the left thigh on the other. The two holes on the thigh, which have been patched, follow the lines of creases or folds formed during use. These two coveralls also showed many lighter coloured lines of damage, which looked like scratch marks. Generally it is easier to observe the course of damage in these less-worn garments.

In **35B(3)** considerable peeling off of fibre layers is shown, in an area between two small holes in the knee. Probably, rubbing action on the knee has led to a scraping of the yarn crowns with fibre layers being pulled away, intensifying any effect caused by laundering. The damage eventually leads to breakage of yarns, and then this loosens the fabric, allowing considerable disturbance and formation of a hole. In other parts of the garment, **35B(4),(5)**, multiple splitting occurs, and this is probably a consequence of bending and twisting of fibres. Even on the inner surface of the fabric, **35B(6)**, fibre ends broken by splitting can be found, although these have probably migrated from the outer surface.

Thus the general mode of failure of these cotton coveralls subject to very extensive use is the multiple splitting found in the laboratory in biaxial rotation fatigue tests (Chapter 13), together with some peeling away of fibre layers.

A third severely worn and extensively damaged coverall was made of Proban-treated cotton, to give flame resistance. There were holes near the trouser hems, which had been shortened by turning up the original hem and stitching in place, and the complete back portion of the right hem was missing. The trouser seat had a T-shaped tear which had been patched and darned. Seams and edges of fabric were heavily abraded, and there were small holes in various places in the garment. Velcro-type fastenings down the front opening, and on pocket flaps and cuff tabs, had been worn considerably, although they still operated. Another Proban-treated coverall, somewhat less damaged, was also examined.

The different pattern of gross wear, for example the lack of holes at the knees, indicates a different pattern of use of these coveralls. However, detailed examination shows a difference in fabric and fibre breakdown, which must be attributed to the effect of the Proban treatment. The general impression of abraded surfaces is that they are not as hairy as untreated cotton fabric surfaces. Abrasion of the fabric surface shows broken fibre ends sitting in the crevices of the weave, **35C(1)**, which is typical of wear in cotton fabrics, but there is less fibrillation and peeling of fibres on the yarn crowns than with the untreated cotton shown in **35B(3)**. There is a general tendency for fibres to split along their length, following the spiral angle of the cellulose fibrils, **35C(2)(3)**, before they fail through breaks which are often sharp and angled, **35C(4)**, although some are more fibrillated, **35C(5a)**. Many broken fibres, either at the weave interstices or at the edges of holes, have rounded ends, **35C(5b)**, due to further wear. Some places, where neatly rounded ends occur, are associated with orange or white discoloration of the navy-blue fabric, suggesting that some chemical damage has occurred.

In the Velcro fastening, **35C(6a,b)**, the ends of some of the hooks had broken off.

Tank crews need garments giving a high degree of thermal protection in case of fire, and a typical coverall is made from a Nomex (meta-aramid) fabric. Three used coveralls showed few signs of wear and tear. Although the fabric surface had been generally abraded, there was no yarn breakage or holes, even in regions where severe wear is common. The worst damage was due to scuffing, particularly in the groin and seat region.

Fibre breakdown is of three types. Firstly, there is peeling leading to fibrillation of fibre surfaces, **35D(1)**, which can cause complete breakage, **35D(2)**, albeit with some coarse splitting. This extensive fibrillation is unusual, and seems to be a special feature of Nomex. Secondly, along scuff lines, **35D(3)**, the fibres are often sharply broken, **35D(4)**. Thirdly, there is the common form of multiple splitting, **35D(5)**, due to bending and twisting.

At the inner trouser seam in the crotch region, a band of webbing has been sewn across the top of the leg. There has been severe abrasion of this seam, and short tufts of broken fibres can be seen in the crevices of the twill weave: these fibres have stubby, bushy ends, **35D(6)**.

Another Nomex coverall had been used as the control in an experimental study of three types of garments containing carbonized fibres to give superior fire resistance. All the garments had been subject to wear trial by completing a specified number of circuits of an assault course.

The Nomex coverall had completed 30 circuits and been laundered six times. There was considerable damage in the knee region, with weft-way slits in the fabric, and there was also a slit near the waistband, where fabric had abraded over a zip fastener. In other parts, where there had been a more gentle rubbing, the fabric had a fuzzy appearance. Wear at seams and edges was not excessive.

In addition to some tangling of surface fibres, the fuzzy appearance is probably due to the characteristic Nomex peeling, which occurs where wear is slight, **35E(1)**, moderate, **35E(2a)**, or severe, **35E(2b)**. However, in the knees, where there is hard rubbing on obstacles of the assault course, there is much more severe damage, **35E(3),(4)**. Fibres have broken after severe scraping, and their ends are curled over in the direction of rubbing.

Fabric covering the zip fastener has been badly abraded, partly because it stands proud of material on either side of the zip, and partly because it is sewn firmly in place over the hard surface of the zip. There is a marked contrast between the material directly over the zip, **35E(5a)** and fabric in an adjacent area, **35E(5b)**. The scraping of fibres on yarn crowns over the zip is shown in **35E(6a)**, whereas the nearby fibres are undamaged, **35E(6b)**.

The first fire-resistant coverall, made of 75% carbonized viscose/25% Nomex, had also undergone 30 circuits of the assault course with six launderings. The damage to the knees was severe and extensive, with a $2\frac{1}{2}$-inch slit of warp yarns in the right knee, allowing weft yarns to fray out. In an area 8×6 inches in the other knee, four holes were darned, with slits extending into the thigh region. There is pilling and matted surface hair in many parts, and a small hole over the zip fastener. The coverall sheds black fragments of carbonized fibre with lengths down to 20 μm.

An inside surface, **35F(1)**, shows a few broken carbonized filaments but otherwise is little damaged. However, the outside region of the worn knee was very disturbed and thin near holes, **35F(2)**. Yarns extracted from this region were found to have thinned down to a few Nomex fibres, **35F(3)**.

The second coverall contained less Nomex (15%) and a different type of carbonized viscose (85%), and showed extensive damage after only 10 circuits of the assault course and one laundering. Many holes in the right knee had been patched, and the left knee had a crescent-shaped hole, $4 \times 3\frac{1}{2}$ inches in size. The areas round the holes were very thin, down into the shin regions. In various places abrasion has caused the surface to become hairy, with some development of pilling. Short carbonized fibre fragments, with lengths from 25 to 110 μm, are easily broken off when handling the garment.

Fabric in the worn knee region shows many yarn breaks, **35F(4)**, with the carbonized viscose fibres showing a sharp, brittle fracture, **35F(5)**. Broken fragments can also be seen in this region, **35F(6)**.

The third coverall contained 15% Nomex, but the 85% carbonized fibre was oxidized PAN (polyacrylonitrile), not viscose; it also showed much more severe damage after 10 circuits and one laundering, than the control 100% Nomex garment after 30 circuits and six launderings. Each leg had crescent-shaped holes about 4×3 inches in size. There were some holes elsewhere, and much of the fabric had become hairy. This garment shed short fragments of the oxidized acrylic fibres.

A lightly worn area on the back of the garment does show some fibre breakage, **35G(1)**. The breaks go sharply across the fibres, **35G(2)**. Over the zip fastener the wear is much more severe, **35G(3)**, and the breaks are sharp and brittle, **35G(4)**. In the most severely worn region of the yarn, the oxidized PAN fibres are broken at each yarn cross-over, and only the few Nomex fibres provide continuity, **35G(5)**.

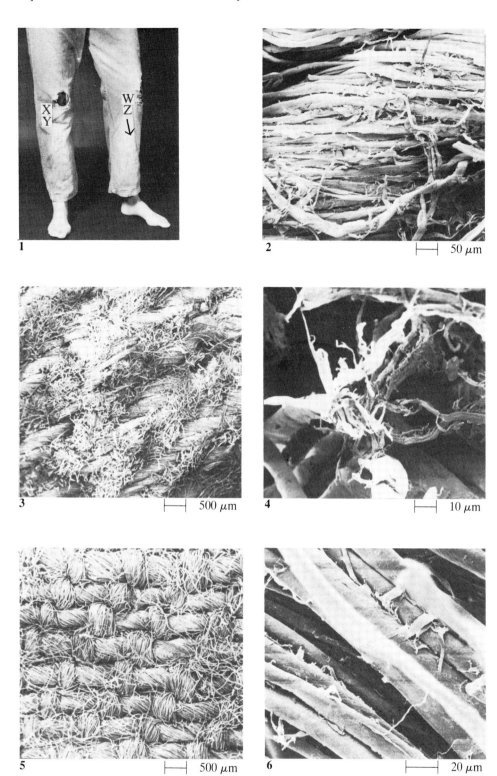

Plate 35A — Severely worn cotton army coverall after extensive use.
(1) General view with identification of regions illustrated in detail. (2)–(4) From region W on **34A(1)**, worn area across edge of a patch on left knee. (5),(6) From region X on **34A(1)**, inside surface of fabric at right knee.

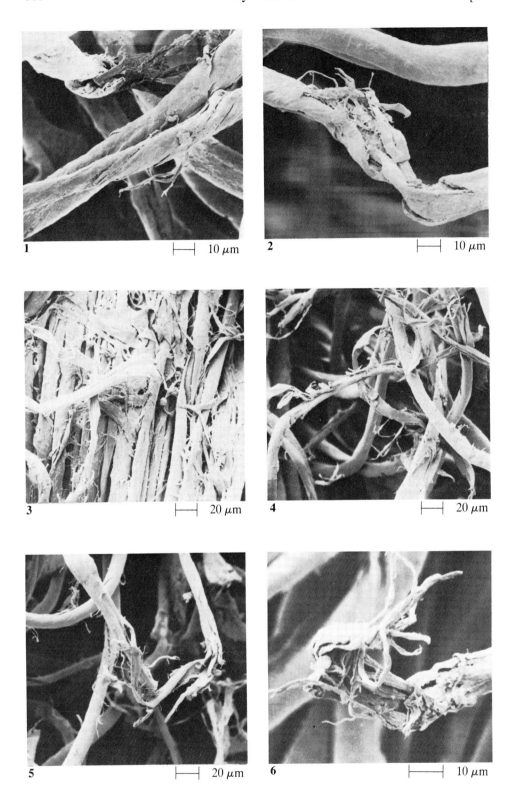

Plate 35B — Severely worn army coverall (continued).
(1) From region Y on **34A(1)**, at a worn crease on the leg. (2) From region Z on **34A(1)**, worn area on left shin.

From less severely worn coverall, used by second corporal.
(3) Between holes on left knee. (4) Outer surface of right shoulder. (5) Across a crease, outer surface just above knee. (6) Near worn left knee, inside surface of fabric.

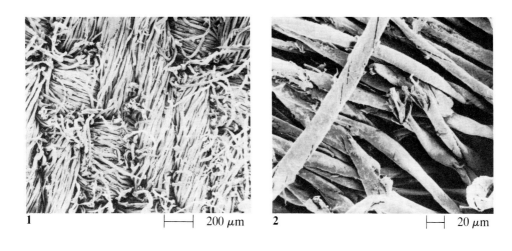

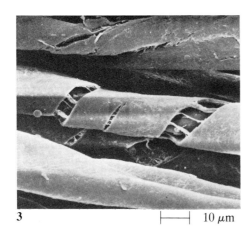

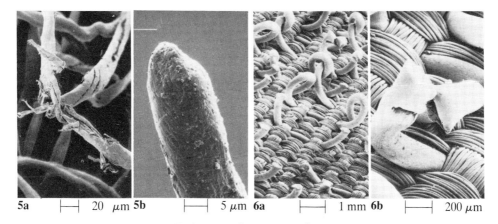

Plate 35C — From a severely worn Proban-treated cotton coverall.
(1) From right knee. (2) From worn pocket flap. (3) From worn seam in crotch. (4) From worn pocket flap.
(5a,b) From edge of right cuff, where there was a complete fabric break. (6a,b) Worn Velcro fastening
from front opening.

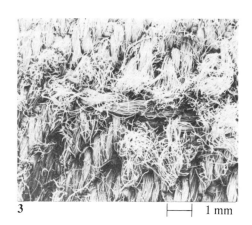

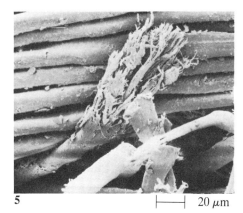

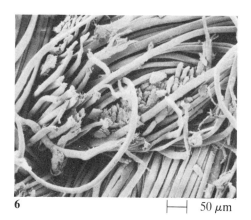

Plate 35D — From Nomex coveralls, used by tank crews.
(1), (2) From mid-shin. (3), (4) From scuff line in seat. (5), (6) From trouser seam.

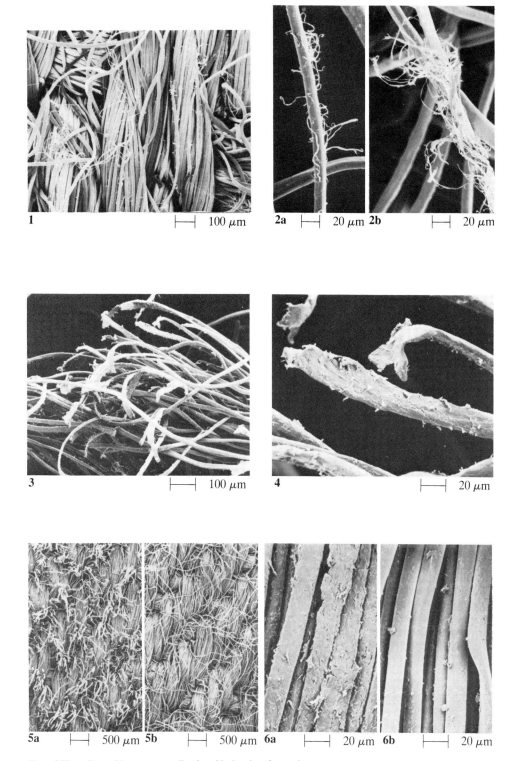

Plate 35E — From Nomex coverall, after 30 circuits of assault course.
(1) From centre back. (2a) From worn area of trouser knee. (2b) From trouser seat. (3), (4) From edge of
hole in left knee. (5a), (6a) Directly over zip fastener. (5b), (6b) Adjacent to zip fastener.

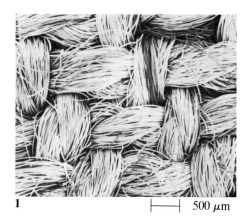

1   500 μm

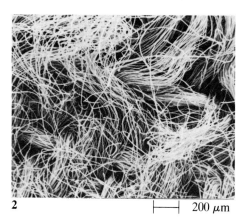

2   200 μm

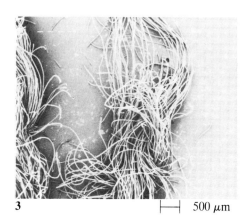

3   500 μm

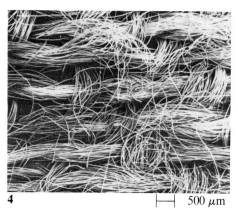

4   500 μm

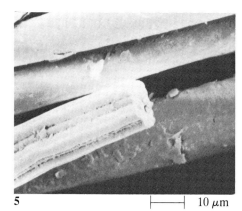

5   10 μm

6   50 μm

Plate 35F — From 78% carbonized viscose/25% Nomex coverall after 30 circuits of assault course.
(1) Inside surface at knee. (2) Thin area of knee near hole. (3) Yarn extracted from knee area.
From 85% carbonized viscose/15% Nomex coverall after 10 circuits of assault course.
(4)–(6) From worn knee region.

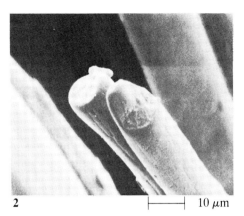

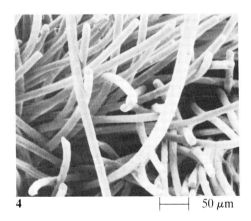

Plate 35G — From 85% oxidized PAN (acrylic)/15% Nomex coverall, after 10 circuits of assault course.
(1), (2) From centre back. (3), (4) From fabric over zip-fastener. (5) Warp yarn taken from left knee.

Part VII
Case studies:
industrial uses

36

INTRODUCTION

In many engineering uses of textiles the mechanical requirements are dominant. The designer is first of all concerned to ensure that the product has adequate strength. However, there are two difficulties; the loading conditions to which the material will be subject are often difficult to estimate, since they depend on the total dynamic response of the system in use; and the loading conditions which the material will withstand are difficult to calculate or measure, since they involve complex stresses acting on the fibres within the product. Consequently, most 'design' has been empirical, based on practical experience. There is, however, a growing need for a proper engineering design approach. A study of the ways in which products fail provides essential background information in this area.

We can revise Fig. 1.1 and present it in more specific form as Fig. 36.1, in order to show the lines of advance which are needed.

But apart from contributing to this long-term goal, the study of failure in textile structures used in industry and engineering is interesting, and can give short-term benefits. It can help the empirical design procedures, since when the mechanisms of failure are shown practical remedies may be suggested: modifying a structure to relieve stresses, adding a lubricant, changing the type of fibre, and so on.

Finally, particularly when there is litigation, it may be necessary to establish the cause of failure, and this can only be done when the different pathology of different causes of damage is established.

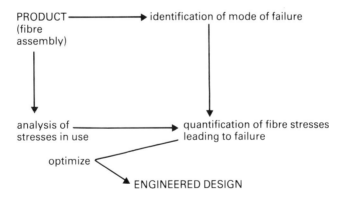

Fig. 36.1 — The route to fabric engineering.

37

AUTOMOBILE SEAT BELTS

I. J. Duerden

Seat belts are a protective device, but, in the rare circumstances in which they break, it is necessary to determine the cause of failure, in order to establish liability and to indicate how to reduce future accidents. However, before examining seat belts which have failed under service conditions, it is necessary to establish the forms of break which occur under controlled laboratory conditions. The following account is taken from work reported by Duerden, *et al.* (1983), of an investigation of new seat-belt webbing material which had been subjected to a variety of destructive loads.

Two types of material were used: nylon 66 and polyester. The construction of the webbing is essentially the same for both materials, although the description which follows is specifically that of the nylon webbing. The webbing, which is 5 cm wide, is made up of 264 warp ends; two of these are single-ply identifying ends, and the remaining 262 are two-ply Z-twist (2.5 t.p.i.) of 940 dtex continuous-filament yarn with 68 filaments per yarn. The filament size is 13.8 dtex or 12 den. Filling (weft) yarn is single-ply 940 dtex zero-twist continuous-filament yarn, again with 68 filaments per yarn. A 2×2 twill pattern is woven on a shuttleless loom with a lock-stitch of 555 dtex yarn, and there are 17 picks/inch. Reversing the twill pattern at intervals across the width produces panels. Use of a 2×2 twill increases the initial stiffness of the fabric over that of a plain weave. After weaving the webbing is dyed. The polyester webbing is also heat-set.

The webbing material used in this study of controlled fracture was first broken in simple tension. These tensile tests showed a mean failure load of 7040 lbf (31.3 kN) for nylon webbing and 6759 lbf (30.0 kN) for polyester webbing.

Samples of each type of webbing were broken in a variety of ways designed to simulate circumstances which might arise in service. It is appropriate to note that the macroscopic appearance of the fractured webbing is affected by the mode of failure, although what will be shown here are the fibre failures.

Webbing was failed in tension using a tensile testing machine fitted with grips which ensured failure due to tension. The approximate length of the webbing under tension was 50 cm and the extension rate was 25 cm/min. In practice seat belts seldom, if ever, fail in simple tension. Nevertheless, the appearance of fibres in webbing extended to failure in this manner provides a useful reference point.

The appearance of a nylon filament found in webbing failed in simple tension, **37A(1)**, is typical of the fibres found in this and similar specimens, and is a mushroom end, such as is found in single-fibre tests at high strain rates. This form results from rapid load transfer after break starts. Polyester fibres have a similar form in webbing failed in simple tension.

In actual service seat belts would be subjected to rapidly applied loads. To simulate this a drop test was devised. The webbing was used to arrest a 273-pound falling weight at decelerations of between 15 g and 35 g. In addition to the simple drop test, modifications were made to the test to simulate the effect of rapid loading on the webbing as it is pulled against either associated restraint system hardware (a D-ring, also known as a B-pillar swivel) or a sheet metal edge. The latter test was designed to assess the effect of intruding damaged vehicle parts on the webbing.

Both nylon, **37A(2)**, and polyester, **37A(3)**, filaments from webbing specimens failed in a drop test have an appearance similar to that of fibres found in specimens failed in simple tension, although the polyester end is somewhat more bulbous than that found on tension failure specimens. In polyester webbing which had been passed through a D-ring and drop tested, other effects of material melting may have occurred, **37A(4)**. At some places in the webbing several filament ends could be seen to have fused together. A nylon filament, which formed part of a webbing specimen broken in a drop test over a sheet metal edge, is shown in **37A(5)**. The appearance of polyester filaments fractured in a drop test over a sheet metal edge, **37A(6)**, is characteristic of this type of failure mode. A similar appearance is found in some nylon fibres in webbing failed in this manner.

In service, seat-belt webbing is subjected to many forms of abuse. Perhaps the two most common are being dragged along outside a vehicle and being repeatedly trapped between the door and the door frame of a vehicle. To simulate this latter phenomenon a sample of webbing was folded perpendicular to its length and then struck repeated blows with a hammer along the fold while lying on an anvil. Beating was continued until the webbing separated. The resulting fibre failure morphology is illustrated in **37B(1)**.

The ageing of webbing samples was simulated, by exposure to ultraviolet radiation, in order to determine the effect of weathering upon mechanical properties. Three levels of exposure were used: 100, 200 and 300 hours. After exposure the webbing was broken in simple tension and the failure morphology examined. The effect of 300 hours exposure upon the failure morphology of nylon is illustrated in **37B(2)**. There is also an effect upon the macroscopic appearance of the webbing, particularly in the case of nylon, where the extent of the damage associated with tensile failure is dramatically reduced by exposure to ultraviolet radiation. The reduction in strength concentrates the zone of damage imposed by the breaking load.

On occasion, webbing is deliberately cut by rescue personnel, and it is important to be able to identify webbing which has been separated in this manner. A polyester filament in a webbing specimen cut with a pocket knife, **37B(3)**, has a fan-like appearance. This effect is typical of a knife cut, and is seen in nylon as well as polyester webbing. A sharply defined flat surface is found in a nylon filament from a webbing cut with a scalpel blade, **37B(4)**. The striations on the surface correspond to fine grinding marks on the blade. The fracture morphology in the nylon fibre, **37B(5)**, is typical of that produced by a scissor cut of webbing. Polyester filaments exhibit a similar form. Not all scissor cuts produced this appearance; some were more like the fibres illustrated in **37B(6)**, which is a polyester filament from a webbing cut by tin snips (metal shears). Note that there is no evidence of pinching, as is seen in other examples of scissor cuts. This form of failure appears in both nylon and polyester webbing.

In the remainder of this Chapter we show illustrations of the failure morphology of webbing which has broken in service. The circumstances surrounding each failure are outlined and conclusions are drawn from examination of the photomicrographs. A report on some of these failures has been given by Duerden and Dance (1984).

The first pictures are from the driver's seat-belt webbing in a vehicle which skidded sideways into the rear of a parked vehicle. Extensive damage occurred to the driver's side (left-hand side) of the skidding vehicle, the maximum intrusion into the vehicle in this region being 38 cm. Examination of both the macroscopic and microscopic appearance of the separated webbing indicated that failure was due to a combination of tensile loading and cutting of the webbing by intruding sheet metal edges which resulted from the collision. Two representative failure morphologies found in the webbing are presented. The fibre shown in **37C(1)** has an appearance typical of many fibres in this sample. Mushroom end failures such as this are a consequence of tensile overload. The failure morphology exhibited by the fibre illustrated in **37C(2)** shows distinct evidence of a shearing action consistent with the fibre being pulled across a metal edge, and may be compared with **37A(5)**.

In the second accident a vehicle was involved in a head-on collision with a small truck. There were five occupants in the vehicle: two in the front who were wearing seat belts and three in the back who were not restrained. The three rear seat occupants were thrown forward upon collision into the back of the front seats. Examination of the webbing and restraint system hardware produced evidence that the webbing had developed a fold along its length and became trapped in the D-ring. Samples of seat-belt fibre were found trapped in the D-ring slot. This hypothesis is also supported by the microscopical appearance of failed fibres found in a narrow zone in the region of the fold crease on the webbing.

The general appearance of fibres in the suspect zone of the webbing shows several different failure modes, **37C(3)**. More detailed examination illustrates a morphology typical of tensile failure accompanied by crushing, **37C(4)**, which is consistent with the fibres having been pulled over a metal edge, **37C(5)**. Two other morphologies can be seen in **37C(6)**, namely a pinching/cutting type and a high-speed tensile failure in the rear.

In another accident the webbing failed in a high-speed roll-over collision in which the passenger wearing the seat belt was ejected from the vehicle. Two separate failure zones were identified by visual examination: one suggested that simple tensile failure had occurred and this was confirmed by scanning electron microscopy; the appearance of the other zone suggested that the material may have been cut. Examination of this region in the SEM showed that the fibres had been crushed. The appearance, **37D(1)**, is consistent with the webbing being damaged prior to the accident by being jammed between the door and the door frame. A comparison may be made with **37B(1)**. It was concluded that failure occurred because of the prior damage in this zone.

It is not too uncommon to find seat belts which have been chewed by a dog. An example from the nylon belts examined in Canada is shown in **37D(2)**. Some other examples from a sample submitted to UMIST by the Royal Aircraft Establishment (RAE) are shown in **37D(3)-(6)**. Most of the seat belt was undamaged, and the nature of the fabric is shown in **37D(3)**. This appearance was typical of most of the material, with little sign of wear and only slight contamination by dirt particles. However, there were two distinct regions of damage, and, in one of these, the belt had been completely severed. The two regions were about 60 mm apart, which is probably the bite width of the dog.

The edge of the severed region, **37D(4a)**, shows the broken fibre ends and the disturbance of the fabric by the chewing. Individual fibre ends, **37(4b),(5)** have flattened wedged-shaped ends, which result from the sharp biting (cutting action). There are signs of damage further back along the filaments, which are distorted and often bent back, **37D(6)**.

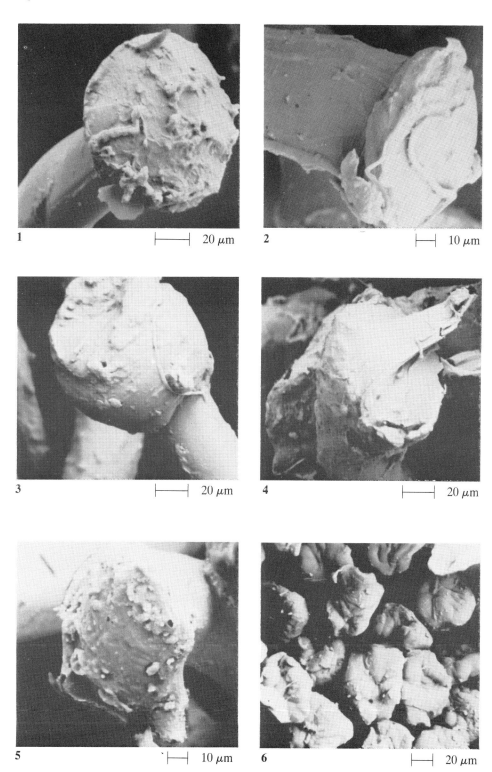

Plate 37A — Laboratory tests of automobile seat-belt webbing.
(1) Nylon filament from webbing broken in tensile test. (2) Nylon filament from webbing broken in drop test. (3) Polyester filament from webbing broken in drop test. (4) Polyester filament from webbing failed in drop test with D-ring. (5) Nylon filament from webbing broken in drop test over sheet metal edge. (6) Polyester filament from webbing broken in drop test over sheet metal edge.

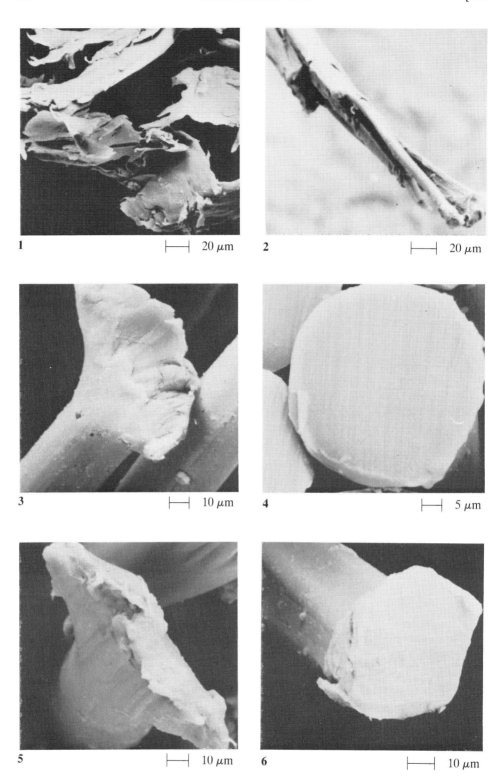

Plate 37B — Laboratory tests of automobile seat-belt webbing (continued).
(1) Nylon filament from webbing broken by repeated pinching between a hammer and an anvil. (2) Nylon filament from webbing failed in tension after 300 hours exposure to ultraviolet radiation. (3) Polyester filament from webbing cut with a knife. (4) Nylon filament from webbing cut with a scalpel. (5) Nylon filament from webbing cut with scissors. (6) Polyester filament from webbing cut with tin snips (metal shears).

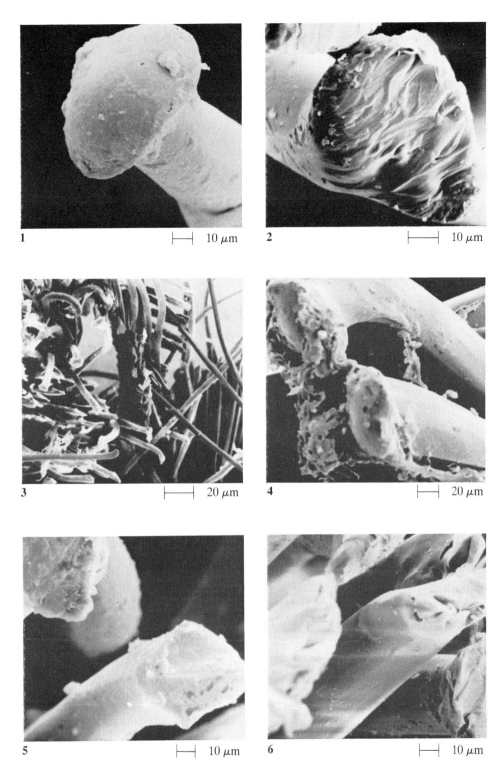

Plate 37C — Seat belt from a skidding accident.
(1),(2) Failed polyester fibres from the broken webbing.
Seat belt from a head-on collision.
(3) Fibres in suspect zone of webbing, at a fold trapped in a D-ring. (4)–(6) Details of fibre failure.

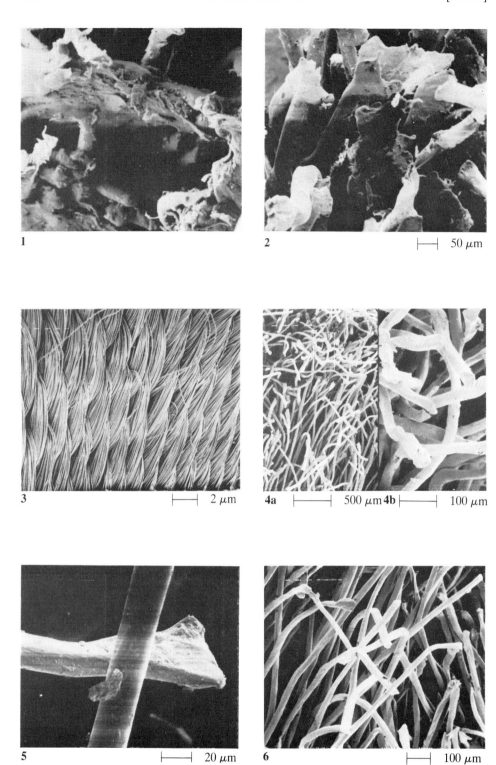

Plate 37D — Seat belt from a high-speed roll-over collision.
(1) Nylon fibre failure.
Seat belt chewed by a dog (Canadian study).
(2) Damaged nylon fibres.
Seat belt (nylon) chewed by a dog (RAE).
(3) Fabric from surface of belt, in a region which has not been chewed. (4a) Fabric at edge, which has been chewed through. (4b) Detail of fibre ends. (5) Severed fibre. (6) Material in chewed region.

38

MILITARY WEBBINGS AND CORDS

Narrow woven and braided materials are used in a wide variety of applications as fasteners and connectors. Depending on their shape and end-use, those of medium size are usually known as webbings or cords. They are found in many places in military equipment, and the case studies reported in this chapter have come from RAE, Farnborough.

The first example is the retraction strap of the ejector seat on a Jaguar aircraft. This was a continuous-filament nylon woven webbing, but had developed a hairy surface owing to projecting loops and broken filaments. Wear was particularly severe on the inner curved edge of the strap, and was probably associated with friction between the strap and the clip through which it slides.

Broken yarns at the badly worn edge of the strap are shown in **38A(1)**. The early stages of filament breakdown suggest a combination of multiple splitting due to flexing and peeling caused by surface shear, **38A(2)**. Some broken fibres have bushy ends, **38A(3)**, which could be due to biaxial rotation fatigue (Chapter 13), but the absence of indications of twist suggest that simple flex fatigue (Chapter 12) may be the cause. Alternatively, other filaments have flattened and mangled ends, **38A(4)**, presumably due to a shearing or cutting under high normal loads from the guide.

Where weft yarns have been severed, they shrink back into the fabric, **38A(5)**. Examination of broken filaments, **38A(6)**, shows flat ends which are sheared in one direction and subdivided into many zones. This suggests that they had first split and then been sheared off or, less likely, been gradually worn smooth.

The second set of examples result from a very different wear environment: they are webbings from arrester harnesses on airfields. Detailed study showed interesting differences between different parts of the webbings, but only some common features will be described here. Surprisingly, webbing which had been used for 12 months on a concrete surface, **38B(1a)**, showed less damage than one which had been used for only 4 days, **38B(1b)**. This presumably reflects the irregular usage of such equipment. The comparatively minor wear after long use occurs on the yarn crowns, where there is a smearing over the surface, **38B(2)**, as a result of material being peeled off, **38B(3)**. The peeling action on the 4-day specimen is more drastic, **38B(4)**, giving a very ragged appearance with some material rolled up. Other regions of the webbing are hairy owing to filament breakage, **38B(5)**, with evidence of multiple splitting, **38B(6)**.

Webbing which had been used for 12 months on a tarmac surface had also been scraped and smeared, **38C(1)**, with some material being rolled up, **38C(2)**. Detailed views show how a substantial part of a filament can split off and roll up, **38C(3),(4)**. Other parts of the webbing showed less severe damage by smearing, **38C(5)**, or an absence of wear but considerable contamination by debris, **38C(6)**.

Equipment used on airfields suffers damage not only through general mechanical wear and environmental degradation and emergency overloads, but also through extraneous causes. For example, **38D(1)** shows an arrester barrier centre marker after it had been damaged by rabbits. The teeth marks can be seen in the plastic-coated fabric. Where the material has been bitten through, **38D(2)**, the fibre ends are flattened into a spatula shape, **38D(3)**. Other fibres show damage over a greater length, **38D(4)**, and were probably broken by abrasive action, rather than being directly cut by the action of the teeth.

In contrast to this a webbing which has been cut with a file shows a different type of damage, **38D(5)**. Adjacent to the severed region, the filament tips are bent over, **38D(6)**, indicating the direction of movement of the file.

The next example is again wear due to normal use, and consists of two nylon border cords from cargo nets used in helicopter supply operations. The cord is a hollow circular braid, which has flattened into a webbing about 8 mm thick. The surface of the cord had been treated with a dark grey coating, which held the filaments together. One sample had only very slight damage, as can be seen from the appearance of the surface, **38E(1)**, although there are a few broken filaments on the edges of the cord, **38E(2),(3)**. The flattened and distorted appearance of the filaments indicates that breakage was due to heavy lateral pressure. The second sample was much more severely worn, and the colour had changed from dark grey to a light silver. The surface of the cord is badly disturbed, **38E(4)**, and on one edge the filaments were twisted and tangled, **38E(5)**. The fibre breakage, **38E(6)**, again appeared to result from the lateral pressure.

The last example is of parachute rigging lines, which are nylon braids. The material had suffered damage which led to the appearance of pairs of small fluffy knobs about 1 cm apart, which were the two broken ends of twofold continuous-filament yarns. A broken end extracted from the fabric is shown in **38F(1)**. The individual filaments give evidence of considerable squashing and splitting, with adhesion by softened material, **38F(2)**, and some broken ends have a complicated mangled shape, **38F(3)**, although others, probably among the last to break when they were taking a high load, are mushroom ends, **38F(4)**, similar to those found in high-speed breaks.

Another rigging line showed more widely distributed fibre damage, **38F(5)**, with evidence of high lateral pressures, **38F(6)**.

Among the samples which came to us from RAE was an interesting example of wear in flax, used in restraint webbings related to ejection by aircrew. The general damage, **38G(1)**, shows the typical wear found at the interstices between yarns in woven and braided fabrics. Squashing, **38G(2)**, possibly by webbing hardware, and fibrillation, **38G(3),(4)**, are two ways in which the flax fibres are damaged. However, a novel feature of deformation in flax fibres, which may also occur in other similar plant fibres, is sharp bending at nodes, as seen in the bottom right corner of **38G(4)**. These open into cracks, **38G(5)–(7)**. The break in the middle of **38G(4)** probably occurred at such a node and was accompanied or followed by some axial splitting.

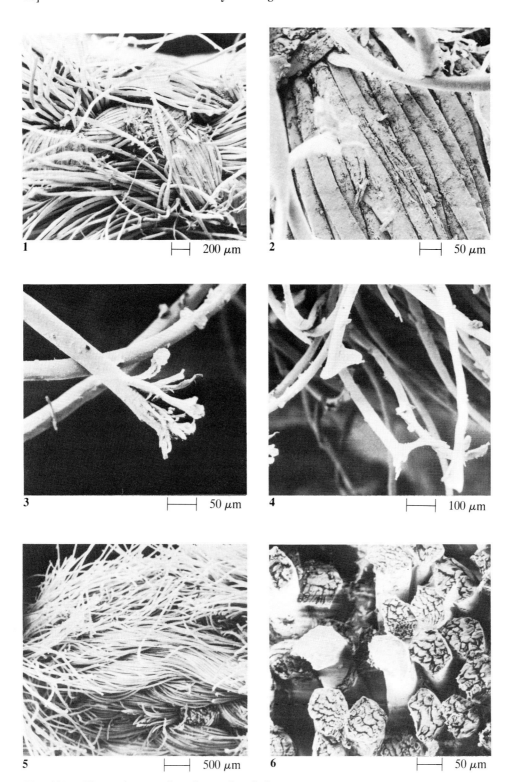

Plate 38A — Worn nylon strap from Jaguar aircraft ejector seat.
(1) Severely worn yarns at edge of strap. (2) Detail of badly worn area. (3),(4) Broken filaments from worn area. (5) Several yarns: note broken end of yarn pulled back into fabric, near bottom right corner. (6) Detail of broken yarn.

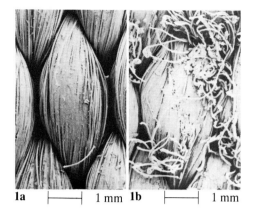

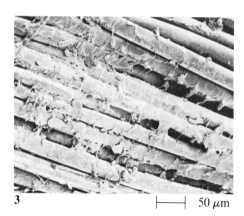

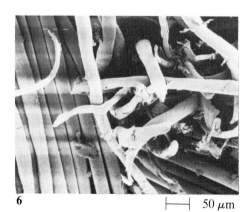

Plate 38B — Nylon arrester harness used on concrete runway.
(1a) Surface after use for 12 months. (1b) Surface of another webbing after use for 4 days. (2), (3) Wear on 12-month specimen. (4) Wear on 4-day specimen. (5),(6) Selvedge region of the 12-month specimen.

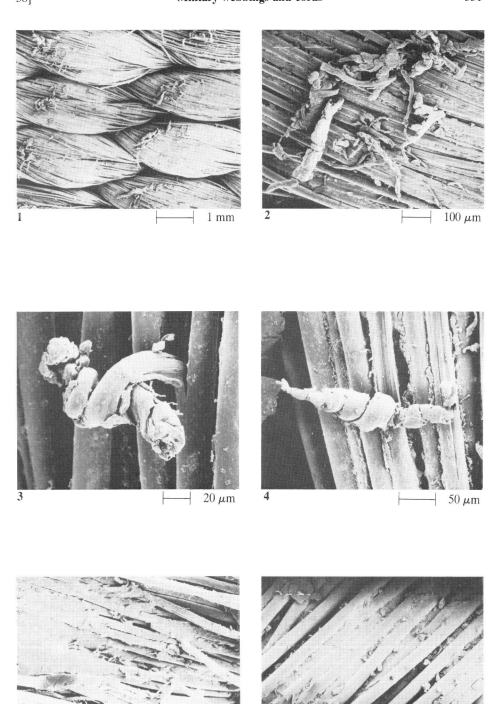

Plate 38C — Nylon arrester harness used on tarmac for 12 months.
(1),(2) Wear on surface. (3),(4) Detail of fibre damage. (5),(6) Other parts of the webbing.

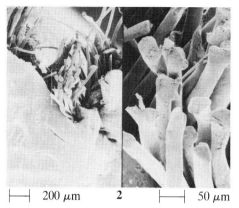

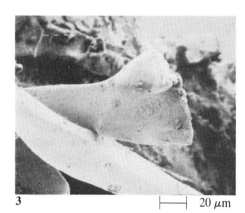

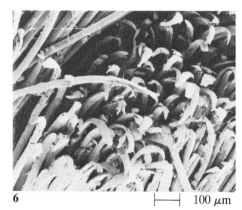

Plate 38D — Nylon arrester barrier centre marker damaged by rabbits.
(1) General view. (2) Detail of bite. (3) Fibre end from bite. (4) Other broken fibres.

Nylon webbing cut with file.
(5) Cut region. (6) Adjacent to cut.

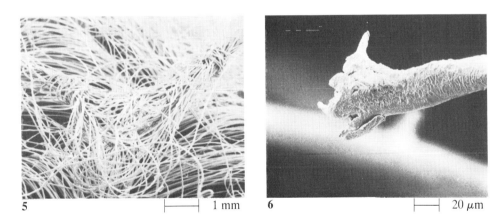

Plate 38E — Nylon border cords from helicopter cargo nets.
(1) Main surface of little-worn cord. (2) Edge of little-worn cord. (3) Detail of fibre breakage. (4) Main surface of severely worn cord. (5) Edge of severely worn cord. (6) Detail of fibre breakage.

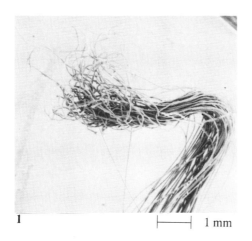

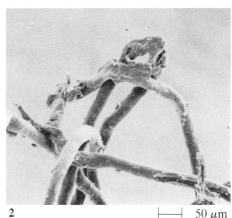

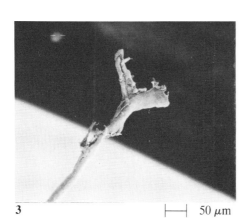

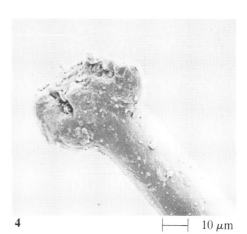

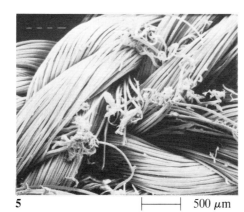

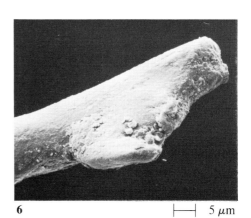

Plate 38F — Nylon parachute rigging line with localized yarn breakage.
(1) Broken yarn end, removed from the braid. (2) Fibre damage. (3),(4) Broken fibre ends.

Another rigging line.

(5) General view of wear. (6) Broken fibre.

Plate 38G — Wear of flax restraint webbings.
(1) General damage. (2) Squashed fibre end. (3) Fibrillation. (4) A variety of forms of damage. (5)–(7) Cracks from bends at nodes.

39

ROPES

Ropes provide examples of interesting forms of fibre failure, particularly associated with abrasive wear both on the surface of the rope and between fibres within the rope.

It has recently been found in studies by a number of investigators that nylon ropes break down rather rapidly when exposed to repeated loading in a marine environment. Although nylon products usually show excellent durability in dry conditions, moisture absorption causes changes in properties. In some single-fibre fatigue tests, involving bending and twisting, this may increase life; but in surface abrasion resistance the performance of wet nylon is poorer than of dry nylon. In contrast to this, polyester fibres with much lower moisture absorption show good abrasion resistance, both wet and dry.

As an example, taken from OCIMF tests, **39A(1)** shows a large nylon rope which had broken after being cycled wet between a low load and 50% of its single-test breaking load: under the repetitive loading the rope lasted for only 970 cycles. A polyester rope would have lasted for millions of cycles at this loading, and would have needed a loading of around 90% of its normal breaking load to fail in less than 1000 cycles. In the nylon rope the original continuous filaments had broken down into short fibres, as a result of a splitting and peeling shown for a collection of fibres in **39A(2)** and in more detail in **39A(3)–(6)**.

The next example comes from a failure in use of a rope, which was then subjected to very detailed examination. The rope was a nylon rope which was used to tie the skirt of a hovercraft onto the vessel itself. Failure was occurring after a short period of use, and it was found that the problem was that a cheaper nylon rope had been substituted for the previously used polyester rope. A return to the earlier practice solved the problem.

A length of the nylon rope, removed because it was near to failing, is shown in **39B(1)**. Some parts of the rope were almost undamaged, **39B(2a)**, apart from slight surface rubbing, a few broken filaments and some contamination with dirt; other parts showed clear evidence of moderate wear, **39B(2b)**; and near the knot in the eyelet of the skirt, the rope had almost broken, **39B(2c)**.

SEM study of the least worn rope surface, **39B(3)**, shows the contaminating particles jammed between the filaments, together with some rubbing wear on the fibre surfaces. The broken filaments have bushy ends, **39B(4)**, in which the fibrillar splitting has developed from one side, presumably due to layers peeling away as a result of abrasion. In the moderately damaged region there are many more broken filaments, but these have the same form of break, **39B(5)**. Close examination of this picture shows that the breakdown into the fine fibrils of bushy ends follows the peeling away of successive layers of material. An early stage of peeling from a fibre surface is shown in **39B(6)**, which also illustrates the varied nature of the contamination with some rounded granular particles and some cubic salt crystals.

In the badly damaged region near the knot, the structure is extremely complicated, **39C(1)**: there are some fibrillated broken ends, but other fibres show contortions characteristic of snap-back after rapid break, and there are some signs of lateral pressure.

Surface wear can be expected in a rope and, if the main body of the rope is intact, does not cause serious weakening. It is therefore necessary to look for damage within the rope. Because of the great confusion in the region of worst damage, it is best to start with the least damaged part, **39C(2)**, and look for wear between the strands. There is clearly appreciable damage to the fibres, **39C(3)**, where the strands rub against one another under the lateral pressures within the twisted rope. Layers are peeled off the filaments, and tend to group together to give a tide-mark, **39C(4)**. Examination of individual fibres shows the peeling of separate layers, **39C(5)**, which then fibrillate. The peeling goes progressively deeper into the fibre until there is breakage, as shown respectively by the upper and lower fibres in **39C(6)**.

The mechanics of failure can be expressed in the following terms. The rope as a whole will undergo repetitive tensioning, owing to the interaction of the vessel, the skirt and the sea, with possibly some bending or twisting. Within the rope there will be inter-strand movement and lateral pressure, which will generate frictional forces between the fibres. Material near the fibre surface will thus be subject to reciprocal shear forces, which cause the layers to peel away, as discussed in Chapter 14. Further rubbing will split the layers into fibrils.

Having established the essential nature of inter-strand wear, it is necessary to examine more severely worn regions to confirm that the same mechanisms are operative. In the moderately damaged region the effects are seen to be similar, but much more pronounced, in **39D(1)**. The yarn surfaces near the centre of the picture have been worn smooth; further out the tide-marks of peeling can be seen; and, at the outer edges of the picture, there are the fibrillated ends of broken filaments. Near the surface of the rope individual broken filaments show the characteristic bushy end, in an extreme 'mop-head' form, **39D(2)**. Within the strand there is considerable filament splitting, **39D(3)**, but the ends are not as bushy and often taper owing to the removal of successive layers.

Examination of the most severely damaged region is difficult because there is so much fibre breakage that the rope structure has become much looser. However, the same form of fibrillated breaks can be observed, and the peeling may occur over long fibre lengths, **39D(4)**. Some fibre breaks are typical tapering ends, **39D(5)**, but others show little sign of peeling and are probably caused by one filament cutting through another, **39D(6)**, under high lateral pressure.

Although the rope and strand surfaces are the most prone to damage, wear also occurs deeper within the rope structure. In the least damaged region inner ply yarns showed kinking and the beginning of surface peeling, **39E(1)**. A closer view of the peeling on the inner surfaces of singles yarns within the rope is shown in **39E(2)**. The surface appearance of the fibre in **39E(3)** suggests that inter-fibre adhesion may have occurred, with the subsequent separation loosening minute portions of the fibre surface. This would later lead to peeling, **39E(4)**.

Fairly severe peeling can be found in the inner surfaces of outer ply yarns, **39E(5)**. In a more severely damaged part of the rope, **39E(6)**, there is kinking of filaments and contamination with particles, which are probably salt crystals.

The next example comes from an extremely demanding rope use: a kinetic energy recovery rope (KERR). In this application the nylon rope is used to pull out a military vehicle when it has got bogged down. The extension of the rope stores up elastic energy, which is then available to continue the pull once the vehicle starts to move. During the operation much heat is generated, and it is usual to find that the ropes will not survive many periods of use.

Strands removed from positions away from the point of break in two failed ropes were examined, and are illustrated in **39F(1)**. The fusing due to heat is very clear, but is strictly localized, with a sharp transition to regions where the fibres are little damaged. Some individual breaks show projecting tails, **39F(2)**, perhaps where softened material has been pulled away, but others show a multiplicity of such effects, **39F(3)**. Particularly in regions of snap-back, heat has caused the fibre surface to become wrinkled, **39F(4)**. The appearance of the interior, seen through cracks in the skin, suggests that the inside of the fibre had become molten, **39F(5)**, and, in a severe case, the molten material had burst through the skin, **39F(6)**.

In the KERR and many other rope uses, damage may be caused either by the action of heat or by mechanical action. In order to investigate the effect of heat alone, H and T Marlow Ropes carried out a set of tests in which they heated ropes to 160°C for 33 hours, and then carried out a tensile test either at the elevated temperature or after cooling down to room temperature.

As a result of the heating, polyester ropes had changed to a pale cream colour. Failure of one rope in the tensile test had resulted in two strands breaking, with considerable recoiling on one side of the break but not on the other; the third strand had not broken. Heat is generated by the sudden release of energy on rupture, and fusing was apparent in many places, including the surface of both the unbroken strand, **39G(1)**, and the broken strands, **39G(2)**. There is also fusing at broken yarn ends. For inner yarns the fusing is moderate in extent, **39G(3)**. The ends of broken filaments tend to thicken towards their tips, in some instances, then tails project from the broken ends, **39G(4)**, presumably of softened material drawn out at the moment of rupture. Where the filaments are free, and not fused, mushroom-shaped ends characteristic of high-speed break are observed, **39G(5)**. In outer yarns the fusing is more severe, **39G(6)**.

Nylon 6 ropes show greater discoloration as a consequence of the heat treatment: they are generally deep gold, but near the break are dark brown or almost black. In the rope subsequently tested at room temperature all three strands had broken, but in the rope tested hot only two of the strands had broken, and the broken yarn ends showed no entanglement.

In the rope broken hot, rupture had occurred by breaks which run sharply across the yarn, **39H(1)**, although closer examination shows that this is in separate steps, **39H(2)**. The main cores of the fibres show a granular break, **39H(3)**, which is typical of the effect of thermal degradation on nylon. There appears to be a different skin layer; and on each fibre the break fans out from a point of contact with a neighbouring fibre. This suggests that there is fusion or very intimate contact between fibres, so that the break can propagate from one fibre to another.

In the rope broken at room temperature the form of breakage is similar in inner plies,

39H(4), except that the individual fibre failure commonly occurs in V-shaped ridges, **39H(5)**, which are along the lines of high shear stress. In outer plies, where lateral pressures have been less, the filaments break at different locations across a yarn, and different forms of fibre break are seen. Sometimes there are transverse granular breaks, but where filaments have fused together failure may be at an angle owing to shear stress, **39H(6)**.

Finally, we show some miscellaneous examples of rope failures. In a Kevlar yachting rope, used for a Genoa sheet, the failure shows the fibrillation which is characteristic of Kevlar, **39I(1)**, but the damage is clearly intensified by the presence of salt crystals, **39I(2),(3)**.

The other examples were supplied by RAE, and show rather different forms of damage.

The first was a nylon rope used in mountain rescue work. The rope had been badly scuffed and abraded, and was covered with fibre fragments and other debris. There was damage to filaments by peeling, and broken ends were flattened and distorted, **39I(4)**, owing to the scraping and abrading in use.

The second, a polypropylene rope used to tow gliders, was also severely worn. The filaments had split into ribbons of various widths, **39I(5)**, so that the appearance at different magnifications was similar. Where the damage had led to filament breakage this was in the form of long multiple splits, similar to those found in flex fatigue.

The third was a nylon aircraft barrier rope. This showed localized damage, with severing of yarns in places along the rope, presumably due to high impact loading. Broken filament ends were ragged, smeared and misshapen. There was fusing together of filaments, **39I(6)**, indicating considerable frictional heating.

39J(1),(2) shows the breakage of a filament taken from the outer surface of a 3-strand nylon mooring line, which had been manufactured at Boston Naval Shipyard and used for several years. This comes from an extensive research programme at MIT, reported by Backer *et al* (1983), Backer and Seo (1985) and Seo (1988), which included the pathology of worn US Navy and Coastguard ropes. The rupture has the stake-and-socket form, which has previously been observed in degraded polyester fibres, **16D(1),(2)** and **34G(4)–(6)**, and in hair, **19D(2)–(4)**. In this example it is a consequence of photo-degradation of nylon.

Natural fibres are now less used in ropes, but **39J(3)–(6)** and **39K** are pictures of a flax rope from RAE. This is interesting as showing how flax fibres break. A laboratory break test of an unused rope is relatively sharp, though it divides into separate bundle breaks and there is some structural disturbance. Some fibre breaks appear to be granular tensile breaks, **39J(3)**, but others show more splitting, **39J(4)**. There is also evidence of breaks at kink-bands, **39J(5)**, which may result from snap-back and may be at nodes similar to those seen in **38G(6),(7)**. Higher magnification views of the detail of failure within a flax fibre show the separate rupture of the ribbons of cellulose fibrils, **39J(6)**. In a used rope, wear seen on the rope surface in **39K(1)** comes from external abrasion. As it becomes more severe, strands break irregularly, **39K(2)**. There is considerable disarrangement of fibres, **39K(3)**, which in places form entangled balls. Breakage of individual flax fibres shows splitting and smearing, **39K(4),(5)**, though some individual breaks, **39K(6)**, are cleaner and more like those found in the laboratory break test.

Polyester ropes last longer under fatigue in wet conditions than nylon ropes which fail by internal abrasion as illustrated in **39A–E**. Even where two polyester strands come together within a sub-rope in a multi-strand rope, there is very little sign of wear, **39L(1)**. This is where a nylon rope would show severe wear, as in **39C(3)**. Although the polyester had been out in a sea-trial of a mooring for a considerable time, hardly any damage is apparent. In some parts of the rope, **39L(2)**, there was an accumulation of debris, which has led to slight peeling. The occasional broken ends have failed along kink-bands, **39L(3)**, sometimes associated with axial splitting, **39L(4)**, or have been squashed, **39L(5)**. The jacket suffers more damage due to external abrasion, as shown in **39L(6)**, with a higher magnification picture as **39L(7)**, showing a fibre bent at two or three places, leading to cracks and then rupture at the most severe damage location.

A large number of small ropes made from different high-performance fibres in a variety of rope constructions were subject to tension–tension fatigue in a joint industry study, FIBRE TETHERS 2000 (1995), aimed at the evaluation of fibre ropes for deepwater moorings. The commonest damage mechanism was axial compression fatigue, which occurs when a rope component goes into compression, even though the rope as a whole is always under tension. The compression results either from twisting of the rope or from differential lengths of rope components.

An analysis by Hobbs *et al* (1997) showed that the mechanics of deformation was similar to that in the buckling of heated pipelines on the seabed. The factors involved are: (i) development of an axial compressive stress along a length of the component; (ii) axial slip against friction; (iii) buckling giving a lateral displacement opposed by restraining forces. The analysis showed that this results in groups of buckles at intervals along the component, separated by straight portions. In the elastic deformation of the pipeline, the buckles are sinusoidal waves, but, in the elastic–plastic mode of yarn or fibre bending, sharp kinks

Fig. 39.1 — Predicted form of buckling of fibres and yarns due to axial compression within a rope. There is a set of solutions with different numbers of kinks between straight parts.

develop. The form of the predicted kinking is shown in Fig. 39.1. The kinks then lead to the yarn breaking into short pieces, separated by longer unbroken lengths.

This theoretical prediction is confirmed by fatigue tests on small ropes with a break load of about 5 tonnes. **39M** shows pictures from the central yarn of a Kevlar aramid rope after tension–tension cycling, in which this yarn will have been subject to axial compression, As the core strand is opened up, **39M(1)**, broken pieces can be seen and, back along the yarn from the last break, a kink, which has started to fail. Similar effects are visible in the extracted length of yarn, **39M(2)**. Examination of fibres by optical microscopy shows kink-bands within straight portions of fibres due to axial compression, **39M(3),(4)**, which can join in pairs and project out of the fibre surface. The axial slippage referred to above can lead to a concentration of kink-bands, **39M(5)**. A sharp kink in a whole fibre and some axial splitting is shown in **39M(6)**. Break at a kink in a fibre is seen in **39M(7)**. Fibre breaks usually run across the fibres at 45°, **39M(8),(9)**, but sometimes with complications, **39M(10)**. An unusual break at a sharp angle across the fibre is shown in **39M(11)**. A more usual break along the typical angle of approximately 45° of a kink-band is shown in the SEM picture, **39M(12)**. A break at a kink in a fibre with some axial splitting is shown in **39M(13)**, which has similarities to **39M(7)**. In **39M(14)**, several pairs of kink-bands can be seen projecting at 90° to each other on the surface of the fibre below the break. The break itself has developed by cracks along a similar pair of kink-bands, with some disturbance due to splitting. A longer axial split is associated with the break in **39M(15)**.

Kinks at intervals along a yarn in a Twaron aramid rope after tension–tension fatigue are seen in **39N(1)**. Optical micrographs show similar effects to those in the Kevlar rope. There are kink-bands on the inside of a bend in **39N(2)** and also axial splitting due to variable curvature, **39N(3)**. The curve from the straight portion into a kink and then to break at the next kink can be seen in **39N(4),(5)**, and the complications of rupture with fibrillation in **39N(6)**. Two kinks in a fibre in a parallel yarn Twaron rope are seen in **39N(7)** and an anomalous double kink in **39N(8)**. Extensive fibrillation is found in another Twaron rope, **39N(9)**, with detail of peeling in **39N(10)**. An unusual example of damage due to squashing is shown in **39N(11),(12)**.

The pictures in **39O** show that appreciable damage can develop in a thousand cycles of severe tension–tension fatigue in a Dyneema HMPE rope. Indications of axial splitting are seen in **39O(1)**. Kink-bands within fibres, **39O(2)**, develop to project out of the fibre surface, **39O(3)**, and finally into a contorted fibre path, **39O(4)**. These forms are essentially in 'straight' fibre paths. Kinks in whole fibres are seen in **39O(5)–(8)** and accompanied by break in **39O(9)**.

A similar development of kink-bands is seen in a Spectra HMPE rope after a million cycles of less severe tension–tension cycling, **39P(1)–(3)**, with slight fibrillation in **39P(3)**.

Polyester ropes are much less damaged by axial compression fatigue. A million cycles of fairly severe tension–tension cycling leads to the appearance of kink-bands within fibres, **39P(4)**, but these have not developed into cracks that would weaken the fibres. There are some gross kinks, such as **39P(5)**. However, as with the effects of yarn buckling shown in **12M(8),(9)**, the main 'damage' consists of a plethora of transverse lines and some axial cracks, **39P(6)**. Examples of squashing due to lateral pressure are seen in **39P(7),(8)**.

In general, the damage occurring in these ropes, which have been subject to tension-tension cycling is similar to that found from the yarn buckling tests shown in **12L** and **12M**. Axial compression fatigue is a potential hazard in fibre ropes used in deepwater moorings. It can be avoided by maintaining a *high enough minimum* tension on the lines. This is a concept unfamiliar to engineers used to metals, including steel wire ropes, where the concerns are with *maximum* loads, which may cause breakage, and *load ranges*, which lead to metal fatigue.

BEND-OVER-SHEAVE TESTING OF AERIAL CABLES
by Petru Petrina and Leigh Phoenix

At its White Sands Missile Range in New Mexico, the US Army has suspended an aerial cable between two mountain peaks with a span of about 5 kilometers. The cable is equipped with a variable tensioning device to permit moving loads of weights up to 10 tonnes, which may be targets on trolleys for air defence and drop testing, at speeds up to 1000 km/hr. The trolleys are supported on the cable by sheaves (pulleys) and moderate braking forces are applied to stop the trolleys. The cable must have a high strength and low mass per unit length, producing a high transverse wave speed. The jacket must be durable and the cable interior must be tolerant to abrasive actions due to rolling sheaves.

As part of the initial testing which led to the selection of the cable, tests were carried out on readily available, but slightly smaller cables, of similar construction. The basic design was a 36-strand, co-laid aramid Kevlar 29 rope in layers of 6, 12 and 18 strands in a nested helical configuration. The helix angles in all layers was 15°. The inner strands were jacketed in polyester and the outer strands in Kevlar, in order to give greater dimensional stability and resistance to surface wear under the sheaves. This rope had a diameter of 5.3 cm and a break strength of at least 200 tonnes. The actual cable used at White Sands had a diameter of 6.25 cm. The cables were in two forms: unlubricated and lubricated. The former contained the usual fibre finishes, but no additional finishes or lubricants. For the lubricated cable, the Kevlar yarns of the outer braided jacket were impregnated with a blend of silicone and Teflon, which was then cured to a flexible dry state.

The 5.3 cm cables were tested in a cyclic reverse-bend-over-sheave apparatus (CRBOS) in the laboratories of Tension Member Technology, under a cable tension of 50 tonnes and with transverse sheave loads of 2.5 tonnes or less. The sheave to rope diameter ratio was about 6 to 1, giving a short nominal contact length of about 1 cm. Cycle lifetimes ranged from 1300 to 22 000 machine cycles. The upper end of this range was regarded as acceptable for the application. The cable with a lubricated jacket had a much longer life than one that was not lubricated. It was also found that rope lifetime varied inversely approximately with the cube of the sheave load, so that it was advantageous to use more sheaves with lower loads.

Portions of these ropes were dissected and analysed at Cornell University. **39Q(1)** shows a strand where the CRBOS test has worn through the jacket. Within the strands the aramid fibres are fibrillated due to internal abrasion, **39Q(2),(3)**. The wear is greater in the outer strands than in the inner strands. Flattening and fibrillation due to high transverse pressures is also seen, **39Q(4)**. The general conclusion was that damage accumulation was primarily due to local sheave contact and was concentrated most strongly in the outer yarn layers of the outer strands. In unlubricated ropes, fibres were severely crushed and abraded under the jackets, but in the lubricated ropes there appeared to be a cushioning effect. The strength degradation in the fibres of the outer strands was roughly linear with number of cycles. The failure progressed in outer strands by weakening and collapse of the outer layer of yarn, thus overloading inner yarn layers through load transfer until the whole outer strand collapsed.

Unlike typical rope/sheave applications, inter-fibre and inter-yarn abrasion due to relative motion caused by rope bending was not by itself a noticeable cause of abrasive damage under these test conditions.

For further studies on rope wear due to passage over sheaves under tension, a Cornell Cable Wear Testing machine (CCWT) was built and has been reported on by Phoenix et al (1993) and Petrina et al (1994,1995). These are fundamental studies, but are relevant not only to the White Sands application, but also to plans by the US Navy to replace the 2.5 cm steel wire rope used in the Underway Replenishment System by a fibre rope. In order to simulate the loading of an outer strand in a rope subject to tensile force and lateral loading from a sheave, the strands in the test machine are under tension and experience lateral compression between a Teflon backplate and a rolling sheave. The backplate provides radial constraint to simulate the radial support provided by the inner layer strands to the outer layer strands. The groove of the sheave is about two strands diameter deep and provides lateral constraint to simulate the contact from adjacent strands in the same layer. For a 1 cm strand, the tensile force was about 1.2 tonnes. In order to establish an empirical relation between lateral force and strand lifetime, the lateral sheave load varied with the test from 0.25 to 0.625 tonnes.

In one study, strands were made of 1500 denier Kevlar 29 yarns, with a braided jacket of alternating Kevlar 29 and polyester yarns, which was similar to that used for the outer strands of the White Sands rope. The strands consisted of 3 core yarns, surrounded by a middle layer of 6 yarns and an outer layer of 12 yarns. **39Q(5),(6)** show how fibres can be bent or kinked as a result of the testing. Fibrillation is starting to develop at regions of high strain. Broken fibres, **39Q(7)**, show fibrils which start to coalesce as a result of friction and transverse pressure. This is most severe in the unjacketed strand, when the contact forces cause the fibres to flatten and sinter together forming a structure similar to one obtained by crushing fibres at about the melting temperature. These studies indicate that the forms of failure are influenced by the jacketing. The rolling and sliding of fibres is amplified when no jacket exists and becomes more localized for strands with damaged jackets.

As part of a synthetic fibre rope technology development programme, a series of 1 cm strands with braided polyester jackets was made from different fibres and tested on CCWT. **39R(1)** shows the tendency of Kevlar fibres to split longitudinally and fibrillate as a result of machine cycling. A tensile fracture occurring in such a test, **39R(2)**, also shows a split form, though, as mentioned below, this is not the common form of failure for Kevlar. Splitting and peeling also occur in Vectran fibres, **39R(3)–(5)**, but is less severe than in aramid fibres. Vectran strands lasted over 10 times longer than Kevlar strands. The studies indicated that Vectran and Technora fibres ultimately fail largely in a tensile mode, despite considerable fibrillation due to transverse stresses, whereas Kevlar fibres are more prone to shear failure by splitting along the fibre length producing a wandering crack from one side to the other over many fibre diameters. The Vectran tensile fracture, **39R(6)**, is similar to **39R(2)** for Kevlar.

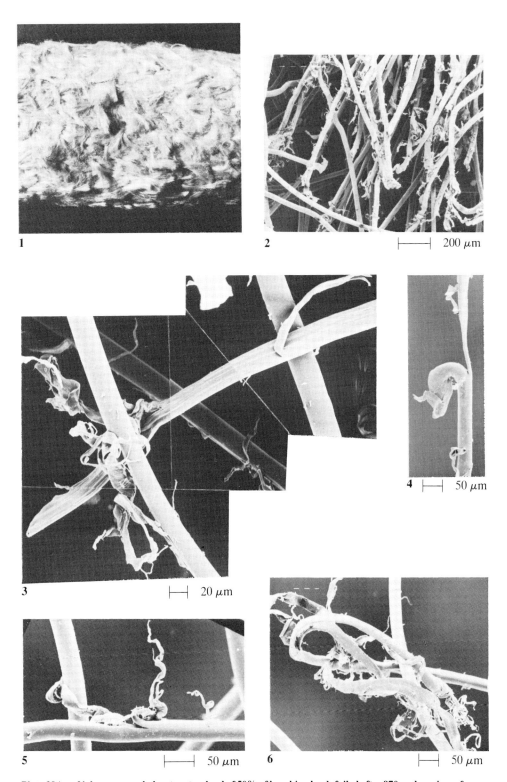

Plate 39A — Nylon rope cycled wet up to a load of 50% of breaking load, failed after 970 cycles: piece of rope away from break (from OCIMF test).
(1) General view of rope. Note that the rope before testing had a smooth continuous-filament surface, whereas the cycled rope shows many fibre ends sticking out. (2) Broken fibres in inner core region. (3) Detail of fibre splitting and peeling in inner core. (4)–(6) Detail of splitting and peeling in fibres from outer braid.

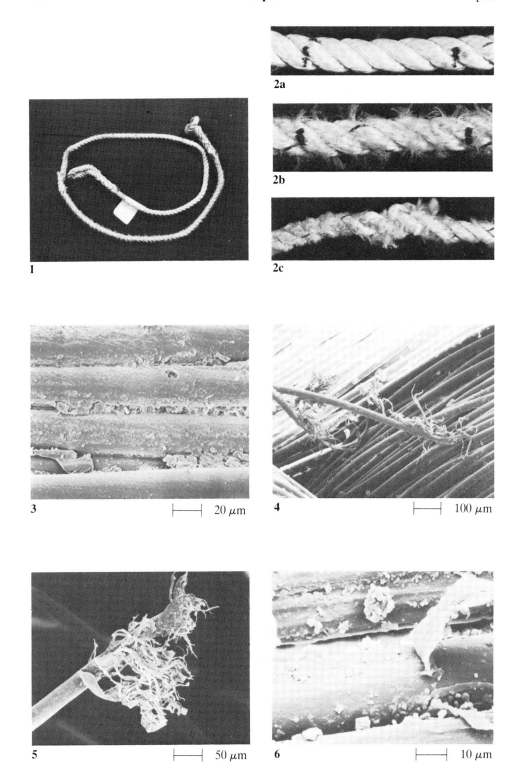

Plate 39B — Nylon rope after use in securing the skirt to the bow of a hovercraft.
(1) The complete rope assembly, which was knotted through an eyelet at the region of severe damage in the middle of the rope on the left. (2) Rope surface in a macrophotograph: (a) region of little damage; (b) region of moderate damage; (c) knot region, with severe damage. (3) Surface of rope in region of little damage. (4) Broken filament in region of little damage. (5) Broken filament in region of moderate damage. (6) Start of fibre peeling in region of moderate damage.

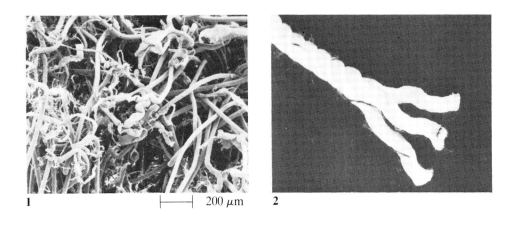

1 ├──────┤ 200 μm   2

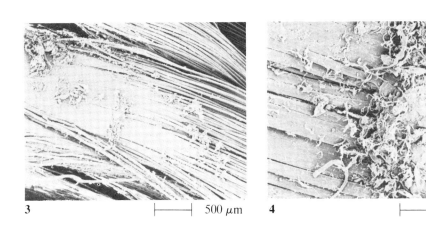

3 ├──────┤ 500 μm   4 ├──────┤ 100 μm

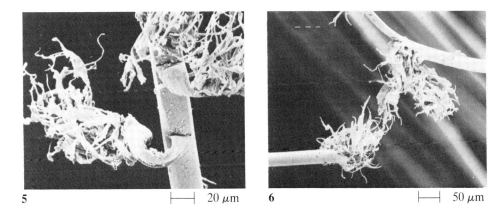

5 ├──────┤ 20 μm   6 ├──────┤ 50 μm

Plate 39C — Hovercraft rope (continued).
(1) Surface of rope in region of severe damage. (2) Rope in least-damaged region, opened out for examination of inter-strand damage. (3)–(6) Detail of the inter-strand damage in this region.

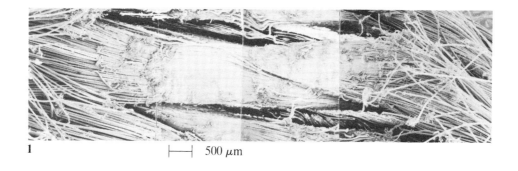

1   ⊢——⊣ 500 μm

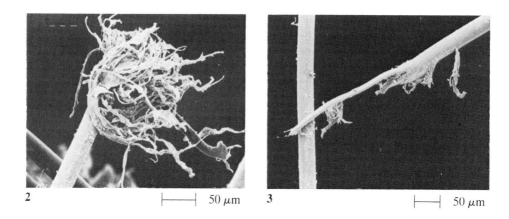

2   ⊢——⊣ 50 μm   3   ⊢——⊣ 50 μm

4   ⊢——⊣ 50 μm

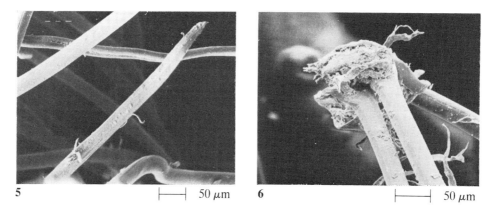

5   ⊢——⊣ 50 μm   6   ⊢——⊣ 50 μm

Plate 39D — Hovercraft rope (continued).
(1) Inter-strand wear in region of moderate damage. (2),(3) Detail of fibre breakdown in this region.
(4)–(6) Fibres from region of severe damage near the knot.

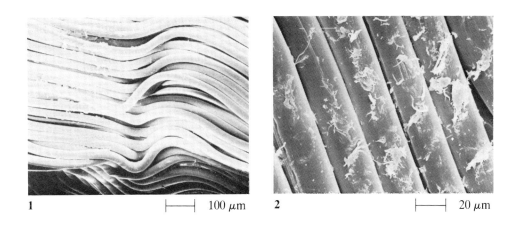

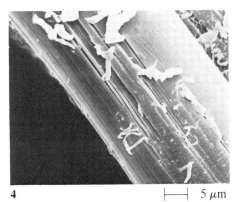

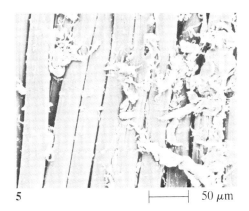

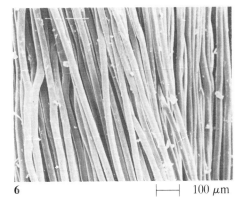

Plate 39E — Hovercraft rope (continued).
(1) Inner ply yarn from least-damaged region. (2)–(4) Detail of fibre damage in this yarn. (5) Outer ply yarn from least-damaged region. (6) Inner ply yarn in a region of more severe damage.

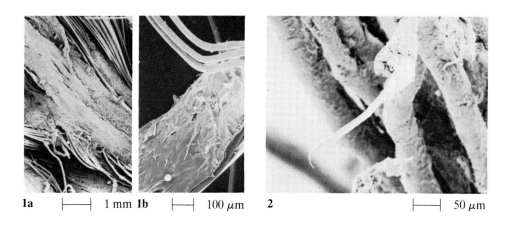

1a |——| 1 mm 1b |——| 100 μm   2 |——| 50 μm

3 |——| 50 μm

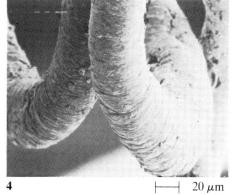

4 |——| 20 μm

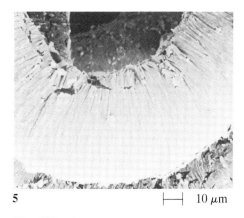

5 |——| 10 μm

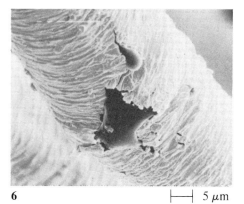

6 |——| 5 μm

Plate 39F — Broken kinetic energy recovery rope (KERR).
(1) Strands away from break in two ropes. (2), (3) Fibre breaks in first rope. (4), (5) Fibre surface changes in first rope. (6) Fibre from second rope.

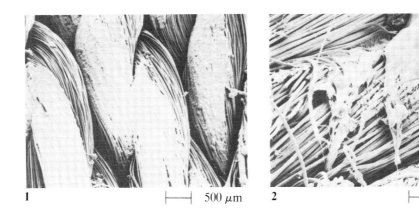

1 ├──┤ 500 μm   2 ├──┤ 200 μm

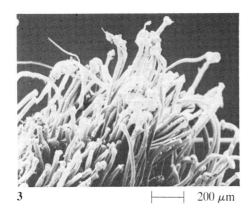

3 ├──┤ 200 μm   4 ├──┤ 20 μm

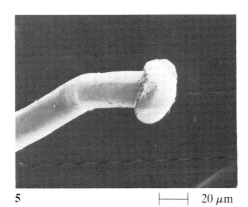

5 ├──┤ 20 μm   6 ├──┤ 50 μm

Plate 39G — Polyester rope heated to 160°C for 33 hours and then broken in a tensile test.
(1) Surface of unbroken strand in rope broken at room temperature. (2) Surface of broken strand in rope broken at room temperature. (3) Broken inner yarn from rope broken at room temperature. (4) Fibre end from this broken yarn. (5) Fibre end from broken yarn in rope broken hot. (6) From broken outer yarn in rope broken at room temperature.

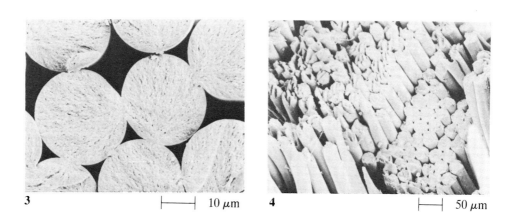

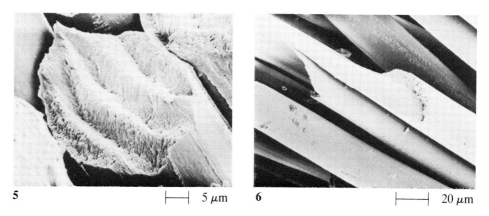

Plate 39H — Nylon 6 rope heated to 160°C for 33 hours and then broken in a tensile test.
(1)–(3) Yarn from rope broken hot. (4),(5) Inner yarn from rope broken at room temperature.
(6) Filament from outer yarn in rope broken at room temperature.

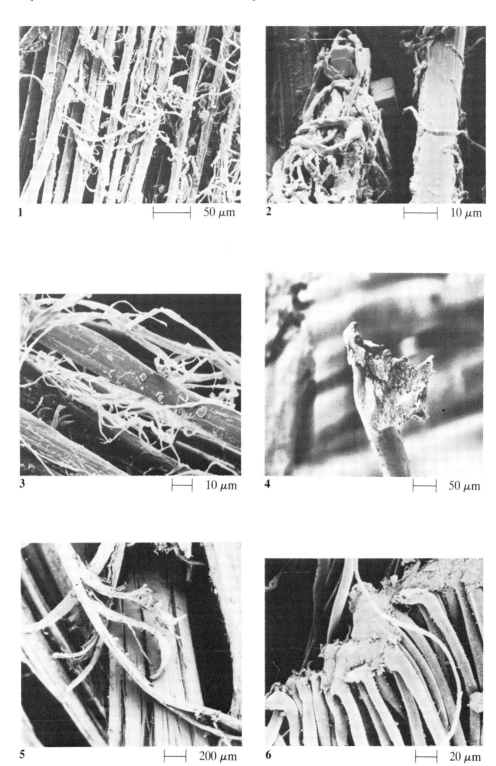

Plate 39I — Kevlar rope used for Genoa sheet.
  (1) Fibre breakdown by fibrillation. (2),(3) Detail of damage, showing presence of salt crystals.
Used ropes from RAE.
  (4) Nylon mountain rescue rope. (5) Polypropylene glider tow rope. (6) Aircraft barrier rope.

Plate 39J — Damage in a nylon mooring line, by courtesy of S. Backer, MIT.
  (1),(2) Opposite ends of a broken filament from a 3-strand nylon rope after several years of use.
Laboratory break test at RAE of unused flax rope.
(3),(4) Flax fibre rupture. (5) Kink-band failure, possibly on snap-back. (6) Detail of internal rupture of flax fibre.

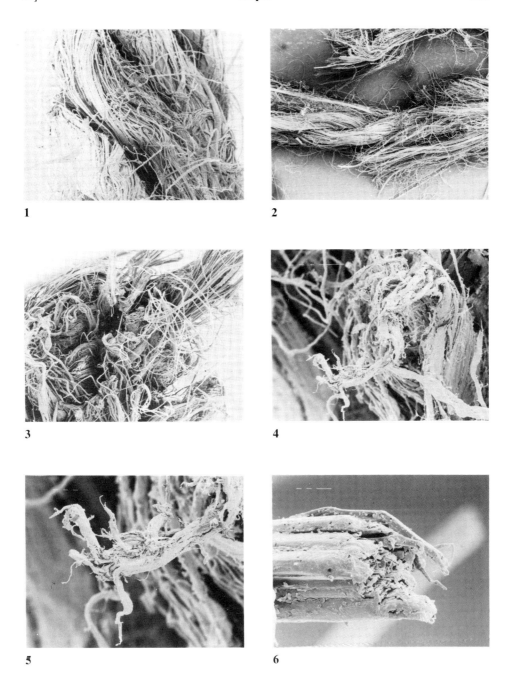

Plate 39K — A worn flax rope from RAE.
(1) External abrasion. (2) Separate strand breaks. (3) Disturbance at strand break. (4)–(6) Fibre breaks.

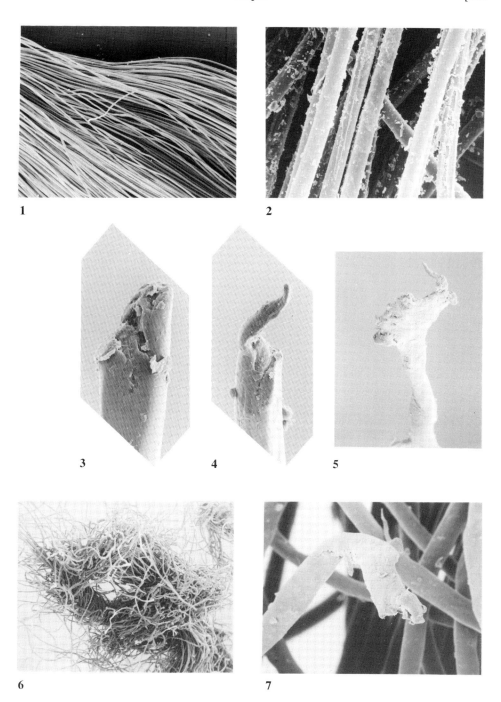

Plate 39L — A polyester multi-strand rope after use in a trial mooring for two years.
(1) Mating point between strands within rope with little damage. (2) Slight peeling and accumulation of debris. (3)–(5) Broken fibre ends. (6) Damaged region of rope jacket. (7) Broken fibre from jacket.

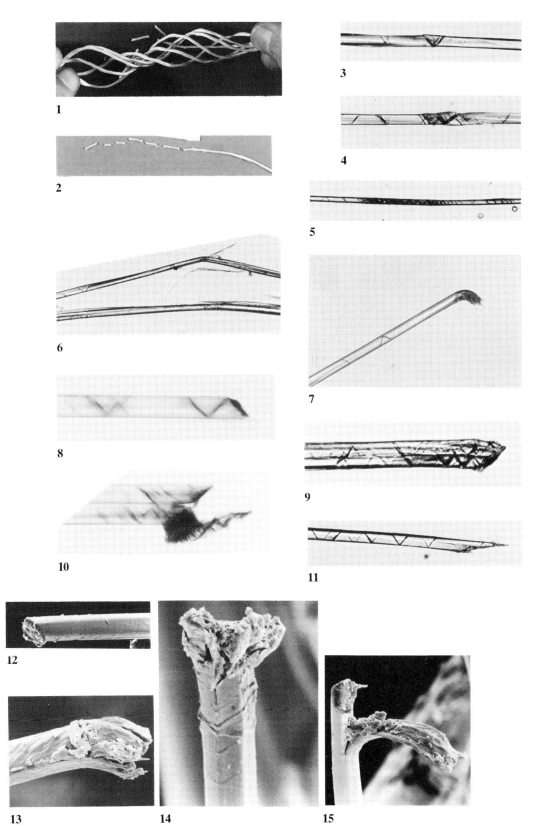

Plate 39M — Five tonne aramid Kevlar 129 rope in a 6-round-1 twisted wire-rope construction after 1 000 000 cycles tension–tension fatigue between about 10 and 50% of break load (60 kN), centre yarn of core strand.

(1),(2) Broken into pieces between 2.5 and 5 mm long. (3)–(5) Kink-bands within fibres. (6) Kink of whole fibre. (7) Break at kink in fibre. (8)–(15) Fibre breaks, along kink-bands and with axial splitting.

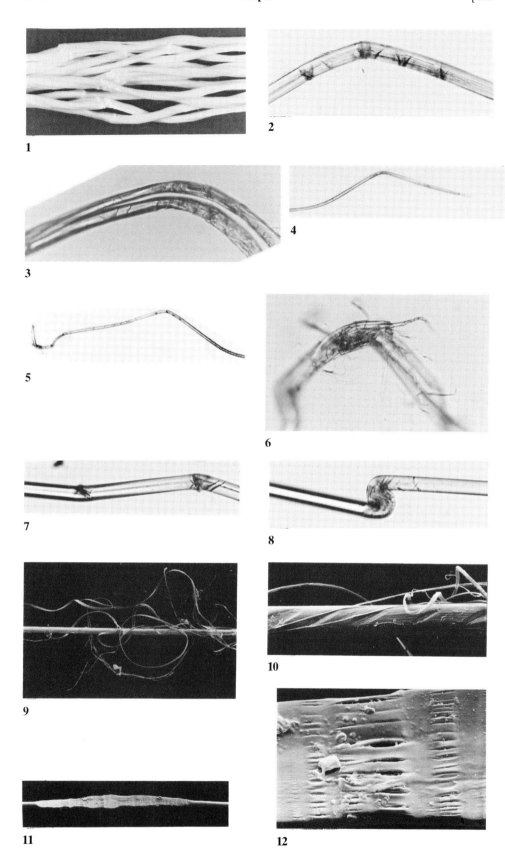

Plate 39N — Five tonne aramid Twaron 1000 ropes, after 1 000 000 or more cycles of tension–tension cycling between about 20 and 60% of break load.
(1) Kinks in yarns at interval along a yarn. (2) Kink-bands in a fibre. (3) Axial split. (4),(5) Bending into a kink and then to a break. (6) Break with fibrillation. (7) Two kinks. (8) Double kink. (9),(10) Fibrillation and peeling from fibre surface. (11),(12) Squashing of fibre.
Note: (1),(9)–(12) Simulated 36-strand torque-balanced rope. (2)–(6) 6-round-1 wire rope construction. (7),(8) Parallel yarn rope.

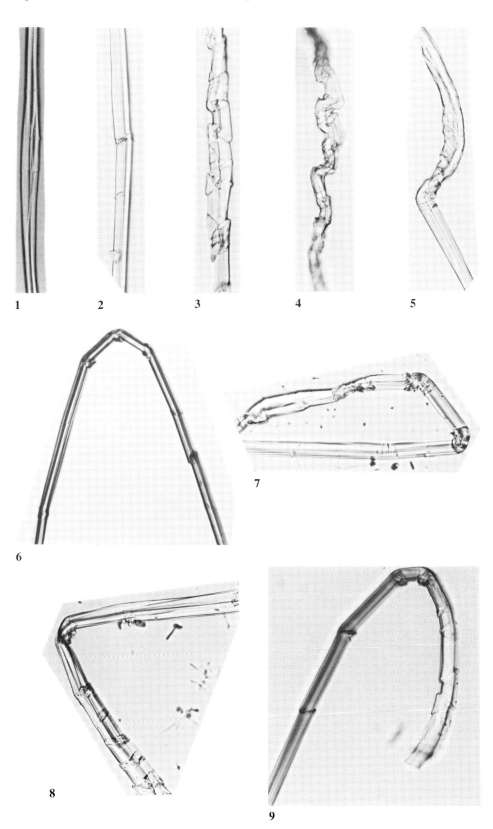

1   2   3   4   5

6

7

8

9

Plate 39O — Five tonne HMPE Dyneema SK60 6-round-1 wire-rope construction after 1080 cycles of tension–tension fatigue between about 10 and 70% of break load had led to failure at the termination.
(1) Indications of axial splitting. (2)–(4) Localised kink-bands in the fibre of increasing severity. (5)–(8) Severe kinks in the whole fibre. (9) Kink and break.

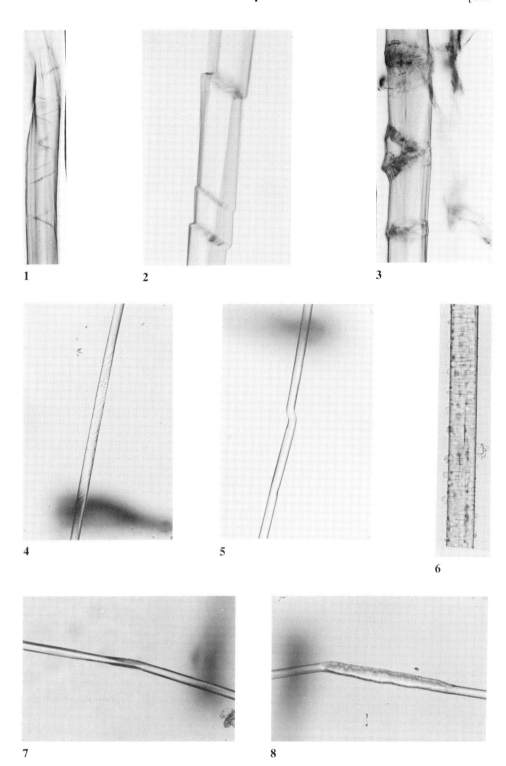

**Plate 39P — Five tonne HMPE Spectra 900 6-round-1 wire-rope construction after 1 000 000 cycles of
tension-tension fatigue between about 25 and 55% of break load.**
(1)–(3) Kink-bands and slight fibrillation.
**Five tonne polyester 6-round-1 wire-rope construction after 1 000 000 cycles of tension–tension fatigue
between 17.7 and 62.5% of break load.**
(4) Faint internal kink-bands. (5) A gross kink in a fibre. (6) Transverse lines and axial cracks. (7),(8)
Squashing due to lateral pressure between fibres.

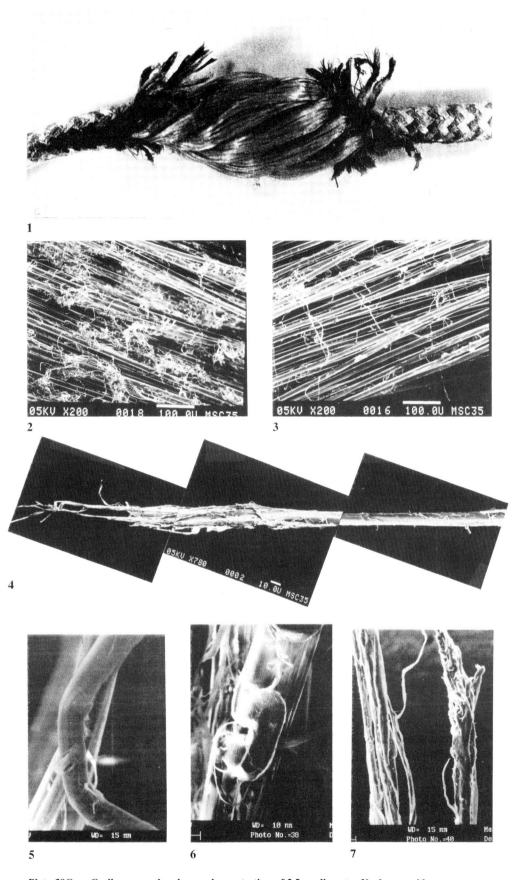

Plate 39Q — Cyclic reverse-bend-over-sheave testing of 2.5 cm diameter Kevlar aramid rope.
(1) A strand with broken jacket. (2) Fibres from an outer layer strand. (3) Fibres from an inner layer
    strand. (4) Flattening and fibrillation due to excessive transverse pressure.
Kevlar strand tests on the Cornell cable wear tester.
(5) From half-jacketed strand tested to failure at 225 kg. (6) From a fully jacketed strand failed at 320 kg.
                  (7) From an unjacketed strand failed at 320 kg.

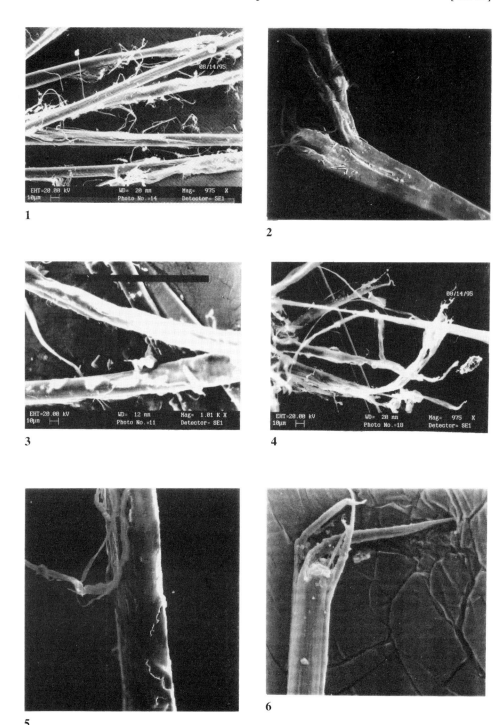

Plate 39R — Strand tests on the Cornell cable wear tester.
(1),(2) From Kevlar strands. (3)–(6) From Vectran strands.

40

OTHER INDUSTRIAL PRODUCTS

This last chapter on industrial uses contains a miscellany of products used for a variety of purposes.

The first example is a faded orange-coloured coated fabric from a wind-cone (wind-sock), which had been taken out of service after the usual period of about six months use on an airfield to show wind direction. In manufacture, the nylon 66 base fabric had been primed with isocyanate and then coated with a thin layer of white Hypalon (chloro-sulphurated polyethylene) followed by a thicker coating on the outside of fluorescent orange Hypalon.

The exposure to the mechanical action of wind forces and to ultraviolet radiation and other environmental effects had caused the coating to break along lines both along and across both faces of the sample. The nylon fabric was visible under the cracks, some of which were straight and followed the line of warp or weft yarns. Other cracks meandered irregularly across the fabric, but curved roughly in a direction which was presumably a result of the particular usage. The coating at the edge of the cracks was loose and easily removed. The most severe damage was a slit, 9.5 cm long, where all the warp yarns had broken, together with two or three weft yarns. In other bare areas there was partial breakage, with some warp yarns but no weft yarns severed.

In relatively undamaged material, shown in cross-section in **40A(1)**, there is good contact between the nylon yarn and the plastic coating; but the first sign of damage is a separation of coating from the yarn crowns, **40A(2)**. The separations can join up to give long lengths of coating which are no longer bonded to the nylon, although this may not be apparent on superficial examination of the material. Once the coating is loose, it is no longer supported by the strong nylon yarns, and it begins to break up and lift off the surface, **40A(3)**. If the fabric specimen is bent into a V along a line of wear, the depth of separation of coating from the fabric is clearly shown, **40A(4)**.

Examination of the face of the coated fabric reveals a pattern of small cracks in the thick Hypalon layer as the first indication of breakdown, with the nylon fabric (highlighted by charging in the electron beam) beginning to show through, **40A(5)**. On the reverse side, with a thin Hypalon layer, the damage was much more severe, **40A(6)**.

Along worn lines, where the exposed underlying fabric had not broken completely, some filament breaks can be seen, **40B(1)**, together with long splits penetrating down into the filaments, **40B(2)**. Surface peeling of filaments can also be seen, **40B(3)**.

Complete breakage of yarns is shown in **40B(4a)**, and often runs straight across a line of yarns, **40B(4b)**. Damage to the filaments can be detected some way back from the broken end, **40B(5)**, and is characterized by splitting. A long central split at the end of a broken filament can be seen in **40B(6)**, although the ends become rounded by further wear. This form of filament failure probably results from repeated bending, as shown in laboratory flex fatigue (Chapter 12).

The major cause of damage of the material is wind flagellation, possibly accelerated by other environmental degradation. The complex rippling motion, due to wind forces, first causes the coating to separate from the fabric. It is only after this has occurred that the fabric begins to be damaged, because bending becomes concentrated where there are cracks in the coating. Ultimately this leads to the breakage of fibres by flex fatigue.

The effect of natural weathering can be seen in a polyester fabric, which had been used in the form of two burgees flying at the mast-head of sailing dinghies for two or three seasons. The first burgee had almost disintegrated, and was torn at the edges, **40C(1)**. The filament break, **40C(2)**, is typical of light-degraded material. Away from the breaks the fibre surfaces have a rough and pitted appearance, **40C(3)**, and loss of material, **40C(4)**. At high magnification it can be seen that, in addition to the holes, there are many fine cracks or crazes perpendicular to the fibre axis, **40C(5)**. The other burgee was not quite as badly damaged, and it shows holes, but no fine crazes, on the fibre surfaces, **40C(6)**. The general impression is of weakening of the fibre due to the influence of light and perhaps some chemical action, with a few broken filaments on the yarn crowns, but most of the filament breakage concentrated in the tears along both warp and weft directions.

The next four examples are from sporting uses. The breakage of the shaft of a carbon fibre/epoxy badminton racket illustrates in real use the type of failure found in composite testing, as illustrated in Chapter 26. The shaft had broken near the point at which it joined the head of the racket. A general view of the fractured end is shown in the photograph, **40D(1)**. The shaft, which is made by winding angled plies of carbon fibre, is hollow; but some glue filling the central hole can be seen in the SEM view, **40D(2)**. The breakage of the several layers of carbon yarns is clearly apparent in **40D(2),(3)**. The effects observed and shown in **40D(4)** include fibre breakage, fibre pull-out and delamination. The detail of individual fibre breaks can be seen in **40D(5)**, and of delamination in **40D(6)**.

The second example is a string from a badminton racket, **40E(1)**, which has broken where strings cross, **40E(2)**. At the centre of the racket there is general wear, shown by peeling away of fibrils from the surface of the string, **40E(3)**. The actual break, which is a single sudden event, shows a complicated morphology, **40E(4)–(6)**.

The break of a tennis racket string (Donnay GLM 640) is shown in **40F(1)**. Near the break a line of cracks can be seen on the outer sheath, **40F(2)**. The fibre breaks, **40F(3),(4)**, appear to consist of some cutting of ends or damage by high localized pressures and some bulbous high-speed breaks, which probably occur during final failure of the string after it has been previously weakened by the other forms of damage.

The third sporting example is a bowstring used in archery, in which one strand had broken, **40F(5),(6)**. There is considerable peeling of filaments, which presumably caused weakness, and the broken ends are then high-speed breaks, with some pulling out of molten tails.

The photographs in **40G** and **40H** illustrate some other forms of damage to fibres. The first, **40G(1),(2)**, is a thick nylon 6 monofil used to transport sheets through an automatic laundry. The form of damage is extensive peeling of the monofilament surface, which was very severe even after only 10 hours service. Nylon 610 performed better, and showed a similar level of damage after 50 hours, and nylon 11 was very much better, with only a small amount of peeling after 257 hours service.

Another type of break is shown in **40G(3)**. This is a very clear example of failure by the kinkband mode, discussed in Chapter 12. It is an optical micrograph by the late S. C. Simmens of the Shirley Institute of nylon filaments from the inner layers of a tyre, which had been worn by simulated use on a rotating drum in a test bed. The damage may be due to compression on the inside of a bend, but could be due to compression of the whole fibre within the rubber matrix. It is true to say that this fibre failed by breakage at a kinkband under compression; but this is not necessarily, or even probably, the cause of failure of the tyre. The real cause may have been a loss of adhesion between the nylon and the rubber, which then allowed the fibre to suffer the deformation leading to its rupture.

A break which appears to be a tensile fatigue break of nylon, as described in Chapter 11, is shown in **40G(4)**. The characteristic tail of the fatigue break can be seen. This break was a fibre from a cord of a parachute, which had been deployed to assist braking of fighter aircraft on landing. The parachute flutters at about 50 Hz and each deployment lasts about 2 minutes, so that the fibres would experience over 100 000 cycles of tensile stress in 20 deployments. The tensile fatigue mechanism is thus a likely cause of damage to the fibre, provided the stress on a fibre falls to zero in each cycle. This is possible since there may well be an alternation of tension and compression waves.

An industrial example of the cutting of filaments is shown in **40G(5)**. This is a small hole in a polypropylene fabric, which is used to filter china clay. The filaments are sheared off at an angle, **40G(6)**, and the breaks were probably caused by a small hard object being forced through the fabric.

The effects of severe light degradation in polypropylene are illustrated in **40H(1)**, in a study reported by Barish (1989). The filaments are from the top of the back cushion of a car seat, where the material had been exposed to light but not to much mechanical wear for about five years. It is known that polypropylene shrinks under these conditions, and, after straightening out crimp, would generate tension. In the weak degraded material on the filament surface, the tensile stress is relieved by the formation of transverse cracks. The boundary between degraded and regular material can be clearly seen in a fibre which has been split open, **40H(2)**.

We next turn from heavy industrial use to personal hygiene. In an unused toothbrush the bristle has an end with a shape determined by the cutting, **40H(3)**; but, after use, this has become somewhat rounded, **40H(4)**, and the surface of the bristle is beginning to wear by surface peeling.

The effects of high-speed ballistic impact are shown in studies by C. Cork at UMIST. A bullet fired at 514 m/s at a nylon fabric penetrates the material and breaks the nylon fibres, **40H(5)**, with the typical high-speed rupture form, shown in Chapter 6. Nowadays, Kevlar has replaced nylon as the material used in 'bullet-proof vests'. Under ballistic impact it fails by axial splitting, **40H(6a,b)**, in a manner similar to a slow-speed tensile break.

An example of failure caused by direct surface wear is shown in **40I**. This is a shoe-lace, which had broken after comparatively little use, as a result of rubbing against the metal eyelets. The outer cover of the lace is a viscose rayon braid, **40I(1)**, in which the fibre surfaces had been worn flat, **40I(2)**, and then had eventually broken. The broken ends can be seen in the interstices of the braid, **40I(3a,b)**. An unusual broken rayon fibre, from close to the place where the braid had broken, is shown in **40I(4)**. The exposed cotton yarns of the core, **40I(5)**, had then broken by multiple splitting, **40I(6)**. A higher-quality lace would be of stronger fibres, such as nylon.

Papermakers' felts are subject to severe forces in hot wet conditions as they are driven through the machines and compressed between rollers. M. A. Wilding and C. Cork at UMIST have examined the damage which can be seen in used felts. Only limited details of the construction and use of the felts, which are made of nylon fibres, were available. **40J(1)** shows the surface of an unused felt, including a bulbous end, probably resulting from some heat treatment. Bulbous ends are most common on the paper side. Cut ends are also found. After use, **40J(2)**, there is considerable surface wear of the fibres, flattening and multiple split breaks. These effects are also seen in part-used felts. The felt in **40J(3)** contains a mixture of coarse (44 dtex) and finer (17 dtex) fibres and shows severe wear after 2 days use, with considerable splitting and peeling of fibres. The damage appears to be less severe after 14 and 16 days respectively in **40J(4),(5)**, which are made of finer fibres (3.3 and 6.7 dtex, respectively), though there is considerable flattening and some splitting. **40J(6)** shows clear examples of multiple splitting failures.

Filter fabrics can suffer damage from chemical and thermal attack as well as mechanical action. The pictures in **40K** are of a degraded polypropylene filter after use at 110°C. The yarn break in **40K(1)** can be seen in **40K(2)** to include a mixture of stake-and-socket breaks, which, as shown in **16D(1)**, are common after chemical attack, and axial splitting. In other places, **40K(3)**, the fibres break into short pieces. Earlier stages of the thermal and chemical attack are shown in **40K(4)–(6)**. Extensive transverse cracking leads to portions of material breaking away.

The rupture of nylon and Kevlar fibres as a result of ballistic impact are shown in **40H(5),(6)**, but in view of the limited information on the form of damage by bullets, as referred to in Chapters 44 and 46, it is worth including more pictures from the laboratory studies of C. Cork at UMIST. When a bullet impacts at high speed on a woven fabric, a pyramidal out-of-plane wave propagates across the specimen, **40L(1)**, preceded by an in-plane wave, which contributes to the increase in length. In a more isotropic knit fabric, **40L(2)**, the deformation is conical. After a nylon fabric has been penetrated at the comparatively slow speed of 254 m/s, **40L(3)**, there is extensive local damage around the hole made by the bullet and the effects of the in-plane extension along the warp and weft directions can be seen reaching to the edge of the fabric specimen. At the higher speed of 514 m/s, **40L(4)**, the hole is much sharper and the warp and weft deformation lines do not extend as far, because penetration has been completed before the wave reaches the edge of the circle. Similar differences are seen in a Kevlar fabric, **40L(5),(6)**.

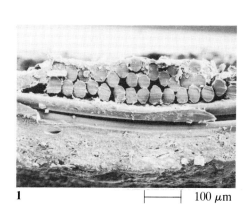

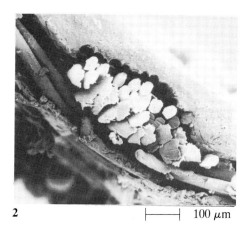

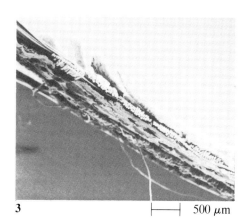

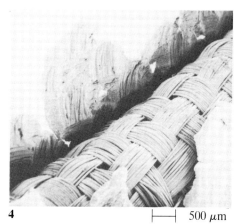

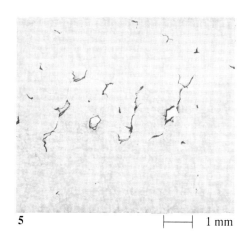

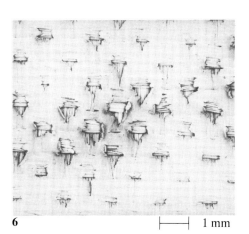

Plate 40A — Examination of worn wind-cone.
(1) Cross-section of relatively undamaged material. (2) Cross-section showing initial indication of damage. (3) First signs of break-up of coating. (4) Specimen folded into a V-bend, showing large areas of loss of bonding between fabric and coating. (5) First signs of damage on the face of the material. (6) More severe damage on the reverse side.

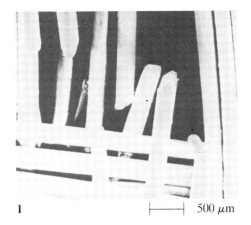

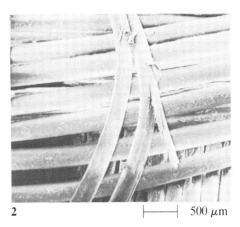

1 |⊢———⊣ 500 μm   2 |⊢———⊣ 500 μm

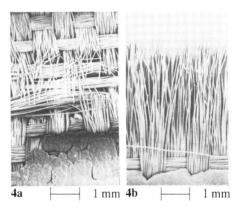

3 |⊢———⊣ 500 μm   4a |⊢——⊣ 1 mm 4b |⊢——⊣ 1 mm

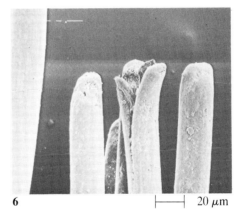

5 |⊢———⊣ 500 μm   6 |⊢———⊣ 20 μm

Plate 40B — Examination of wind-cone (continued).
(1) From region where fabric is exposed, but yarns are not broken, except for a few filaments.
(2),(3) Development of filament damage. (4a, b) Regions of more severe wear. (5) Filaments some
distance back from a yarn break. (6) Filaments at a yarn break.

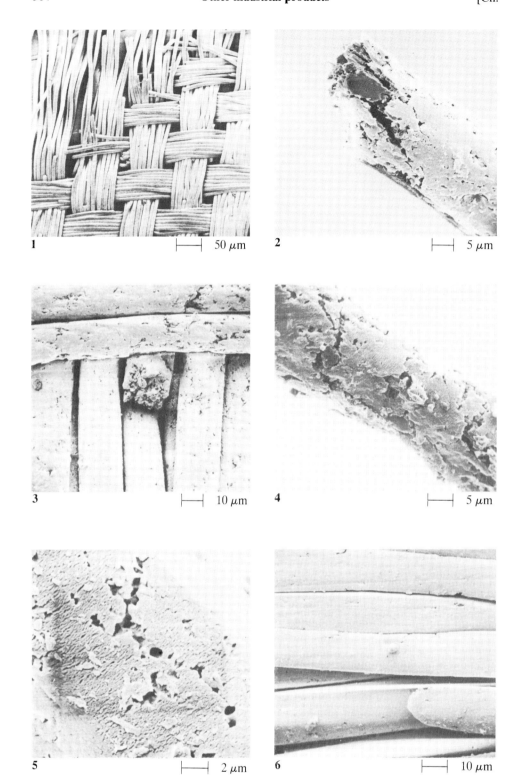

Plate 40C — Natural weathering of burgees on a sailing dinghy.
(1) Torn region at edge of fabric. (2) Broken end of filament. (3) Surface of yarn crowns, with one broken filament. (4) Part of fibre away from broken end, with thinning due to loss of material. (5) Crazing of filament surface. (6) Second burgee, not as severely damaged.

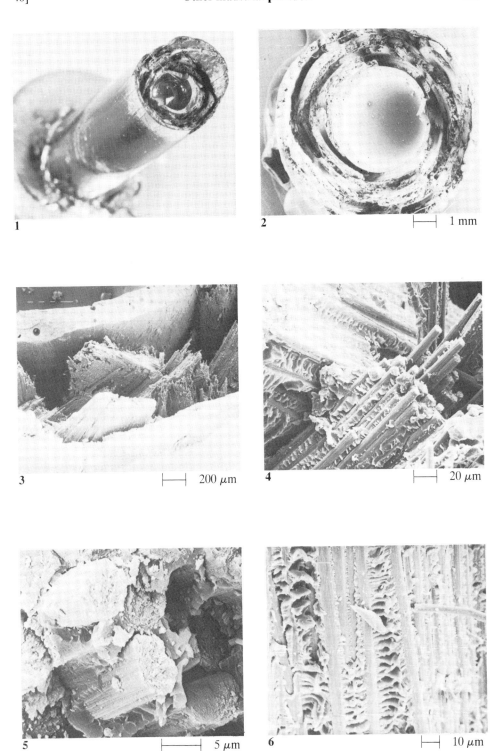

Plate 40D — Breakage of the shaft of a badminton racket made from a carbon fibre composite.
(1) Photograph of fracture. (2) End-on view of fracture. (3)–(6) Detail of breakage.

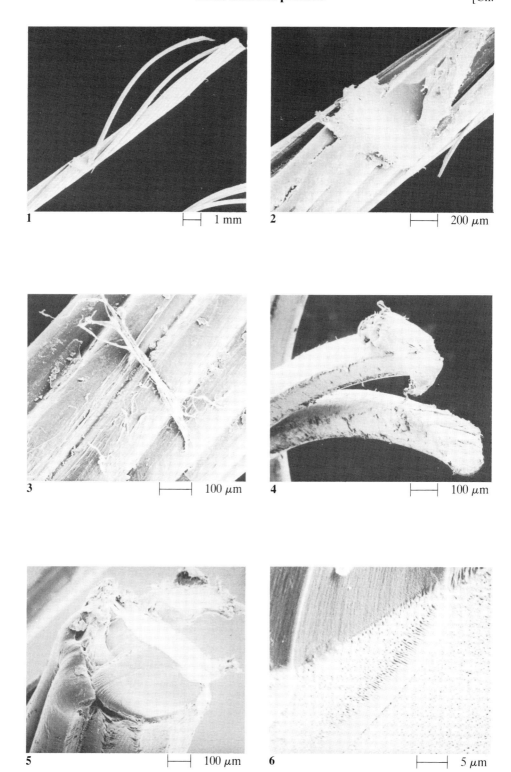

Plate 40E — Broken string of badminton racket.
(1) General view of break. (2) Cross-over point of strings. (3) Peeling on string surface, due to general wear at centre of racket. (4)–(6) Detail of break.

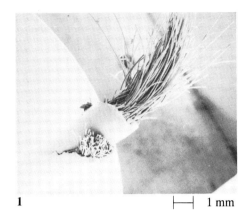

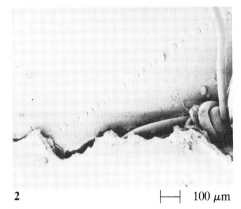

1   ├─┤ 1 mm   2   ├─┤ 100 μm

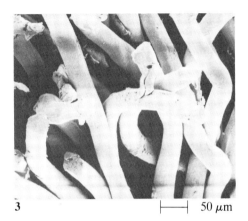

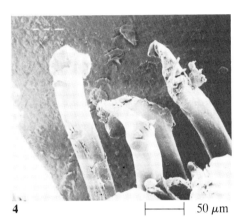

3   ├─┤ 50 μm   4   ├─┤ 50 μm

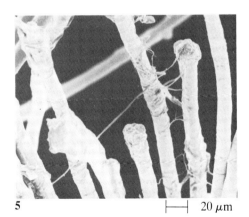

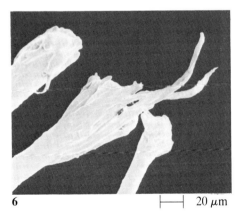

5   ├─┤ 20 μm   6   ├─┤ 20 μm

Plate 40F — Break of tennis racket string.
(1) General view of break. (2) Detail of damage to sheath. (3),(4) Fibre breaks.

Break of one strand of archery bowstring.
(5) Fibres near break. (6) Detail of breakage.

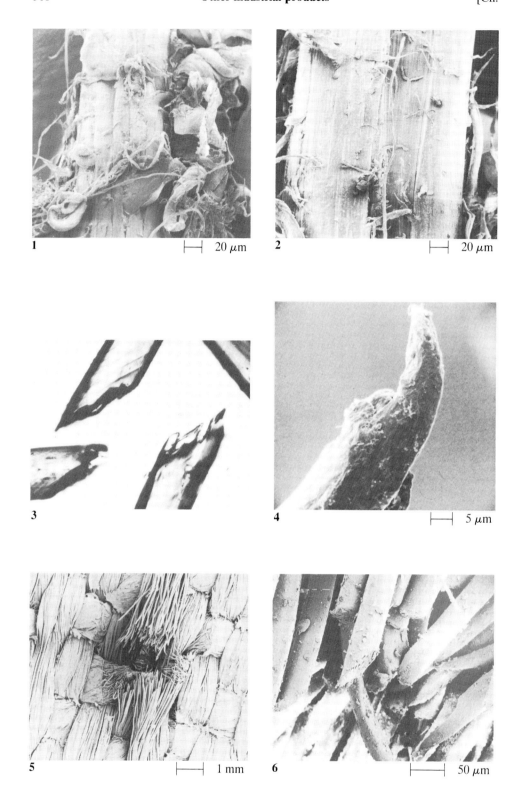

Plate 40G — Four forms of fibre damage.
(1),(2) Nylon 6 monofil used to transport sheets through an automatic laundry, after 10 hours service.
(3) Nylon filament removed from worn tyre; optical micrograph, from S. C. Simmens. (4) Break in nylon
filament from a parachute cord, deployed during aircraft landing. (5),(6) Hole in polypropylene fabric,
which had been used to filter china clay.

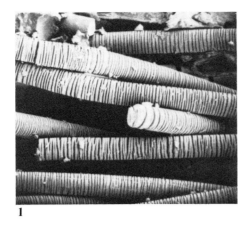
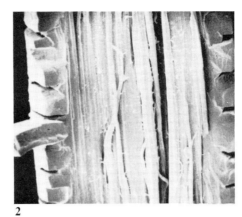

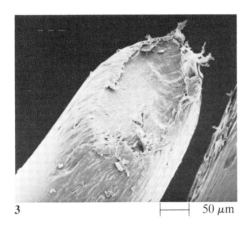

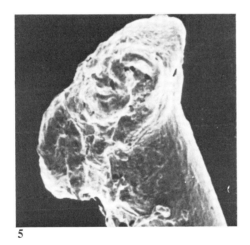
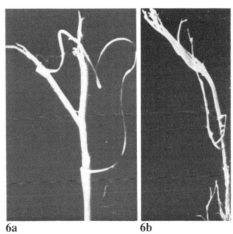

Plate 40H — Polypropylene fabric after five years exposure to light in a cover on the back seat of a car.
(1) Filaments with transverse cracks. (2) Internal appearance after splitting open.

Toothbrush, Oral 30B.
(3) Bristle in new brush. (4) Bristle in used brush.

Ballistic impact resistance.
(5) From nylon fabric, impacted by bullet at 514 m/s. (6) From Kevlar fabric, impacted at: (a) 240 m/s;
(b) 523 m/s.

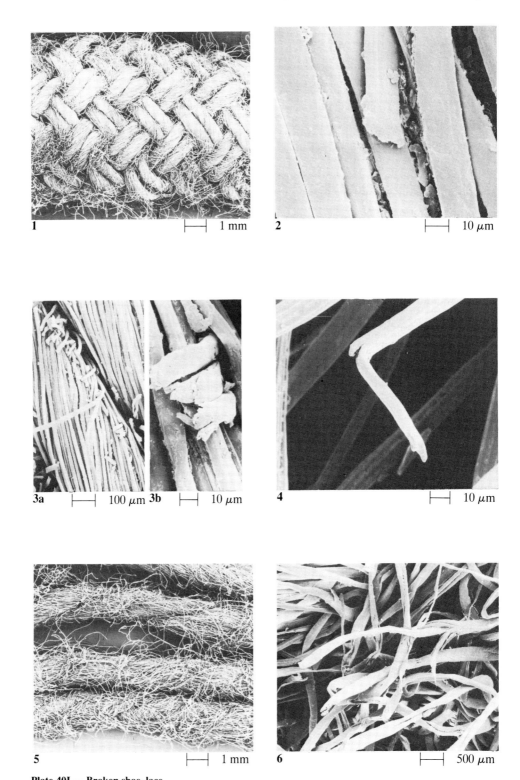

Plate 40I — Broken shoe–lace.
(1) Surface of outer braid, near the break. (2) Surface wear and break of rayon fibres in braid.
(3a, b) Broken fibres in braid. (4) Rayon break, close to break in lace. (5) Exposed cotton yarns from core.
(6) Detail of cotton fibre damage.

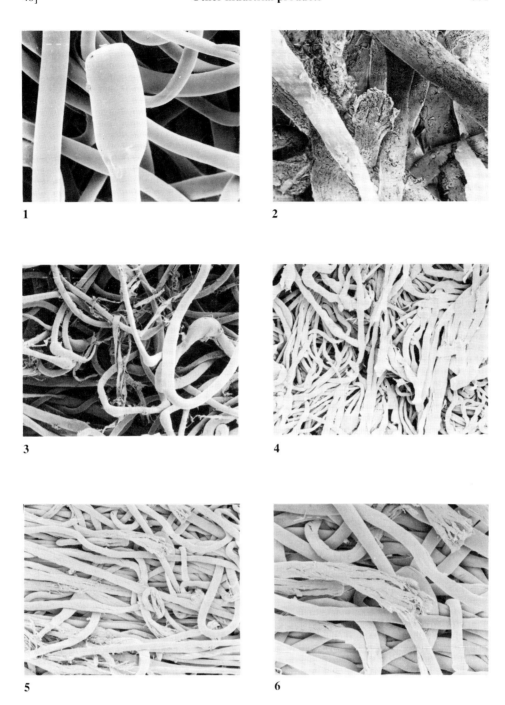

Plate 40J — Wear in papermakers' felts.
(1) Unused felt. (2) Used felt. (3)–(6) Part used felts.

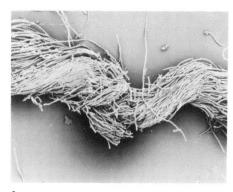

1

2

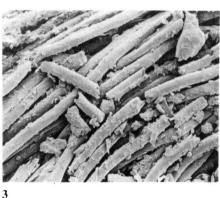

3

4

5

6

Plate 40K — A degraded polypropylene filter.
(1) Yarn breaking. (2) Details of fibre breaks. (3) General damage. (4) Transverse cracks and pieces breaking off.

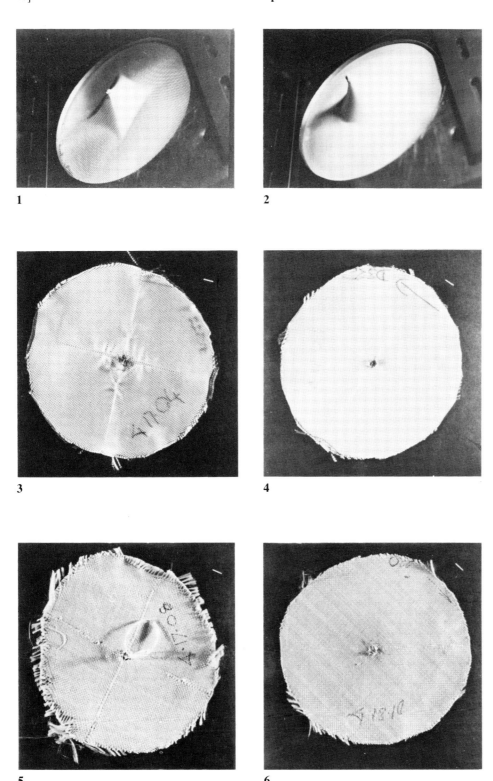

Plate 40L — Ballistic impact.
(1) Penetration of woven nylon fabric. (2) Penetration of knit nylon fabric. (3) Residual damage after impact on nylon fabric at 254 m/s. (4) After impact at 514 m/s. (5) Kevlar fabric after impact at 240 m/s. (6) After impact at 554 m/s.

Part VIII
Fibre archaeology and textile conservation

41

INTRODUCTION

The emphasis of this book has naturally been on the relatively short-term durability of textiles: the seconds or minutes of a laboratory test; the minutes or hours of processing; the days or months, or at most a few years, of use. But in the context of archaeological and museum studies it is necessary to examine materials which are hundreds or thousands of years old (the earliest known constructed textiles date from 8000 B.C.).

Natural textile fibres are vulnerable to damage and degradation. As an illustration, only three weeks' burial in moist biologically active soil at 20°C is sufficient to reduce cotton, linen or wool fabrics to such a tender state that they disintegrate under their own weight; similar damage occurs in other cellulose and protein fibres. The more resistant synthetic polymer fibres have a history of no more than 50 years. It is not surprising that in many archaeological contexts textiles do not survive, other than as impressions in clay, on bricks or pottery, as pseudo-morphs following mineral replacement, or when charred or carbonized. Fortunately attack by the majority of biological antagonists is virtually eliminated if one or more of the following conditions persists: (1) absence of water; (2) temperature less than 5°C; (3) absence of air. Most of the significant collections of archaeological textiles have been preserved by such conditions.

The vast majority of textiles have been produced for use. In most ages and civilizations this use was severe and continuous, with items being handed down and modified, as well as patched and repaired until mechanical damage reduced them to a collection of rags. Such textiles suffered considerable fibre damage and disruption prior to being discarded, and were therefore especially vulnerable to rapid biological degradation. As a consequence textiles rarely survive their useful lifetime. Those that do are often found in collections and museums, where despite careful attention they continue to degrade. On display they suffer photo-degradation, and in today's cities they become progressively more acidic, so that cellulosic materials suffer acid-catalysed oxidation. They may even suffer insect damage.

Many of these differing forms of damage produce changes in morphology which are recognizable when examined with the SEM, and work at UMIST is continuing with the aim of generating an atlas of the morphology of long-term fibre degradation. The completion of this work will make it easier in many cases to identify the cause and sequence of the wear and damage, and in particular to distinguish damage occurring during the production and use of the material from damage occurring before or after discovery by the archaeologist.

FIBRE DAMAGE DURING GROWTH AND MANUFACTURE

Characteristic abnormalities occur in both wool and cotton fibres if their growth is interfered with. An illness in the animal or an absence of suitable grazing produces thin or tender places, and many ancient fleece types produce breaks naturally, prior to shedding. An interruption in the growth of the cotton hair, such as drought or premature harvesting, can prevent the secondary cell wall from developing, and produce an immature fibre. In silk, the tensile strength, fibre cross-section and cocoon length are all influenced by the condition and diet of the silk larva, and problems with either of these result in short cocoons and 'thin' filaments.

Cotton and bast fibres are protected from light during their growth, but wool is not, and in sunny countries the tips of the fleece exposed to ultra-violet radiation are often consider-ably degraded prior to clipping. This damage shows up as increased dye uptake, and the fibre tips may even break off during subsequent processing. Cotton is subject to the attentions of the boll weevil, and the results of such attack are well documented.

While all these forms of damage exist in ancient textiles, it is doubtful whether there is a justification in searching for them in a routine examination. However, it is possible to envisage situations where the identification of breaks, or sun-damaged tips, could provide useful evidence of fleece type or origin.

FIBRE DAMAGE DURING MANUFACTURE

The majority of hand methods of manufacture treat the fibre material gently, and cause little mechanical damage. Exceptions to this general rule include the extraction of bast fibres from the plant stem. With flax the process of retting initiates the bio-breakdown of the cellulose and lignin in the stem. Linen fibres are highly crystalline, and they resist this attack and hence remain intact while the 'wood' is tendered. Excessive retting results in damage to the fibre bundles, with the binder suffering first. Such fibres are likely to split down to the ultimates in subsequent processing and use.

Probably the most extensively damaging form of treatment was the use of certain mordants, such as iron, in order to fix dyes on protein fibres. There is no doubt that iron mordanting speeds up degradation; its use was proscribed by the eighteenth-century Flemish tapestry weavers' guild, because it was known to cause rapid deterioration. The mechanism of such damage is complex, involving the catalysis of protein hydrolysis and the acceleration of photodegradation, as shown and discussed in relation to **43A(2),(3)**, so that it is difficult to relate this process to specific morphological changes. The presence of iron can be proved in many ways, and so far our work involves the 'collection' and recording of the type of damage which results. At this stage there is no clear-cut pattern of morphological change which can be specifically linked to mordant-induced damage.

The use of tin compounds to weight silk causes the rapid photochemical oxidation of the silk protein. Egerton (1948) has shown a reduction in strength of 74% in just 4 weeks' exposure to summer sun in Manchester. Fibre breaks show variable morphology, from simple perpendicular brittle fractures to angular crack propagation, as shown in **43A(5),(6)**.

The processes of felting, fulling, raising and shearing all produce gross changes in the fibre arrangement within a fabric, and these changes usually remain detectable even after burial. However, wool textiles also felt in normal use, and it is often impossible to establish whether felting is deliberate or accidental, unless a relatively large sample survives. In a similar way, polishing or rubbing of linen fabrics smears and smooths the surface ultimates. An example, **41A(1)**, is from the shroud in Tutankhamun's tomb described in the next chapter. This structural change does not survive extensive wear and wet treatment, and is a useful indication of the extent of the use of the object. Again certain types of wear produce a similar effect, and careful examination of a large sample is necessary for positive identification.

FIBRE DAMAGE DURING USE

The useful life of most textiles is characterized by progressive changes. The record of the alteration of fabric structure and appearance, and of fibre damage, due to mechanical wear is the major theme of Parts VI and VII of this book. Work at UMIST reported by Cooke and Lomas (1987) and subsequent publications by Peacock (1988a,b), Mannering (1994), Cork *et al* (1997), and Rast-Eicher (1996), together with investigations by others, have demonstrated that these changes survive extended periods of burial, and can be detected with confidence with the use of the SEM.

Another change associated with wear is creasing. Creases may vary from soft recoverable folds in the fabric, with little associated fibre deformation, to sharp permanent features associated with fibre yield deformation, or mechanical conditioning. With garments, a crease pattern develops which is specific to the relationship between the garment's cut and fit and the anatomy of the wearer. Crease patterns have been shown to survive burial, in reports by Cooke (1988a) and Granger-Taylor (1988), and may contribute to the identification of the function of the object.

Textiles suffer many other types of attack during normal use. They may simply hang in a window, subject to photodegradation. Even in the UK this process is rapid, with a 90% strength loss for an undyed cotton twill after 6 years in a south-facing window in Manchester. In a damp unventilated environment mould growth can develop and there may be damage by the activities of insect pests, as shown in **43C(1)–(5)**. Even in apparently safe places there may be damage by fire and flooding. Many of these forms of damage survive in ancient and archaeological textiles even after burial, and they can often be identified with the aid of a stereo-optical microscope, once the characteristic morphologies have been learnt from the study of high resolution SEM photomicrographs.

DAMAGE DURING BURIAL

Burial in soil often provides the ideal environment for the complete destruction of cellulosic and proteinaceous textile fibres. The soil ecosystem operates by breaking down cellulose and protein macromolecules into smaller, more readily accessible units. Fungi and bacteria produce synergistic combinations of enzymes, cellulose and proteolase, which are capable of attacking both the amorphous and crystalline zones of the fibre. *Fusarium oxysporum*, *sporotrichumpruinosum* and *pencilliumfuniculosum* are specific for cellulose and common in soil. Tests involving the soaking of 2-inch strips of untreated cotton fabric in a culture of *fusarium oxysporum* and the subsequent storage with a moisture content of 24% at 30°C ± 2°C produced strength reductions of 45% after 4 days, 92% after 7 days and 100% after 14 days.

Wool seems to be more prone to bacterial attack, for example by *Bacillus mesentericus*, *B. subtilis*, *B. cereus* and *B. putrificus*, but also supports the rapid growth of *penicillium*, *aspergillus* and *actinomyces*. The attack often initiates in contaminants such as soaps, sizes or suint, and then spreads to the wool fibres. The weak point in the undamaged wool fibre is the distal scale edge, and damage at this point in the otherwise resistant exocuticle provides access to the less resistant endocuticle and cortex. The common breakdown mechanism is hydrolysis of the peptide link, caused by trypsin, and an enterokinase activator. When completed such breakdown removes 10% by weight of the fibre, in the form of cell membranes, nuclear remnants, cytoplasmic debris and endocuticle, and results in a 90–95% loss in strength. The keratin of the cortical cells is not attacked, unless the disulphide bonds are broken (Lewis, 1975).

It is fortunate for textile archaeology that fungal and bacterial activity is temperature- and pH-sensitive, and water-dependent. The most rapid attack occurs under the following conditions; 25–40°C, pH 6.5–8.5 and r.h. >95%. Acidic and alkaline conditions inhibit the process, and temperatures less than 5°C prevent active attack.

A number of different archaeological contexts have consistently yielded well-preserved textiles. The most extensive finds have come from desert conditions, such as Egypt and the Sudan, where the virtual absence of water has prevented biological attack. The Northern European acid peat bog has preserved many organic remains, including wood, animal and human cadavers, and textiles. Unfortunately acid conditions lead to acid-catalysed hydrolysis of cellulose, which eventually dissolves; consequently wool and silk survive, while linen, cotton, nettle and jute vanish. A contributing factor to the preservation environment of the peat bog is the development of anaerobic conditions, which also often develop in other waterlogged sites. A number of these have produced extensive textile remains, for example Viking Dublin, Viking York and Vindolanda on Hadrian's Wall (see Chapter 42).

The permafrost layer has the potential to preserve most organic remains, without the selective acid removal of cellulose or alkaline hydrolysis of wool and silk. However, such finds pose considerable problems to conservators, as the return to normal temperatures initiates rapid decomposition. There is evidence that freeze-drying will provide the solution to this problem, Peacock (1988). Certain metal salts, such as the corrosion products of iron, copper and bronze, also inhibit biodegradation, but they catalyse the hydrolysis of both cellulose and protein. In this context the formation of negative casts or positive pseudomorphs may preserve much of the surface and structural detail of the fibre, Janaway (1983), despite the almost total destruction of the textile itself. In one respect the oxidation of cellulose can be advantageous. Under the right conditions cellulose will oxidize in a controlled manner, without total disruption of the fibre structure. This charring is assumed to involve slow combustion, with a limited supply of oxygen. In much the same way that charcoal retains many of the structural features of wood, carbonized textile fibres have recently been shown to retain their structural features, Cooke (1988b), as shown and discussed in relation to **43B(1)–(3)**. The oxidation of cellulose also seems to proceed without combustion, at room temperature, albeit very slowly. Such carbonization can be found in very ancient fabrics such as the Tutankhamun Anubis shroud, as shown in **42D(3)**.

In general the problems of identifying damage associated with the microbial attack of burial are considerable. With modern textiles staining methods are used to reveal fungal (mildew) attack, but these are inappropriate for archaeological objects. Ancient textiles show changes in break morphology, the residues of hyphae **41A(2),(3)**, or colonies of bacteria or fruiting bodies **41A(4),(5)**. When these are not present we can only surmise on the causes of damage such as the extreme erosion seen on a Coptic 'rondelle' in the Whitworth Gallery collection **41A(6)**.

POST-EXCAVATION DAMAGE

Ideally post-excavation damage would not occur. In practice there is rarely adequate funding for immediate conservation, or for appropriate long-term storage facilities. Textiles which have survived for very long periods owing to a happy combination of burial conditions almost inevitably face a more destructive environment as a result of excavation. Who can realistically envisage the safe storage of a Coptic textile for a further 1500 years, or a Pharonic linen object for another 3000 years? Perhaps the greatest risk arises between discovery and conservation. Each burial context leaves the textile open to differing risks and therefore dictates a different treatment method. The actions to avoid can be summarized as: wetting dry textiles; drying wet textiles; neutralizing acid or alkaline textiles; aerating anaerobic textiles; warming frozen textiles; and cleaning metal objects on site. The small-finds manager should be made aware of the rapid destruction of textiles in warm, wet, biologically active environments and the particularly damaging effects of storing wet textiles in polythene bags.

Damage during conservation is increasingly less likely, as the present high standards of training and professional awareness continue to be refined. During long-term storage the risks are easy to identify: fluctuations in temperature and humidity; regular handling by scholars and students; exposure to dirt, insects, spores and bacterial contamination; exposure to light on display; exposure to acidic gases, such as oxides of nitrogen and sulphur; and disasters, such as fire and flood. The studies reported here help to identify and explain the possible forms of damage, but the problem is how to prevent the damage at an acceptable cost.

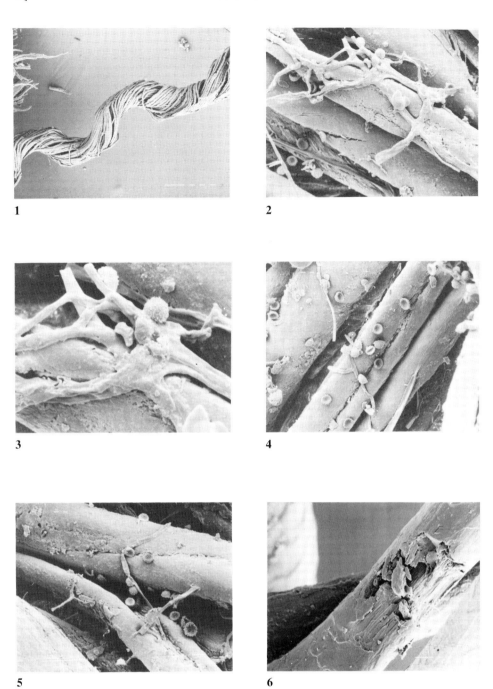

1

2

3

4

5

6

Plate 41A — Shroud from tomb of Tutankhamun.
(1) Yarn from shroud showing crown flattening.
Damage to 19th century tapestry from Victoria and Albert Museum, London.
(2),(3) Residues of hyphae. (4),(5) Colonies of bacteria or fruiting bodies.
Coptic 'rondelle' from Whitworth Gallery, Manchester.
(6) Unspecified damage.

42

MECHANICAL WEAR IN ANCIENT TEXTILES

The combined use of optical and scanning electron microscopy has proved to be a powerful tool in the study of mechanical wear and degradation of ancient textiles. Samples of old fabrics almost invariably shed fibre fragments, and these form the initial target of optical microscopy. Mounted in distilled water, under a coverslip, they are examined for fibrillar breakdown, cracking, splitting, peeling, and for brittle fracture. The fabric surface is then scanned with a stereo-microscope for evidence of macro wear patterns, as well as fractured fibres *in situ*. Once the macro wear pattern is understood, the stereo-microscope is used to identify representative sampling zones for SEM examination. The SEM study provides confirmation of the wear mechanisms, both at the level of structural damage to the yarns as well as in the individual fibres, and produces a permanent photographic or digital record.

Linen fabrics have a reputation for smoothness, coolness and durability. The author had access to a collection of linen sheets and dresses, which had been in regular use for about 40 years, and these were examined using the above methodology. The fabrics were strong and supple, and there was no evidence of fibre shedding. The microscopical examination revealed complex changes in both yarn and fibre structure, and it is probable that these changes, resulting from the effects of washing and wearing, contribute to the feeling of smoothness associated with linen. The changes can be summarized as:

(1) Flattening of yarn crowns, **42A(1)**, which involves the breakdown and fibrillation, often in sheets, of the cellular bundles in the flax fibres, **42A(2),(4)**, and the smearing of surface material, **42A(3)**.
(2) Abrasive action, leading to fibre fatigue and 'brush ends', **42A(5)**.
(3) Extensive wet alkaline treatment producing surface peeling, **42A(6)**, possibly with some separation of cells.
(4) Extended fatigue damage resulting in fibre loss, thin places and broken threads, **42B(1)**.

A similar study of a worn silk dress, ca. 1920, revealed the following pattern of breakdown:

(1) Abrasive action on the yarn crowns produces peeling and fibrillation, **42B(2)–42B(4)**, and ultimately rounded ends in the weave interstices, **42B(5)**, as the fibrils wear off, **42B(6)**.

Excavations at the site of Vindolanda a Roman fort on Hadrian's Wall, have yielded a large number of organic objects dating from the Flavian and Trajanic periods, ca. AD 100. These finds, including writing tablets, leather goods and textiles, were preserved in the anaerobic but not waterlogged conditions of a compacted bracken floor of a wooden military building in the Vicus.

A portion of a leg wrapping was made available for an initial wear study. The loose fibre debris associated with this fabric showed evidence of fibrillation and brittle fracture. The SEM study confirmed the mechanisms of damage as:

(1) Biaxial-type fatigue and fibrillation associated with pilling, similar to that described in Chapter 30, namely multiple fatigue sites along pill anchors, and local clusters of such damage, **42C(1),(2)**. The detailed morphology of this damage is exceptionally well preserved, **42C(3),(4)**, as is the scale structure of the wool, **42C(5)**.
(2) The sample also contained a number of well-defined holes. The fibre breaks associated with this damage were exclusively brittle type, probably resulting from local biodegradation, **42C(6)**.

The extent of the wear damage, together with the very well-preserved longitudinal creases typical of leg wrappings, which were similar to puttees, suggested extensive use.

The success of this initial study led to a three year Leverhulme Trust funded research project at UMIST, completed in the summer of 1995, which has made an extensive examination of the Vindolanda corpus of textile finds made up of more than 750 individual textile fragments. This project has used a range of analytical methods including the use of the SEM and image-analysis to study the fibre, yarn and fabric structures. This examination showed that surprisingly few of the fragments had originated from the same fabric web, and those that did match in terms of weave, twist-angle, fibre diameter, etc, were probably joined prior to excavation. The SEM provided clear low-magnification images of the yarn systems, which facilitated the use of image-analysis to automate twist angle measurement, **42D(1)–(3)**, Cork *et al* (1996), and accelerate the process of fibre diameter measurement.

The majority of the fragments showed the typical signs of very considerable use, and it would appear that clothing was worn until it fell apart, and the remains used for patching other garments, under the rigorous conditions of life for Roman auxiliaries serving on Hadrian's Wall. Even the fabrics found in the officers' quarters, although of higher quality, had the same extensive fibre damage, and the same fragmentary nature. A number of the heavier fabrics showed a definite raised nap, which was visible with a stereo-zoom microscope, as well as with the SEM. The study confirmed the use of diamond twill as the most popular structure on Hadrian's Wall, in common with much of the northern Roman Empire at that time, and recorded 42 different diamond twill constructions.

Museums are another source of historical textiles for examination. For example, **42D(5),(6)** is a silk bodice from the Textile Conservation reference collection at Hampton Court, which is discussed in more detail in the next chapter.

During recent conservation of textiles in the Victoria and Albert Museum, from the tomb of Tutankhamun, a number of small fragments of weave became detached, and these formed the basis of an SEM examination. The fragments came from two textiles, a shawl and a shroud, found draped around the statue of Anubis, which guarded the entrance to the antechamber of the tomb.

The fragment of shawl showed no signs of wear. However, the individual fibres, which appear to be flax, have suffered a concertina-type deformation, **42E(1)**, which seems to be associated with length shrinkage, **42E(2),(3)**. Not all fibres are equally affected, and the cause of this damage is not yet clear.

In contrast, the fabric of the shroud, dated by inscription to the seventh year of the reign of Akhenaten, **42E(4)**, has suffered considerable flattening of yarn crowns, **42E(5)**. The fibres in protected parts of the weave show transverse cracking, often at nodes, but no surface damage, **42E(6)**, whereas fibres in the crowns exhibit fibrillation and smearing, typical of changes associated with wear and wet cleaning, **42F(1),(2)**. The extent of the crown damage suggests considerable mechanical rubbing, together with wet treatment, and may have occurred either through use, or perhaps during a process of wet finishing and smoothing to impart softness prior to use. The shroud is actually a shirt with a neck opening, which would only accommodate the head of a child, and the inscriptional date coincides with the birth of Tutankhamun.

Occasionally the evidence of wear in an archaeological object is visible to the naked eye, and can be recorded using macrophotography. A Coptic child's tunic, recently conserved in the Whitworth Gallery, Manchester, has seen considerable use, and has been darned and repaired frequently, **42F(3),(4)**. In the lower back there is a thinning of the fabric, associated with extensive crown damage resulting from lengthy wear, **42F(5)**. Also of considerable interest is the weave distortion or bagging, which has become a permanent feature of the garment, together with the creases at the side seams formed by the stresses developed during sitting, **42F(6)**.

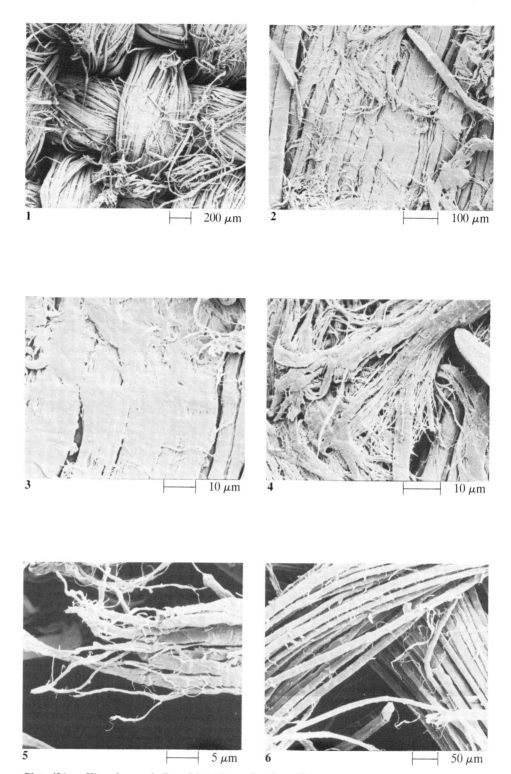

Plate 42A — Wear damage in linen fabrics in use for about 40 years.
(1) Crown damage. (2) Crown damage, fibre flattening and fibrillation. (3) Fibre smearing. (4) Linen
fibrillation. (5) Fatigue fracture. (6) Surface peeling.

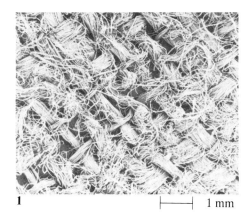

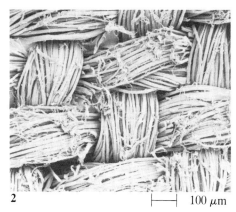

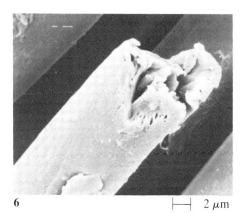

Plate 42B — Wear damage in linen (continued).
(1) Thread breakdown.
Wear damage in silk dress after 60 years.
(2) Crown breakdown. (3) Fatigue breaks in weave. (4) Peeling and fibrillation. (5) Crown breakdown. (6) Fibre fracture and rounding off.

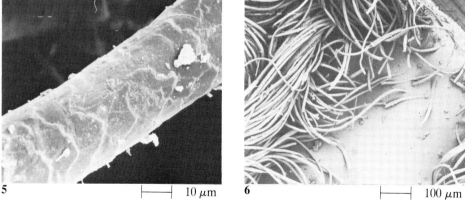

Plate 42C — Fibre damage in Roman leg wrapping from Vindolanda.
(1) Pill site with multiple fatigue breakdown. (2) Pill anchor. (3) Biaxial-type fatigue. (4) Fatigue breakdown. (5) Scale damage. (6) Brittle fracture.

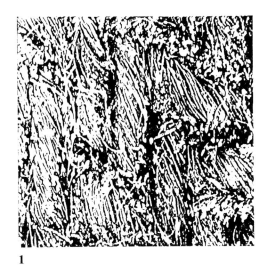

1

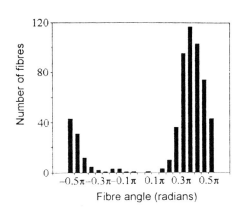

2   **3**

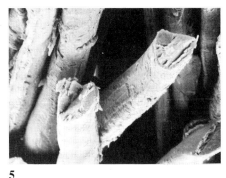

4   **5**

Plate 42D — Roman fabric from Vindolanda.
(1) Frame-grabbed digitally stored SEM image. (2) Final stage of image analysis, showing vector lines for twist-angle measurement. (3) Twist angle distribution histogram generated from image analysis.
Late 19th century silk bodice from Hampton Court
(4) Brittle fractures in warp and weft. (5) Close-up of fractures in warp.

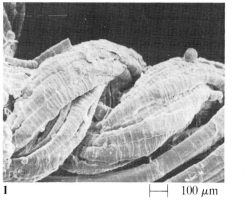

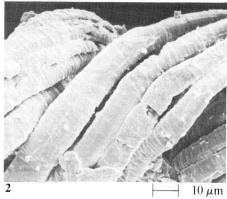

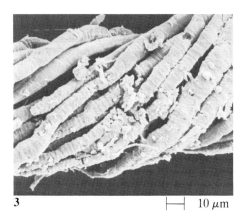

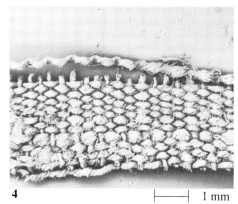

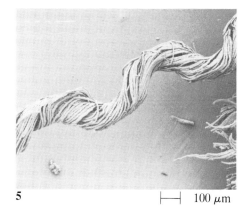

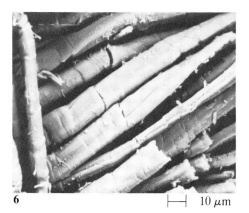

Plate 42E — Textiles from tomb of Tutankhamun.
(1) Flax fibres from shawl. (2) Concertina damage. (3) Concertina damage and shrinkage. (4) Fabric from Anubis shroud. (5) Crown flattening. (6) Transverse cracks.

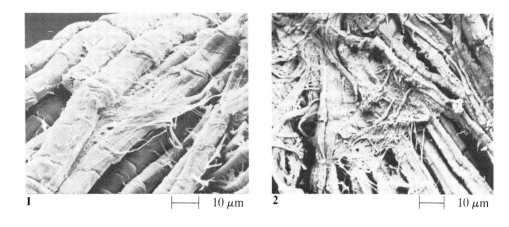

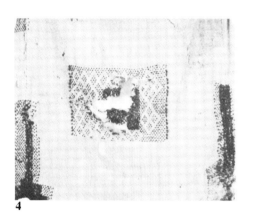

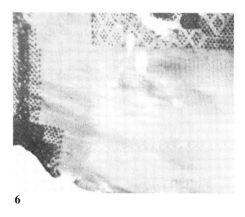

Plate 42F — Textiles from the tomb of Tutankhamun (continued).
(1) Fibre breakdown. (2) Fibre fibrillation.
Coptic child's tunic.
(3) General front view. (4) Darning. (5) Seating in back. (6) Abrasion damage and thinning.

43

ENVIRONMENTAL DAMAGE

Textiles are liable to suffer environmental damage at any stage in their life time, from fibre growth through to storage in a museum. In contrast with the evidence of wear and use discussed in Chapter 42, environmental damage is difficult to identify with precision, and it is often impossible to decide when it occurred. The latter comment is particularly true for archaeological textiles, as the degradation associated with burial is likely to obliterate prior damage.

Textiles designed for display usually suffer photodegradation throughout their lives, and it is only recently that effective means to reduce this damage have been taken in our museums and collections. At greatest risk are objects with a long natural life, such as drapes, carpets, upholstery and, particularly, tapestries. When new, tapestries are immensely strong objects. The typical products of Flanders and northern France were capable of supporting 2000 times their own weight, Howell *et al* (1997), and they are consequently capable of hanging on display for 300–400 years. The first evidence of damage is provided by changes in dye colour. The initial reduction in tensile strength of the fibres, due to molecular chain scission, has little effect on the tapestry owing to its technical over-construction. Nevertheless, 300 years' exposure does cause damage, including the loss of areas of weft and the consequent disruption of the design.

In order to understand the process of breakdown, a survey was carried out, initially using hand lenses and subsequently an arm-mounted stereo-zoom microscope, of the hanging tapestries in Hampton Court Palace. It was apparent that the protected backs of the tapestries were in a much better condition than the fronts. The colours were much brighter and there was less damage. Further examination showed that certain colours were much more subject to damage, dark brown wools being most seriously affected, together with pink, cream and apricot-coloured silks. The study was then restricted to a tapestry which had been taken down for conservation. This was one of the Alexander series, *Alexander with His Horse Bucephalus* (ca. early eighteenth-century, Brussels workshop). An initial tactile survey indicated that even a tapestry in good general condition had large areas which were dry, abrasive and rough to touch. The problem was restricted to the face side, and was more serious in areas of the design dyed dark brown, which had been close to a window since the time of George I (1714–27). A Nikon stereo-zoom microscope, ×2 to ×40, was used to study the tapestry. The back showed little evidence of fibre damage. The weft crowns were intact and in excellent condition, **43A(1)**. The damaged areas of the face revealed the cause of the roughness, an unusual form of fabric degradation. The weft crowns had suffered very considerable brittle fracture, **43A(4)**, which had converted a proportion of the weft face into a sharp 'pile'. These groups of fractured fibres existed in the interstices of the weave on either side of each weft crown, **43A(2),(3)**. This process of crown breakdown is progressive, and will ultimately lead to total breakdown of the tapestry, Cooke and Howell (1988).

In order to understand the selective nature of this light-induced damage, samples of weft from each of the dark colours were subjected to X-ray emission microprobe examination on an SEM. The resulting elemental analysis revealed a significant iron (Fe) peak in the dark brown sample, whereas the other colours were free from traces of iron. It is probable that the damaged wool had been treated with an iron mordant, and the damage had resulted from a combination of iron-induced hydrolysis and photodegradation.

Royal bed hangings are also subject to photodegradation. Queen Charlotte's bed in Hampton Court Palace has been exposed to light since 1715, a similar length of time to the Alexander tapestries, and has suffered considerable fading and photodegradation. Many of the warp fibres are shattered by angular fractures, **43A(5),(6)**, and each movement of the drapes causes more fractures to occur. This damage is not due to tin weighting as this process was only introduced in the late nineteenth-century.

A more recent study of a late c19th century silk bodice from the Textile Conservation Centre reference collection showed catastrophic brittle fracture in both yarn systems, **42D(5),(6)**. EDAX analysis demonstrated the presence of tin (Sn), which had caused the rapid degradation of the fibre, as well as aluminium (Al) and silicon (Si), which probably indicate cleaning with fullers' earth, a process common with silk objects prior to the use of dry cleaning.

Fire is usually the ultimate destructive agency for textile materials, and yet under the right conditions cellulose fibres will oxidize in a controlled manner without total disruption of the fibre structure. In much the same way that charcoal retains many of the structural features of wood, the charring of cellulosic textiles is known to preserve the macrostructure of the fabric. The conversion of cellulose to carbon eliminates most of the risk of biodegradation, and archaeological textiles are often found in a carbonized condition in contexts which would destroy both cellulosic and proteinaceous material. Such objects are difficult to deal with, owing to their fragile state, and their examination has usually been restricted to thread counts and the determination of twist direction. Recent work at UMIST on carbonized textiles from Soba (Sudan ca. AD500) has shown that not only is the yarn structure, twist, etc, preserved in great detail, **43B(1)**, but fibre surface and cross-sectional information is sufficient to allow the positive identification of cotton, and even reveal the maturity of the fibre, **43B(2),(3)**.

A further form of 'oxidation' frequently found in grave goods is the non-specific damage which occurs when textiles are in contact with body fluids, and with certain embalming materials. The examination of an 'oxidized' fragment from a Coptic sprang cap in the Whitworth Gallery collection in Manchester reveals considerable brittle fracture, **43B(4),(5)**, but the surface scale structure of the wool is still intact, and the internal cellular structure is remarkably preserved, **43B(6)**.

The problems of identifying damage associated with the microbial attack of burial are considerable. With modern textiles, staining methods are often used to reveal fungal (mildew) attack, but these are inappropriate for archaeological objects. Ancient textiles show changes in colour and brittleness, and with luck the residues of spores or hyphae, or colonies of bacteria, serve to identify the cause of the damage, as in a Coptic wool textile, **43C(1)**, and a Pharonic linen, 18th Dynasty, **43C(2)**. Extensive fibre damage in a Vindolanda sample would seem to be due to bacterial attack, **43C(3)**, as there are no signs of hyphae or spores.

Many insects are capable of damaging textiles with their mandibles but only relatively few have adapted to using fibre protein as their main source of food. In Europe two orders, Lepidoptera and Coleoptera, pose the most serious threat. The larva of the clothes moth is an avid selective feeder, often choosing particular dyestuffs, for example eating only greens and pinks from a multicoloured embroidery. A study of the debris left after such an attack suggests that the larva eats individual fibres as a child would eat a stick of candy, namely munching down from one end, **43C(4)**. Eaten fibre debris is often found covered with 'moth silk' when the case moth is the cause of the damage, **43C(5)**. This larva extrudes very fine filaments, 0.5–1 µm diameter, to form a protective cocoon during feeding. A further sign of clothes moth activity is the characteristic droppings, **43C(6)**, which often reveal the colour of the wool eaten. The carpet beetle, which feeds on wool both in the larval and adult stages, would seem to bite in a more random manner, often starting half way along a fibre. The droppings are very similar to those of the clothes moth.

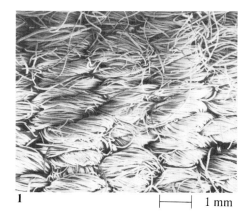

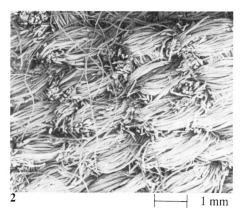

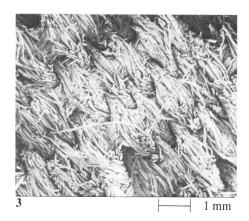

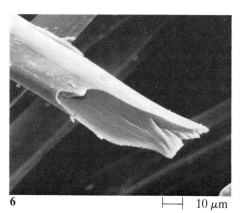

Plate 43A — Tapestry damage, Hampton Court Palace.
(1) Undamaged crowns on back. (2) Face damage. (3) Structure close to breakdown. (4) Fibre fracture in face crown.

Queen Charlotte's bed drapes.
(5) Silk fracture in face. (6) Silk fracture.

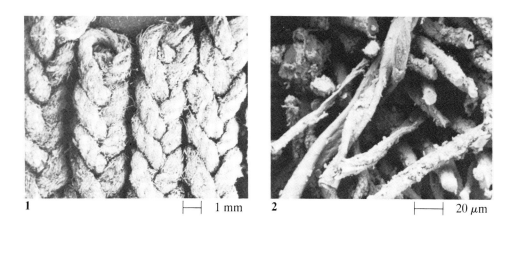

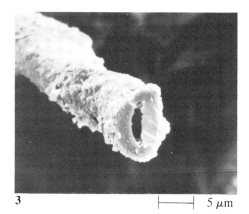

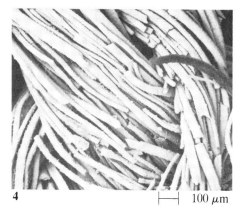

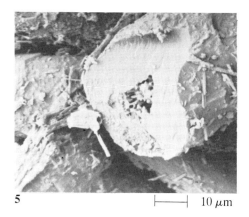

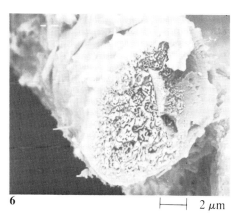

Plate 43B — Carbonized textiles from Soba, about 1500 years old.
(1) Crossed-loop knitted structure. (2) Cotton fibre convolutions. (3) Cotton cross-section in fracture.
Coptic sprang hat in Whitworth Gallery
(4) Fibre fracture in yarn. (5) Fibre fracture, wool. (6) Cortical cell structure in fracture.

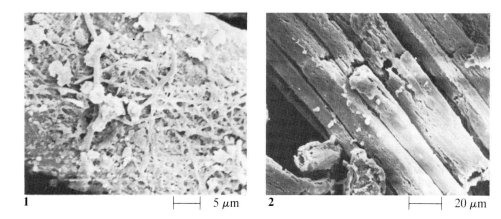

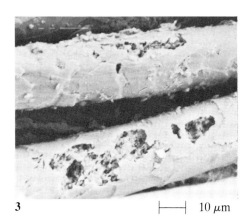

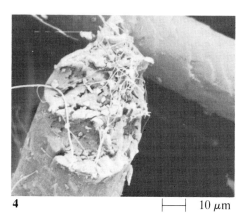

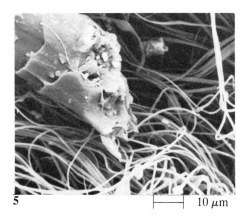

Plate 43C — Biodegradation
(1) Mildew on Coptic wool. (2) Bacterial damage on Middle Kingdom linen. (3) Bacterial damage on Roman wool from Vindolanda.
Case moth damage
(4) Mandible pattern on wool fibre. (5) Wool fragment in moth 'silk'. (6) Moth excreta.

Part IX
Forensic studies

44

TEXTILE DAMAGE IN FORENSIC INVESTIGATIONS

Franz-Peter Adolf and John Hearle

Damage to textiles plays a part in many forensic studies. Civil liability litigation was the motivation for the work described in Chapter 37 on automobile seat belts. The questions to be answered in such cases are: *Did the belt fail due to some fault of the manufacturer? Or was it misuse by the owners of the vehicle? Or was it deliberately cut in order to impute blame to the belt supplier?* There are many similar examples where textiles deteriorate or where their failure leads to other losses and claims are made. For example, if ropes break in towing or mooring, the loss of a vessel or an oil-rig may lead to very large claims. Significant claims, though not as large, may be made when a product, such as a carpet, does not perform in a guaranteed or expected way and its useful life is shorter than it should be. The information spread throughout this book on many types of textile and many applications provides a means of approaching problems of this sort. For example, the evidence in **34G** showed that the breakdown of overalls used in a virology laboratory was due to the autoclaving conditions and not to any fault of the linen-hire company in selecting the fabric.

The use of textile evidence in criminal cases raises more specific questions. Damage to fabrics caused by knives and other more or less sharp objects is significant over and over again in clearing up certain kinds of crimes and forms the principal subject of this chapter and the next. Earlier reports on damage to clothing by cuts and tears include the papers by Monahan and Harding (1990) and by Heuse (1982). In contrast to the effects of stabbing, the damage to textiles caused by bullets in shooting incidents has been much less studied or used, but the results of a recent investigation are given in Chapter 46.

Crimes involving knives or other sharp objects are mostly a bloody matter because they often deal with murder or violence. The leisure jacket shown in **44A(1)**, dressed on a mannequin, is a typical example. It shows three damaged areas, marked by white arrows in order to demonstrate their relation to the injured parts of the body. The one stab only injured the right shoulder but the two others injured the left kidney and the liver respectively. These last stabs were fatal, indicating that the weapon must have been of a certain length. The example also demonstrates that the examination of damage to textiles should generally include information from the medical protocol, if there are injured persons, or the post-mortem protocol, if there are fatalities.

In cases where damage to textiles plays a role the forensic expert has to answer two sets of questions. One set is: What type of damage is present? Why is it angular, **44A(2)**? Was it caused by cutting or by tearing? Or was it caused by other effects, for example by rubbing caused by a fall onto the pavement, as in **44A(3)**? The other set is: What kind of object has caused the damage? Was it a knife or another tool, for example a screwdriver? Was it sharp or blunt? Did it have one edge or two edges? Some possibilities are shown in **44A(4)**.

There is no manual which describes the procedure for the examination of damage to textiles. There is the basic knowledge that two important facts dominate the examination of such damage. The first is that the features of the damage are of a morphological nature; the second is that these morphological features are clearly preserved in different ways in the different parts and the different kinds of textile constructions — the fabric, the threads, the fibres. The morphological characteristics of damaged areas are usually better preserved and more definite in the edges of damaged non-wovens or woven fabrics than in those of knitted fabrics. The characteristics are mostly clearer in fabrics made from non-textured filament yarns than in fabrics made from textured or staple yarns. Concerning the fibres, the characteristics of the fibre ends may show a greater or lesser degree of variance depending on the type of fibre and on the kind of weapon used to cause the damage. This indicates that the development of the morphological features in a damaged area is strongly influenced by the elasticity of the fabric, by the flexibility of the position of the yarns, by the construction of the yarns and by the fibre material itself. This knowledge leads to two simple rules for the general course of action.

The first and the most obvious rule is: *do not alter the damage by stretching the object or by any other kind of moving and manipulation*. The other rule follows the same principle, which dominates crime scene working: *at first you only have to look at the damage in order to get an overview of the potential morphological information*. From the beginning of the examination a stereomicroscope is therefore needed to get a detailed view. Besides this you finally need your eye and your brain to store and to combine the features that you have seen.

Sometimes, as shown by **44A(5)**, which is referred to in more detail later, SEM pictures at higher magnification are useful. However, as discussed in the next chapter, the microscopical features of the fibre ends in textile damage vary over a wide range, so that it is difficult to determine the cause of damage from the observation of single fibre ends. In the present state of the art, this limits the value of SEM observations in forensic examinations.

The next question is: Where are the features located which can be used to characterize damage? There are two areas of interest — the edge lines and the end areas of the damage. Sometimes, the features are not clearly pronounced in the damage to the outer fabric, for example in the case of jackets, overcoats or trousers made from thicker fabric. Then you should try to look at the damage to the lining or to the interlining of the clothing. Further, it must be mentioned that the macroscopic appearance of the edge lines can often be ambiguous. The damage to a knitted cotton T-shirt, **44B(1)**, demonstrates this. The split was caused by a combined cutting and tearing process. This is not indicated by its macroscopic form. It only becomes clear if the yarn ends are examined. A cut yarn with the typical plain end is first seen, **44B(2)**. Then, in contrast to that, a torn yarn with the characteristic formed end like a thin pointed beard or brush is also visible, **44B(3)**.

From this example we can deduce that as a general rule the most important characteristic which is located in the edge lines is the form of the yarn ends. The importance of that knowledge is emphasised by two other pictures. In **44B(4)**, we see a macroscopic view of the damage in a lining made from acetate fibres. It only gives ambiguous information as mentioned above. In **44B(5)**, we see the edge line under the stereomicroscope once more showing the typical plain form of the cut yarn ends.

Now to the other area of interest — the end areas of the damage, which are the most important parts to study. They often have clear contours which indicate whether a cutting tool like a knife was used and if that tool was one-edged or two-edged. **44C(1),(2)** shows the two end areas of the damage to a nylon shirt belonging to a murdered pensioner, which was obviously caused by a one-edged kitchen knife. In **44C(1)**, you see the pointed end caused by the edge, and in **44C(2)**, the other end formed by the back of the blade. That blade must have had a broad angular back in order to form the swallow-tailed end. In this case, the jacket of the murdered pensioner was also pierced by the same stab. **44C(3)** demonstrates the same form of damage to the interlining of the jacket. The damage in these parts of the clothing was more precisely defined than in the outer fabric.

In another murder case the clearest view of the damage was found in the nylon lining of the anorak of the victim. In this case, the macroscopic form, **44C(4)**, already indicates that the tool must have been sharp and one-edged and had a broad but round back. The last can be deduced from the deformed end area at the left (marked by arrows) and is clearly to be seen under higher magnification, **44C(5)**.

From some systematic experiments carried out a few years ago we have noticed that further interesting features may be located in the end areas of damage. Fig. 44.1 demonstrates the finding that the stabbing device can draw the pierced fabric into the body because the fabric and the body are both elastic and can be stretched. So, some part of the drawn-in fabric may come in a certain contact with the edge of the knife. That may cause the smaller area of secondary damage, which is seen in **44D(1)**. The drawing in of the fabric may also result in only a few of the fibres on the surface of the yarns of the fabric being cut, as shown in **44D(2)**.

The same experiments also showed that two edged knives mostly cause more or less bent damage. The reason for this is that in practice the tool does not penetrate the fabric without any twisting movement and in an exact rectangular position. The effects are schematically demonstrated in Fig. 44.2. The first diagram A shows that a twisting movement of the knife has the tendency to cause a double bent damage. B and C demonstrate that stabs tilted at an angle cause damage with only one, more or less sharp bend. **44D(3)** shows a practical example of the double bent type of damage caused by a two edged knife.

A simple, but not everyday, case story, which did not involve a crime, shows the value of forensic examination of clothing. An alcoholic was found bleeding to death on the street. The damage to his shirt, **44D(4)**, and the fatal injury in his chest were congruent in their size and their position. A damaged and bloody plastic bag with some broken bottles in it was collected from near the corpse. The criminal police asked if the injury and the damage in his shirt could have come from a fall on to the plastic bag with the broken bottles in it. Although the damage in the shirt showed an irregular macroscopic form, **44D(5)**, it was seen under the

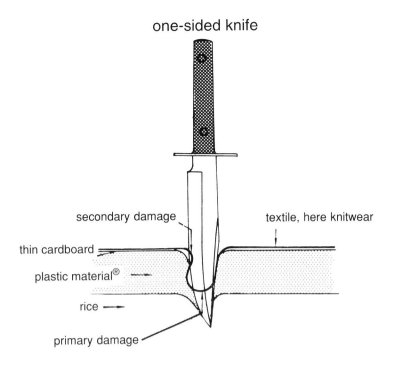

Fig. 44.1 — How secondary damage is caused.

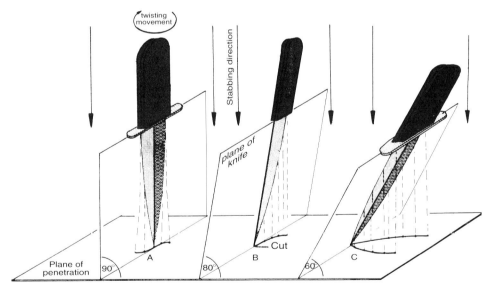

Fig. 44.2 — Principle of the origin of bent damage caused by two-sided knives.

stereomicroscope that all threads in the edge lines were definitely cut by a very sharp object. Even the unusual formed end areas did not show any characteristics of a tearing process. From the examination of the plastic bag it was apparent that there were some smaller cuts and a larger one. The larger cut, **44D(6)**, matched the damage in the shirt of the alcoholic in size and in form. From this and from other information the criminal police concluded that the death of this man must have been an accident.

In addition to the evidence from macroscopic observation, studies of detail in fibre ends can also be of use. The various forms of tensile and fatigue breaks, which may have resulted from normal wear, or, for example in the high-speed breaks in Chapter 6, from violent tearing, are spread through the early chapters of the book. Cutting and burning, which will be more relevant in forensic work, is covered in Chapter 20. In particular, **20F** consists of pictures from Foos (1993) of Bayerischen Landeskriminalamtes, which were taken with forensic applications in mind. **44A(5)** is also from the paper by Foos and shows acrylic fibres welded together in a garment damaged in an offence. The appearance matches fibres from a test made with the suspect blunt knife and shown in **20F(5)**.

Another real application of the examination of fibre ends in the SEM is described in a paper by Stowell and Card (1990). The fibres came from a woman's black nylon nightgown after an alleged sexual assault. The garment was supposed to have been cut into two halves by a knife along the vertical strips of lace on each side of the nightgown and across the shoulder straps. The investigators made experimental studies of the fabric cut by scissors and by a sharp unused scalpel blade and also when torn by hand. Fibre ends were examined in the SEM and characterised according to the type of damage. The scissor cuts were described as *squeezed inward on both sides and flattened* and the illustration is similar to the scissors-cut polyester fibre, **20A(6)**. Fibres from the torn fabric were reported as having *smooth, nonfractured* [sic] *ends with a more or less pronounced 'bulb' formation . . . although, in a few torn fibers, the bulb was seen only in parts of the edges of the fiber end*. This description is slightly misleading. The fibres from the torn fabric are clearly fractured, in the sense of being broken, and are typical of high-speed breaks of nylon fibres. One illustration is similar to **6A(3)**, with a complete mushroom end, and the other is similar to **6A(4)**, with a small V-notch indicating a transition between a high-speed break and a slower ductile break. The ends of fibres from the scalpel *showed a variety of shapes and forms*. Some were *relatively clean cut ends*, including some with *striations across the cut surface*, which were similar to the razor-cut nylon fibre ends in **20A(1),(3)**. Others *varied from elongated and twisted to fractured*, with illustrations reminiscent of the ductile twist breaks of nylon in **17A(2)–(4)**. A few *had indications of a 'bulbous' formation*. The first group will be fibres that were directly cut by the scalpel, while the other two sets of fibres will have failed as a result of secondary tearing and stretching of the fabric in regions away from the blade.

Thirty-six fibres from the nightgown were also examined. The most notable feature of the investigation was the use of a quantitative comparison of the incidence of the forms of break, as shown in Table 44.1. The quantitative evaluation of break types is discussed further by Pelton in the next chapter.

From their analysis, Stowell and Card concluded that, although the visual and macroscopic evidence could not determine the cause of break, the SEM micrographs indicated *that the shoulder straps could have been cut by a knife*. Neither the macroscopic appearance nor the SEM evidence was conclusive for the sides of the gown. The predominance of bulbous ends shows that the fibres had been broken by stretching and not by cutting with a sharp blade. The occurrence of fractured and elongated ends, which were not present in the experimental tears, leads to a suggestion that *an instrument of some kind*, possibly *a knife with a much blunter blade edge than . . . the scalpel . . . but it could also have been an instrument other than a knife* — or it might have been tearing by hand in a different way.

The information in this chapter and elsewhere in the book shows that examination of damage to clothing and other textiles can be valuable in solving crime or providing evidence in litigation. Where civil law-suits are for claims on the textile itself or are the direct result of failure of a product, detailed examination of the textile is clearly essential. As Johnson and Stacey (1991) say:

if a mountaineer is found seriously injured at the bottom of a cliff and his climbing rope is severed in two, it is necessary to determine why the rope failed. It may simply have been too worn or too light for the task, but if it appears to have failed due to a manufacturing fault, the injured party may attempt to sue the manufacturer. On the other hand, the possibility that someone has deliberately damaged the rope must be considered. There is also the possibility that the injured climber's friend has deliberately severed the rope after the fall so that a claim can be made against the manufacturer. It is the task of the forensic textile scientist to determine just which of these possibilities provides the most likely explanation.

Table 44.1 — Incidence of damage

Specimens	Number of fibres	Types of break and approximate percentage
EXPERIMENTAL		
scissor cut	21	squeezed [100%]
tearing	15	bulbous and smooth [100%]
scalpel cut	31	clean cut [45%]/fractured and elongated [22%]/bulbous [33%]
FROM NIGHTGOWN		
shoulder straps	18	clean cut [25%]/fractured and elongated [50%]/bulbous [25%]
sides of gown	18	fractured and elongated [25%]/bulbous [75%]

For greater use in criminal cases, more research is need into characterising damage according to the weapon used and the nature of the attack. Some studies have been made by Johnson and Stacey (1991) at the University of New South Wales (UNSW), who point out that the experimental conditions must be carefully chosen. The backing is important and, since *human volunteers are hard to find*!, a side of pork with the skin on is a good approximation. *Quite different morphologies are created if the knife movement is artificially created at a constant speed, and if the fabric is held tensioned in a mounting frame. An accurate simulation requires the scientist to act out a frenzied stab attack on the fabric draped over a piece of pork.* Nevertheless, at that stage of their research, they stated that *because of the great variety of fabric and weapon types, it is not possible to develop a generalised description of the morphology of stabbed fabrics.* More recent research at UNSW is reported in Chapter 46.

It is perhaps because of these difficulties that fibre/textile damage rated only two paragraphs by Carroll (1992) in a book on *Forensic examination of fibres*. His conclusion was that:

While the forensic specialist is often asked to 'match' a suspect weapon with damage found in a garment, this is rarely possible. The garment can be examined to characterize the age of damage present, for example, 'recent' or 'fresh' versus 'old' in nature; the proviso to distinguish recent from old being that the garment has not been laundered since the damage occurred. The type of damage may be characterized as cut, rip, or seam separation. The suspect weapon may also be used to produce test damage, simply to indicate whether or not it is capable of producing damage consistent with that in the garment.

This is a minimalist approach. The co-operation of skilled microscopists with textile scientists is capable of more.

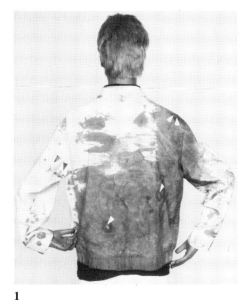

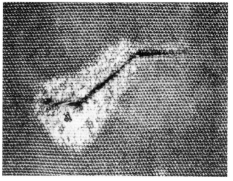

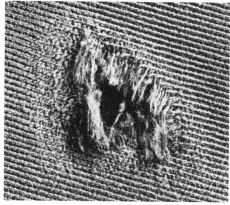

Plate 44A — Damage in fabrics.
(1) A leisure jacket with three stab cuts marked by arrows. (2) An angular cut. (3) A break from rubbing against a pavement. (4) A range of cutting instruments. (5) SEM picture of acrylic fibres welded together after attack with blunt force, from Foos (1993), scale mark = 10 μm.

1

2

3

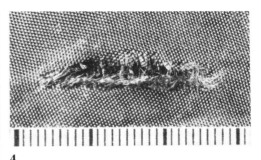

4

5

Plate 44B — Cutting and tearing in a knitted cotton shirt.
(1) General view. (2) Cut yarn end. (3) Torn yarn end.
Damage in an acetate lining.
(4) Macroscopic view. (5) At higher magnification, showing cut ends.

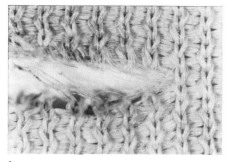

1

2

3

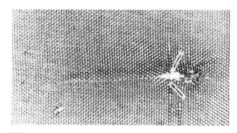

4

5

Plate 44C — Case of a murdered pensioner.
(1),(2) Opposite ends of cut in the nylon shirt. (3) The interlining of the jacket in the same case.
Another murder case.
(4),(5) Cut in nylon lining of anorak.

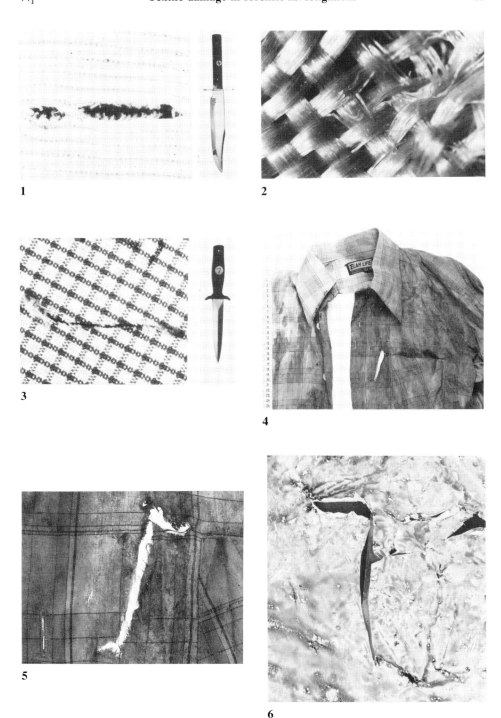

Plate 44D — Experimental knife cuts.
(1),(2) Cut with a one-sided knife. (3) Cut with a two-sided knife.
Fabric damage, but not a crime.
(4) Cuts in a shirt. (5),(6) Matching cuts in the shirt and in a plastic bag full of broken glass.

45

USE OF SEM IN TEXTILE FORENSIC WORK

William Pelton

The field of forensic sciences today utilizes sophisticated laboratory instrumentation and incorporates many disciplines of which textile science is one. Both forensic and textile scientists have been using SEM micrographs of fibre and yarn damage in forensic investigations to:

(1) create a record for evidence;
(2) identify fibre or yarn features associated with different sources of damage, Choudhry (1987), Ishizu *et al.* (1974), Paplauskas (1973), Pelton (1995), Stowell and Card (1990), Wong (1984);
(3) report the results of controlled experimentation into known sources of textile damage, Pelton and Ukpabi (1995), Stowell and Card (1990).

Sources such as glass-fragments, knives, scissors, blunt-instruments, tearing, animal bites, animal claws and abraded hoisting cable strands have been identified as causes of textile damage. There is no established forensic protocol to identify the source of textile damage by observing fibre-end morphology. Scientists *have been using* ad hoc *procedures to present SEM fibre and yarn evidence. . . .* [Investigators have] *presented* [evidence] *. . . with very few micrographs to support their opinion*, Pelton (1995). There has been little comment on the diversity of fibre-end features which could appear in a single source of fabric damage. Scientists have raised issues associated with the use of the SEM to identify unknown sources of textile damage and the forensic interpretation of these SEM micrographs, Crispin (1987), Pelton (1995), Young (1989).

Forensic scientists, such as Choudhry, Ishizu, Paplauskas and Wong, and not textile scientists, first suggested that SEM micrographs of fibre- and yarn-end appearances could be used to distinguish knife-cuts from scissor-cuts, sharp instrument cuts from tearing, and sharp instrument damage from animal bites. Currently, the forensic literature does not have a database of SEM fibre or yarn micrographs, illustrating fibre-end appearances, which could be used to compare unknown to known damage sources. The problem is that much of the published textile SEM data has been based on single fibre experimentation. Textile damage observed in a forensic investigation is normally associated with an assembly of twisted fibres in a fabric structure. Therefore, the source of the damage is not the only variable; fibre type, yarn structure, fabric construction, applied finishes and fabric orientation in garment construction are variables which also influence the appearance of the textile damage observed. Forensic investigators tended to overlook the influence of these critical textile variables when the specific mechanism creating the fabric damage was allegedly identified using SEM micrographs of individual fibres, individual yarns or a montage of yarn ends. A montage is created by scanning and photographing several millimetres of fabric damage. Individual micrographs are then positioned in the correct orientation and glued together forming a record of the damage which could be observed by the naked eye as evidence in court.

In Australia, one criminal case, the trial of Linda Chamberlain for the alleged murder of baby Azaria Chamberlain, has been documented in which SEM micrographs of individual fibres, individual yarns and a montage of yarn ends were introduced as evidence at different points from the initial investigation to the final Royal Commission Inquiry by The Honourable Mr Justice T. R. Morling into wrongful conviction, Crispin (1987), Morling (1987), Young (1989). The Chamberlain Defense argued that a dingo had taken the 9-week old child from the campsite tent, while the Crown alleged that the mother had murdered the child in the car, placed the body in a camera bag and subsequently disposed of the body. The child's body was never found but her damaged garments were discovered by a tourist approximately 4 kilometres from the camp site. At the inquests and trial, the Crown's textile expert produced SEM micrographs of fibres and yarns from the damaged portions of the jumpsuit and indicated that:

- some of the damage portions appeared to be straight;
- the severed fibres in individual yarns terminated in a similar plane;
- the appearance of one fibre within one yarn cluster was identical to that of the classic scissor-cut (i.e., a pinched end with lateral compression), R v Chamberlain and Chamberlain (1982).

The Defense did not use a SEM until the Royal Commission commenced.

The Crown conducted one experiment with a zoo dingo and their experts concluded that the gross morphology of the fabric damage was consistent with tearing. Since the experiment was carried out in wet weather, much of the damage was encased by dried mud. The Crown's textile expert was unable to produce any SEM micrographs of dingo severed fibre-ends. No additional dingo experiments were conducted by the Crown until the Royal Commission commenced. The Crown's forensic experts were familiar only with fabric damage caused by an attacking canid, not with the damage caused by a canid picking up an infant and carrying it away.

Meanwhile two scientists, Chapman and Smith, started conducting experiments in which a domestic dog was offered food sewn into a small knitted fabric bag approximately 20 × 15 cm, Crispin (1987). Bags were produced from fabric identical to that of the missing child's jumpsuit. From their first experiment, Chapman and Smith found 'tufts' of yarn (i.e., short lengths of yarn) floating in the dog's water bowl after it had extracted the food from the bag. Earlier, the Crown had established that yarn tufts along a damaged segment of knitted fabric was a clear indication that fabric was cut and not torn. Chapman and Smith continued to experiment with dogs and dingoes. Their information was used by the Defense in the appeal process. By the end of the Royal Commission, Chapman and Smith had produced 75 specimens of canid damaged fabric. Specimens with similar macro features to scissor damaged fabric were scanned by the SEM. Defense experts felt that the SEM micrographs (100×– 300×) of consecutive damaged yarns did not add any further information to the macro examination (5×–40×).

Using a low power stereo-microscope during the inquiry, the Defense textile expert found 28 points of similarity between canid-damaged fabric and the child's damaged jumpsuit, Crispin (1987). *Of those twenty eight points, he found that twelve would have been difficult to reproduce with scissors even if one had known what kind of damage to cause. Four points of similarity would be very difficult to reproduce, if not impossible, to reproduce with scissors,* Crispin (1987) page 291. A number of the similarities were observed in the exhibit of the Crown's initial dingo experiment. Over the years, the dried mud had disintegrated revealing damage which, it had been assumed, could be produced only by cutting.

During the inquiry, the textile evidence focused on SEM micrographs of individual fibres, individual yarns and, particularly, montages of yarn ends. The Crown introduced evidence suggesting that canids crushed fibre-ends in the severing process, whereas scissor-cuts produced a 'planar array' of consecutive yarn ends. Planar array was defined as a precise *alignment of severed ends of fibres within a number of consecutive yarns,* Crispin (1987) page 284. This was the first time the phrase *planar array* was associated with textiles. The definition of *precise* varied from one Crown expert to another. Crown experts suggested that planar array was visible in SEM micrographs but not visible under low power optical microscopes. No SEM micrographs of individual fibres severed by canids or cut by scissors were submitted as inquiry evidence, but SEM montages of both dingo- and scissor-damaged fabric were introduced by the Crown. Crown experts claimed that the dingo-severed fabric could not achieve the same degree of planar array that scissor-damaged fabric could. Under cross-examination, Crown experts could not agree on the planar array definition nor on the limit of deviation in the planarity. A law court is no place to establish textile definitions. Only established terminology accepted by the scientific community should be used in court evidence.

Defense experts testified that the use of SEM montages to distinguish the 'planar array' was beyond the limitation of the micrograph interpretation, Crispin (1987). They agreed that the montages produced a clear magnified image of the damage, but the two-dimensional SEM montages did not have a reference scale to detect fibre-end deviation measurement within and between the yarn clusters of a three-dimensional structure. No quantitative data were produced by the Crown. When Crown experts were cross-examined about mounting procedures to attach a fabric specimen to the SEM stud, manipulation of specimens around the stud and method of attachment could alter the degree of planarity observed. Since textile materials are flexible, knitted fabric specimens were observed curling towards or away from the stud, Morling Transcript (1987). Again, this fabric curling could influence the degree of planarity observed as the stud is rotated, and moved back and forth to capture the best image. As with terminology, a law court is no place to argue SEM protocol. Only established protocol accepted by the scientific community should be used to prepare court evidence. If necessary, in particular difficult cases, this may involve going back further into the scientific presentation in order to find an acceptable starting point.

In order to establish the validity of the Crown experts' claims that scissor-cuts could be distinguished from dingo-damaged fibre-ends, Raymond, the Commission's forensic scientist, set up a blind experiment for the Crown and Defense textile experts, Young (1989). He prepared a SEM stud with five different short tufts of nylon yarn. This experiment gave scientists the opportunity to demonstrate the significance of fibre-end appearances as a means of distinguishing scissors-cut from dingo-severed fibres. After viewing the five specimens for more than two hours fibre by fibre, no-one volunteered any comment. The characteristics observed in each tuft had many similar overlapping features. Several fibres had an appearance consistent with the micrograph **37D5**, which was published later in the first edition of this book. This micrograph depicts features of a nylon seat-belt fibre severed by a dog. No unique feature such as the classic scissor-cut appearance was noted in any one specimen. Raymond then revealed that three tufts were taken from dingo-damaged jumpsuits, one was produced by a pair of scissors, and the final tuft was the Crown's inquest exhibit from the missing child's jumpsuit. Raymond's experiment made a significant impact in this case on the reliability of SEM fibre-end micrographs to distinguish the cause of fabric damage, Morling (1987).

The inquiry then turned its attention to the SEM montage evidence and called its own textile expert in an attempt to clarify the impasse created by the Crown and Defense textile experts (Crispin, 1987). He testified that textile scientists used the SEM to study features of severed or ruptured fibre-ends; however, they relied on a low-power optical microscope when asked to determine the planarity of a severance line. He felt that the term 'planar array' had been misused. The concept meant that the alignment of fibres must be within extremely fine tolerances produced by very sharp scissors. After observing several SEM montages of yarn-ends, this expert found similar planarity in damaged specimens produced by sharp instruments (i.e., scissors and knife) and by dingo dentition. In his opinion, the planarity of the damaged fabric in the child's jumpsuit had less planarity than that created by sharp scissors. He stated that one could not expect the fabric structure (i.e., single jersey pile fabric) in the child's jumpsuit to produce 'planar array' as defined by Crown experts.

The reliability of the SEM montage procedure was also challenged in the cross-examination of the Crown's SEM expert. One particular specimen, which the expert had cut with a scalpel, had produced two different SEM montages of the same damage. The only difference was the way the specimen had been mounted to the stud: one montage had the fabric face against the stud, while the second had the fabric back against the stud. The expert testified that the one montage illustrated the concept of planar array while the other montage contained no planar array. Specimen mounting techniques changed the textile interpretation of the SEM montages. This type of SEM observation was unreliable.

By the end of the Royal Commission, the validity and reliability of the SEM micrograph interpretations were both seriously challenged. Although the SEM has been a useful diagnostic instrument of fibre failure morphology, the problem was that no appropriate database of SEM micrographs existed. This was certainly true prior to the publication of the first edition of this book in 1989, and, even now, the database is limited in its relevance to forensic investigations. At the time of the 1987 inquiry, no scientist was able to quote references to any published SEM research distinguishing scissor-cut from tears, or scissor-cut from dingo-severed fabric. Research into features to distinguish one source of damage from another was conducted throughout the course of the criminal proceedings. Initial speculation on the appearance and features of dingo-damaged fabrics was proved incorrect by the conclusion of the inquiry. SEM micrographs, however, contributed to the inquiry by illustrating that canids could produce features similar to those caused by sharp instruments. The micro-analysis, therefore, supported the macro-analysis.

In order to move towards the establishment of a database, Ukpabi and Pelton (1995) first reviewed the use of the SEM to identify the cause of fibre damage in forensic investigations and found only six journal articles reporting some aspect of SEM usage to document evidence, to describe unique features, and to report the results of known forensic textile damage. Scientists have expressed different opinions on the usefulness of the SEM to distinguish the cause of textile damage and, in some cases, have not agreed on the interpretation of SEM micrographs. Included in the review was the quantitative investigation by Stowell and Card (1990), which was referred to in Chapter 44.

Reports of a more extensive quantitative study have been published by Pelton (1995) and Pelton and Ukpabi (1995). A plain woven untextured multifilament nylon fabric was damaged in three ways: cutting with sharp 21 cm dressmaker's shears (scissors cut); cutting with a sharp carving knife with a 20 cm blade (knife cut); and rupture on an Elmendorf tear tester (impact tear). For the cuts, one person held the fabric under minimal tension, with no supporting substrate, while another cut the fabric with scissors or slashed it with the knife. The prior expectation was that the shearing action of the scissors would cause lateral compression, the knife would give a clean cut, and the high-speed tear would give the characteristic mushroom cap. These forms are illustrated in Fig. 45.1.

Pelton examined over 600 damaged fibre ends in the SEM, either in yarn clusters or as individual fibres, and assigned 322, which showed clearly defined features, to the above

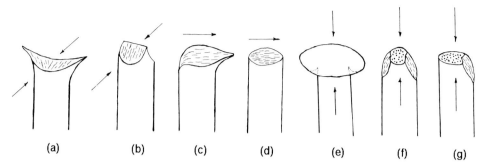

Fig. 45.1 — Classification of fibre breaks.

SCISSOR (shearing action):   (a) pinched, lateral distortion; (b) pinched
KNIFE (slashing action):   (c) flat top with lip; (d) flat top
IMPACT TEAR: (e) mushroom, bulbous
LOW STRAIN RATE TEAR: (f) double ductile fracture; (g) single ductile fracture.

Table 45.1 — Assigned causes of fibre damage, Pelton (1995)
(Reprinted, with permission, copyright ASTM)

Actual source	Assigned descriptors				
	Lateral compression	Clean cut	Mushroom cap	Undefinable	Total for source
scissor	6	89	—	8	103
knife	14	60	6	25	105
tear	—	5	92	17	114

categories, plus an additional 'undefinable' category for those which did not conform to any of the three types. The results are shown in Table 45.1.

Pelton and Ukpabi (1995) then asked a panel consisting of a textile technician, a textile professor, a textile graduate student and 11 undergraduate clothing and textile students to view 117 micrographs. They were given written instructions on how to view the unknown pictures, and how to assess damage based on the theoretical models, shown in Fig. 45.1, and SEM micrographs of known scissor cut, knife cut and impact tear micrographs. In addition to the three options and the undefined category already mentioned, a slow-speed ductile fracture, also shown in Fig. 45.1, was included among the available choices. Appropriate statistical procedures were used to randomise the presentation and determine *recognition probabilities* for assigning correct cause; a value of 0.2 or less would result from a random allocation of cause, and 1 would show perfect assignment. The result of the study was that the recognition probabilities for individual panellists ranged from 0.35 to 0.46. The mean percentages of correctly assigned causes were 15% of the 39 scissor cuts, 37% of the 42 knife cuts and 72% of the 36 impact tears. Table 45.2 shows how the causes were assigned.

Finally, Pelton and Ukpabi attempted to assign 248 fibre end micrographs to descriptors developed from those used by Stowell and Card (1990). Their results are shown in Table 45.3. Comment on the meaning of the descriptors is included in the discussion of the micrographs in **45A,B** below.

The experimentation reported by Stowell and Card (1990), as described in the previous chapter, and by Pelton and Ukpabi (1995) has been significant in starting to establish an SEM database for forensic textile applications. However, in total, the studies were limited to four sources of damage, three fabrics, and one fibre type. The two studies reported results using different quantitative approaches and neither relied solely on qualitative observations. Pelton and Ukpabi replicated the damage created by Stowell and Card but used a different fabric structure. Stowell and Card reported unique features distinguishing cuts from tears and scalpel-cuts from scissor-cuts. Pelton and Ukpabi's experiment found that subjects could identify correctly most of the torn fibre-end specimens but were mostly unable to identify either the knife-cut or the scissor-cut specimens. The two studies reported completely different conclusions using similar sources to create the textile damage. Their conclusions are significant because the results suggest that fabric structure, amount of fabric cover and/or amount of yarn twist could influence what is seen in SEM micrographs. The fabric used by Pelton and Ukpabi was a tightly woven structure with a compact multifilament yarn; whereas, Stowell and Card described one component as lace which is normally an open structure.

When Pelton and Ukpabi's scissor-cut and knife-cut results are compared to those reported in the Chamberlain investigation, the compact woven fabric structure may have

Table 45.2 — Actual and assigned causes, Pelton and Ukpabi (1995)
(Reprinted, with permission, copyright Canadian Society of Forensic Science)

Actual cause	Total	Average assigned cause				
		Knife	Scissor	Tear	Ductile	Undefined
scissor	39	17	6	3	7	6
knife	42	16	8	5	5	8
tear	36	1.4	0.6	26	4	4

Table 45.3 — Distribution of fibre end forms, Pelton and Ukpabi (1995)
(Reprinted, with permission, copyright Canadian Society of Forensic Science)

Descriptors	Actual cause		
	Scissors	Knife	Tear
pinched appearance* (with or without lateral distortion)	0	14	9
rivet head	16	25	0
mushroom cap (bulbous and smooth*)	0	0	48
smooth cylindrical end* (flat top)	16	28	9
smooth cylindrical end with lip	15	39	0
cylindrical concave end	4	0	0
ductile fracture	0	0	2
undefinable (fractures, elongated*)	3	18	2
Total	54	124	70

*Denotes descriptors similar to those reported by Stowell and Card (1990).

created more variation in the fibre-end appearances than those from an open knitted structure. Pelton (1995) has suggested that SEM research associated with forensic textile investigations is creating more questions than it is currently solving.

When researchers have experimented with thermoplastic fibres, they have expected a clean-cut fibre end from a sharp knife or razor, though more squashing from a blunter knife, lateral compression from scissors, a mushroom cap from an impact (high velocity) tear and a ductile fracture from very low velocity rupture. Pelton (1995) used a series of SEM micrographs to document why the subjects in Pelton and Ukpabi's experiment (1995) had difficulty in distinguishing between scissor-cut and knife-cut fabric damage. Examples of the appearance of fibre ends, similar to those published by Pelton (1995), are shown in the following SEM micrographs. They are all from a tightly woven, multifilament nylon fabric and show overlapping features. The pictures are arranged with scissor cuts on the left, **45A,B 1(a)–1(f)**, knife-cuts in the centre, **45A,B 2(a)–2(f)**, and impact tears on the right, **45(A),(B) 3(a)–3(f)**.

Both the knife cut, **1(a)**, and the scissors cut, **2(a)**, have similar pinched ends with lateral distortion, which might have been anticipated only with the shearing action of scissors. Lateral distortion is also visible in **1(b)** for scissor-cut and in **2(b)** and **2(c)** for knife-cut. Striations are apparent on the surfaces cut by both scissor, **1(b)(d)(e)**, and knife, **2(b)(d)(e)**. In **1(c)**, which shows a double-cut fibre end, both blades of the scissors have sheared the fibre; this feature was unique to fibres cut by sharp scissors and could possibly distinguish scissor-cut from knife-cut yarns, Pelton and Ukpabi (1995). A clean-cut with lip is evident in **1(e)** for scissor-cut and in **2(e)** for knife-cut; the lip and striations indicate the direction of severance. Shown in **3(a)–(f)** are a variety of fibre-end features for impact tears. Descriptors such as mushroom, **3(a)**, rivet-head, **3(b)**, inverted mushroom, **3(c)**, globular **3(d)**, double-ductile with catastrophic fracture, **3(e)**, and globular with initial crack, **3(f)**, appearance could describe the different shapes of the impact tear specimens. According to theories of fibre rupture, ductile fractures should not be associated with impact tearing. Fabric and yarn variables may be causing ductile fractures to occur. The scissor-cut, knife-cut and torn fabric fibre-ends in **1(f),2(f)**, and **3(f)**, respectively, all have similar features which are associated with tensile fractures. The appearances of **1(f)** and **2(f)** suggest that some fibres are fractured in fabrics before the blade can make contact with the fibre, as a result of pressure on the part of the fabric that is in contact. The high magnification allows detailed surface features to be observed in several of the micrographs, in addition to the general forms described above.

Scissors are essentially two pivoting knives contacting and cutting a material in opposite directions. In theory, the effect of the two blades coming together could be visible on a single fibre-end (i.e., a pinch-end with lateral compression). This observation should not be anticipated in damaged fabric since the fabric is made from yarns which are usually twisted bundles of fibres. Fibre cross-sectional shapes could also influence fibre-end appearances of thermoplastic fibres. Yarns from the same set of experiments are shown in **45C(1)–(4)**. The scissor-cut in **45(C)(1)(2)** suggests that the fibres could remain 'tightly clustered' with 'all fibre ends terminating in a similar plane'. The sharp scissors causing the damage have created different fibre-end appearances (i.e., clean cut, cut clean with lips, double-cut). No pinched-end with lateral compression examples are visible in this yarn. The compact fabric and yarn structure may account for this feature not being detected. Many individual fibre-ends within the yarn cluster of **45(C)(1)** have features similar to those within the yarn cluster of **45(C)(3)**, a knife-cut specimen. All yarn micrographs were produced at 250× magnification.

Fibres within the knife-cut yarn, **45(C)(3)** are 'loosely clustered' with 'all fibre ends terminating in a similar plane'. The majority of fibre-end appearances in this micrograph could be described as clean-cut. A closer examination of the fibre ends in the yarn cluster reveals a series of parallel striations suggesting a direction of severance. The grooves in the knife-cut striations are more prominent than those of scissor-cut striations. The depth of grooves could be associated with the method of sharpening and could possibly distinguish one sharp instrument from another.

Impact tearing, **45(C)(4)**, exhibited 'no clustering' of the fibres. Fibre ends 'terminated in different planes' with a 'random orientation'. Descriptors such as bulbous or mushroom shaped would characterize most fibre ends. The range and distribution of impact tear features was quite different from those observed for either knife- or scissor-cut fibres. Neither the knife-cut nor impact tear fabric samples exhibited the tightly clustering features of **45C(1)** or the fibre-end feature indicated in **45A(1c)**.

As illustrated in these micrographs, fibres ruptured within a fabric instead of fractured individually have increased the range of fibre-end appearances observed in a given source of damage. The difficulties of making valid interpretations of fibre breaks are emphasised by the scissor-cut, knife-cut and torn fabric fibre-end appearances in **45(B)(4a,b,c)**, respectively. These illustrate some of the unexpected features which Pelton and Ukpabi (1995) catalogued as undefinable. The enlarged diameter and inverted-cone shaped end in the scissor-cut **45B(4a)** could be associated with an impact tear force; the elongated fracture in **45B(4b)** resembles an insect damaged fibre-end appearance; and the enlarged twisted fibre end in **45B(4c)** may show the influence of twist on yarn rupture. Fibres, therefore, can be fractured by tensile or shearing forces before a severing instrument makes contact. The effect of fabric structure, yarn type, and different generic fibres on fibre-end appearances must be studied further.

Forensic protocols exist for fundamental textile analyses such as fibre identification. In situations such as identifying the cause of fabric damage, however, there is no established protocol or substantial body of published research to draw upon. Interpretation of the damage depends on the scientist's knowledge of textile garment, fabric, yarn, and fibre properties, and the possible interactions of these properties. In identifying the cause of fabric damage, the investigation should be systematic — moving from the macro to the micro (i.e. from garment to fabric to yarn to fibre). SEM micrographs, as illustrated above, could be associated with the latter two components. A system similar to the concept used in fingerprint comparison in which 'points of similarities and differences' are established could give more useful information about the cause of textile damage than assessing the damage against a set of known criteria. This approach has been illustrated in discussing features observed in **45A,B(1,2,3a–f)**.

SEM yarn or yarn montages should not be used to measure the planarity of the three-dimensional image. **45C(1)** and **45C(2)** are micrographs of the same scissor-cut yarn, but viewed from different perspectives. Another typical scissor cut yarn from the test fabric is shown in side view in **45C(5)** and in end view in **45C(6)** at different magnifications. These micrographs illustrate some advantages of the SEM to record and document fibre-end features. **45C(2)** gives a clearer indication of the fibre-end appearances — fibres at the bottom of the micrograph having the clean-cut lips pointing up to the yarn centre, while fibres at the top have their lips pointing down. In another yarn example, the perspective presented in **45C(5)** shows that the severed fibres within the yarn all terminated in a similar plane. If the stud were rotated and tilted to give a perspective similar to Fig. 4(b) from Pelton (1995), reproduced as **45C(6)**, fibres in the yarn centre may be seen to have been severed by both blades (i.e., double-cut) with a similar appearance to **45A(1c)**. As the stud is rotated and the angle changed from one position to another along the damage, better images of the features can be recorded. The degree of planarity among the fibres or the deviations from a common cutting plane, however, could not be viewed in either **45C(1)** or **45C(2)**.

The SEM can be a useful diagnostic instrument for fibre morphology in forensic textile investigations. SEM micrographs document evidence which the scientist sees under the microscope and present a clear image of what the scientist is attempting to explain. Judges, lawyers and jury members are not experts and visual documentation assists in technical explanations of textile evidence. Because of the many variables involved when damage is created, scientists need to document the range of features which were observed under the SEM. To date, scientists have viewed damaged fibre features on the SEM but have not presented courts with sufficient documentation about the range of features that one could expect.

With the current limited SEM information published on damaged fibre end appearances associated with yarns and fabric, assessing the cause of damaged fabrics using a set of criteria could be problematic because another source may produce similar criteria, as demonstrated in the case against the Chamberlains. The concept of 'similar and different' observed features provides the scientist with more concrete evidence. As well as looking for differences in the SEM micrograph features, the scientist should also be asking the question 'what other sources could cause similar features to these observed in the SEM micrographs?'. Often forensic laboratories are not associated with the primary investigation. The investigator presents the scientist with the damaged textiles and asks the question 'Could a given source (e.g., glass fragment, knife, screw driver, etc.) cause this damage?' Scientists have been responding that the damage *is consistent* with the identified source without stating in their written documentation that the damage *could also* be produced by other sources based on macroscopic examination. Written statements can be misinterpreted in the jury room to mean that the identified source is the only possible source. SEM micrograph features could confirm a particular source (e.g. screw driver) or could suggest a number of possible sources (i.e. glass fragment, knife and scissors).

A final question which should be asked is 'what are the limits on the interpretation of SEM micrographs in forensic textile investigations?' At present, scientists do not have enough knowledge and understanding to give a full answer to such a question. However, several points do come to mind. The first limitation on forensic textile interpretation is our inadequate knowledge of SEM features caused by different rupturing sources. The second is associated with comparing the damaged fibre features taken from fabrics to those established for single fibre fractures. The third is related to the limited number of SEM observations, which are currently being used in court to support an informed opinion. The fourth is the use of SEM montages to establish the degree of planarity observed along the severance. The fifth is the fact that some fabric structures give similar overlapping features for sharp instruments.

In conclusion, no SEM protocol exists to distinguish the cause of textile damage in forensic investigations. A SEM protocol should require scientists to examine a number of adjacent damaged yarns, the clustering of fibres within yarns and individual fibre-end features in a systematic manner. The systematic analysis should be documented by a series of SEM micrographs.

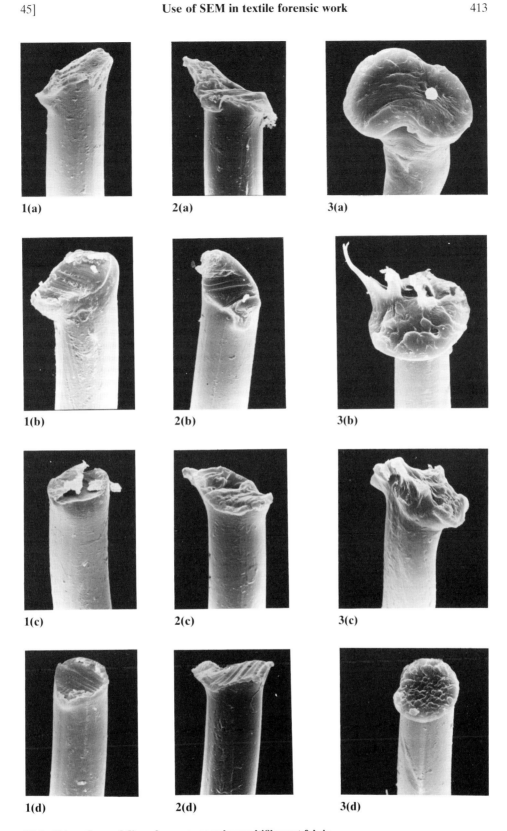

Plate 45A — Severed fibres from woven nylon multifilament fabric.
1 (a–d) Scissors cut. 2 (a–d) Knife cut. 3 (a–d) Impact tear.

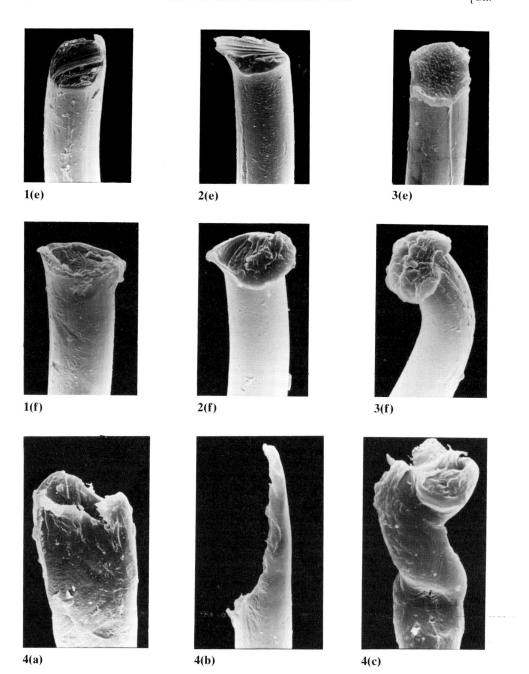

1(e)   2(e)   3(e)

1(f)   2(f)   3(f)

4(a)   4(b)   4(c)

Plate 45B — Severed fibres from woven nylon multifilament fabric (continued).
1 (e–f) Scissors cut. 2 (e–f) Knife cut. 3 (e–f) Impact tear.
Severed fibres showing unusual features.
4 (a) Scissors cut. 4 (b) Knife cut. 4 (c) Impact tear.
(Reprinted, with permission, copyright ASTM)

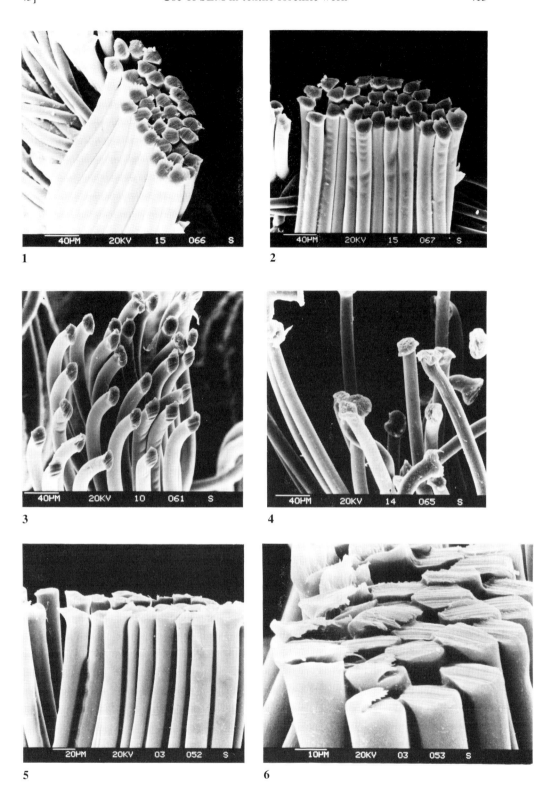

1

2

3

4

5

6

Plate 45C — Severed yarns from the fabric of 45A,B.
(1) Scissors cut. (2) Scissors cut viewed from another angle. (3) Knife cut. (4) Impact tear.
(Reprinted, with permission, copyright Canadian Society of Forensic Science)
Another scissors cut yarn.
(5),(6) Viewed from different angles at different magnifications.

46

COMPARISON OF BULLET AND KNIFE DAMAGE

Fran Poole and Michael Pailthorpe

NOTE: This chapter consists of edited extracts from the report by Fran Poole (1996).

The examination of clothing in murder cases, where the body has decomposed, often calls for expert interpretation concerning the cause of death. A key aspect often relied upon is the type of damage to the clothing, and an opinion as to what caused it. Chapters 44 and 45 discuss damage due to knife cuts and related causes. However, apart from laboratory testing of aramid and nylon fabrics for 'bullet-proof' garments, as shown in **40H(5)(6)** and **40L**, and the few early observations by Paplauskas (1973), there has been extremely little research to examine the relationship between the mode of damage and the appearance of damaged ends associated with ballistic impact. Consequently, limited information is available on the differences in the appearance of fibre ends damaged by bullets versus severance by knives. As mentioned in the last chapter, the lack of information on the causes of damage to textiles has, in cases like the death of Azaria Chamberlain, left textile expert witnesses wide open to criticisms concerning the validity of their judgments.

In July 1996, Ivan Millat was convicted of the murders of seven backpackers. Five of the bodies were skeletonised, which made it impossible for the pathologist to establish the cause of death, but partial remains of clothing had the potential for assisting forensic examiners with their determinations. Textile experts were consulted and were able to comment on the severance damage in the fabric, but they were unable to offer any opinion on the ballistic damage to the clothing because of the lack of research available to support opinions concerning ballistic impact.

The aim of the present study was to examine the differences in failure morphology caused by ballistic impact versus severance, as seen at three levels of resolution, namely macroscopic, microscopic and SEM. A simulated human torso was created by securing a large thick piece of pork belly to a canite backboard mounted on a metal frame. The test garment was fitted over the pseudo-torso. Five attacks were made on different parts of the torso. A common murder weapon, a .22 calibre rifle, was used to fire two types of bullet through the fabric at two distances, contact and 20 cm; and a knife was used to attack each garment with a downward stabbing motion. Details of the five test garments, differing in fibre and fabric type, and of the weapons, which are shown in **46A(1)–(3)**, are given in Table 46.1. The test rig, with and without a garment on the torso, is shown in **46A(4),(5)**.

After the five attacks, the damage was assessed. First, for the macroscopic examination, the shape of the damaged areas and the separated edges were examined visually and photographed. Then a stereo-microscope with a magnification range of 6×–32× was used for the microscopic examination. At low magnification the whole damaged area could be seen and photographed. Finally, the penetration areas were cut out and mounted for SEM examination. A selection of the observations is presented here.

Except for the wool knit, which is discussed below, the fabrics exhibited irregularly shaped, but roughly circular, holes for the bullet damage, and linear severance lines in 'planar array' for the knife damage. The five macroscopic views of the woven polyester/cotton fabric, shown in **46B(1)–(5)**, are a typical set. The knife cut was also sharply defined for the tight nylon warp knit, but was less sharp for the knit cotton, **46B(6)**, and the fluffy knit polyester. The difference between bullet and knife was clear from the macroscopic examination of all four fabrics, excluding the loose wool knit. There are indications that the hollow-point bullets

Table 46.1 — Garments, fabrics and weapons

Garment	Fibre type	Fabric construction
T-shirt	cotton	single jersey weft-knit
vest	polyester	single jersey weft-knit
business shirt	polyester/cotton	plain weave
slip	nylon monofilament	warp-knit
jumper	wool	single jersey weft-knit

Weapons

.22 calibre *Lithgow* single shot rifle:		serrated edged steak knife:	
overall length	1000 mm	overall length	215 mm
barrel length	610 mm	blade length	120 mm
diameter of bore	5.61 mm	blade width	15 mm

ammunition:
   [A] Winchester Superspeed .22 long rifle high velocity **solid-point** bullets at 350 m/s
   [B] Winchester Superspeed .22 long rifle high velocity **hollow-point** bullets at 400 m/s

produce sharper edges to the holes than the solid-point. Both contact and non-contact ballistic damaged areas on all five fabrics displayed a darkened ring of gunshot residue around the perimeter of each bullet hole. This was most easily visible on the lighter coloured garments and was more difficult to see on the maroon woollen jumper.

The microscopic appearance of the first four fabrics, shown in **46C(1)–(6)** and **46D(1),(2)** support the macroscopic observations in showing a clear difference between the bullet and knife damage. Gunshot residues were clearly seen around the bullet holes on all five fabrics. In addition, yellow fragments of canite from the backboard were also visible in the ballistic damaged areas and obscured the view of the fibres. Improvements are needed in the mounting of the simulated torso, in order to avoid contamination by the backing.

Microscopic and macroscopic views of the wool knit are shown in **46D(3)–(6)**. There seems to be no significant difference in the appearance of the holes caused by bullets and by the knife, making it virtually impossible to determine visually what caused the damage. The lack of differentiation is due to the resilience of wool and the coarse, loose knit structure of the jumper. Since the transverse dimensions of both the bullet and the knife are comparable to the stitch size of the fabric, their difference in shape is poorly resolved. This is further obscured by the lack of clear stitch definition in the material and by the spring back of the wool due to its good elastic recovery. The structural differences are clearly seen by comparing the holes in the tight, fine nylon warp-knit, **46D(1),(2)**, with those in the wool knit, **46D(3),(4)**.

A number of fibres from each of fifteen samples (omitting the hollow-point damage) were viewed with the SEM. As expected, there were many differences between the fibre end appearances in the different fabrics. However, a more remarkable observation was that a large range of different fibre end appearances were recorded within the same area of fabric damage. Examples of the fibre appearances are shown in **46F.–G**

The cotton fibre ends, illustrated in **46E(1)–(4)**, displayed no clear pattern for either ballistic or severance, and were a mixture of what would be expected from tensile breaks and from flexing and transverse pressure. The ballistic fibre ends all exhibited signs of distortion or rupture giving a generally ragged appearance. Some knife severed ends showed elongated projections of the tips, while others showed fibrillation and spitting of the ends or jagged edges cut away from the sides of the tips similar to the result of a sawing action. There were no unique features that enabled bullet damage to be distinguished from knife damage.

Some of the fibre ends in the polyester fabric, which show a mushroom cap, may result from the brushing and napping of the fabric, but they may also result from high-speed rupture. Although the polyester fibre ends, **46E(5)–(10)**, exhibit clearer appearances than was seen in cotton, there was little to distinguish bullet damage accurately from knife damage. The polyester and cotton fibre ends in the woven blend fabric, **46F(1),(2)**, gave the same problems as the 100% cotton and polyester fabrics. The individual fibre ends could be distinguished from each other, but the cause of damage could not be determined.

For the nylon monofilament warp-knit, the difference between ballistic and severance damage at the fibre level was clear. The fibre ends from ballistic impact, **46F(3)–(5)**, showed the classic features of mushroom caps caused by fibres broken at high speed or due to other forms of heating. A similar form is shown in **40H(5)**. Fibre ends from stabbing damage, **46F(6),(7)**, show the characteristic squashed or more sharply cut forms noted in Chapter 20. At a larger scale, the knife damaged ends remained in pockets of yarn clusters clearly showing the line of severance of fibres on a similar plane, whereas the bullet damaged ends were separate from each other.

The wool fibre end appearances of the contact damage, **46G(1),(2)**, were not distinguishable from those of the non-contact damage, **46G(3),(4)**. However, there was a subtle difference between bullet and knife damaged fibres. On several of the fibres from ballistic impact, portions of the cuticle (scales) of the wool had split off the tips of the fibres and had lodged further down the shaft of the fibre. This was not seen in any of the fibres following stabbing, **46G(5),(6)**. Other ballistic damaged fibres showed total rupture of the fibre, displaying the internal structure of the fibre end, or gouged out areas along the fibre path. The severance damaged fibre ends were clean and the edges were rounded.

In summary, the macroscopic and microscopic examinations of damaged clothing made it possible to distinguish between bullet penetration and knife severance in four of the five fabrics: weft-knit cotton, weft-knit polyester, woven polyester/cotton, and warp-knit nylon. SEM studies of fibre ends were successful in distinguishing between bullet and knife damage in only two of the fabrics: warp-knit nylon and weft-knit wool.

Observations at all three levels of resolution are thus of potential value in forensic studies. However, the nature of the damage and the possibility of identifying the cause is highly dependent on the type of fibre and possibly even more on the form of the fabric. It is doubtless also dependent on the details of the gun and the bullet or of the knife and on the particular form of attack. There is also the question of when and where the clothing is recovered. The examination of clothing may be particularly important where the remains are not found until some time after the attack, so that the pathological examination of the body is less revealing. However, changes may then have occurred due to environmental damage, which must be taken into account.

For the full potential of the examination of damage to fibres and fabrics to be realised in forensic studies, an extensive database needs to be established covering the enormous variety of fabrics and causes of damage due to criminal acts or other causes. Until such information is available, the only satisfactory option for forensic scientists is to compare observed damage with appropriate test specimens, which are realistically prepared on similar fabrics subject to relevant forms of simulated attack.

If opinions are going to be sought and given as to the cause of damage to clothing in cases similar to the Lindy Chamberlain or Ivan Millat trials, the textile expert must have knowledge of the range of variables pertinent to the case, which may influence or change the appearances of the fabric damage and the fibre end morphologies. Experts have got it wrong in the past. Further research may prevent this from happening again.

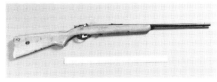

1

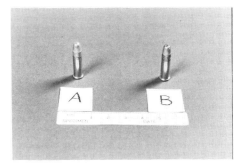

2

3

4

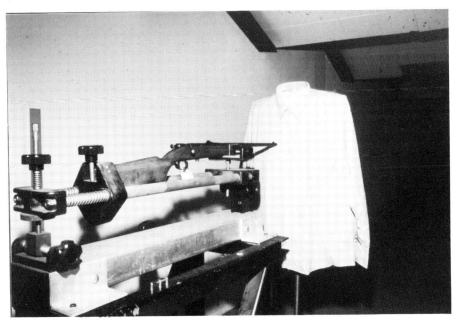

5

46A — Experimental facilities.
(1) *Lithgow* single shot .22 rifle. (2) Winchester superspeed bullets: A — solid point; B — hollow point.
(3) Serrated edged steak knife. (4) Rifle secured in the test rig with the simulated torso in front of the
muzzle. (5) One garment fitted over the simulated human torso.

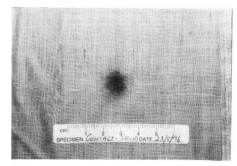

1

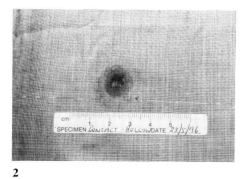

2

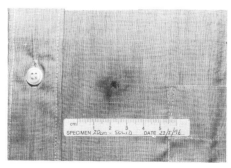

3

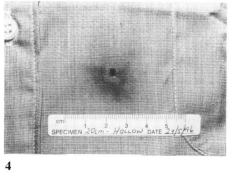

4

5

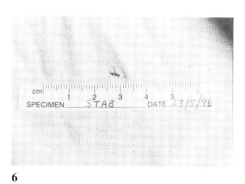

6

46B — Macroscopic observation of damage to woven cotton/polyester.
(1) Contact, solid-point bullet. (2) Contact, hollow-point bullet. (3) 20 cm, solid-point bullet. (4) 20 cm, hollow-point bullet. (5) Knife stab.
Macroscopic observation of damage to cotton weft-knit.
(6) Knife stab.   [Scale numbers are in cm.]

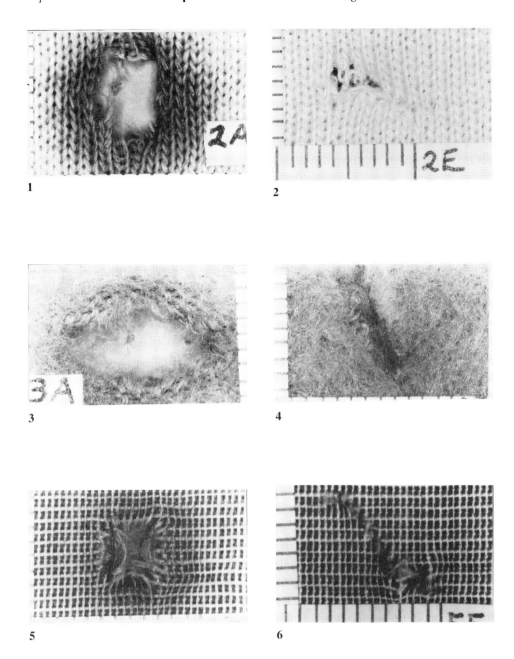

46C — Microscopic observation of damage.
(1) Cotton weft-knit: contact, solid-point bullet. (2) Cotton weft-knit: knife stab. (3) Polyester weft-knit: contact, solid-point bullet. (4) Polyester weft-knit: knife stab. (5) Polyester/cotton woven: contact, solid point bullet. (6) Polyester/cotton woven: knife stab.   [Scale marks in mm.]

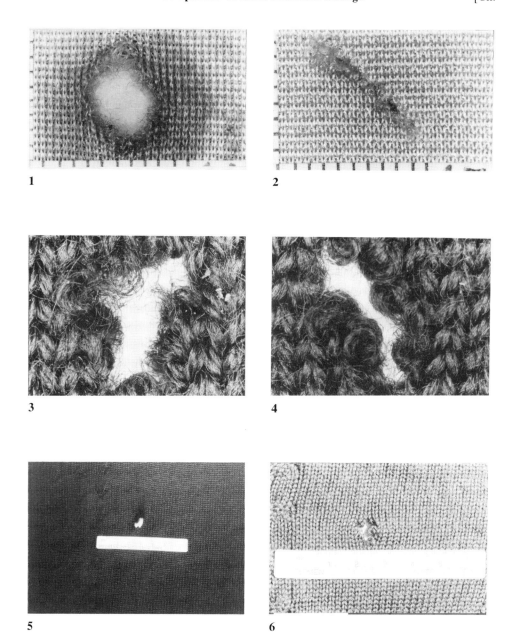

46D — Microscopic observation of damage (continued).
(1) Nylon warp-knit: contact, solid-point bullet. (2) Nylon warp-knit: knife stab. (3) Wool weft-knit: contact, solid-point bullet. (4) Wool weft-knit: knife stab.
Macroscopic observation of damage to wool weft-knit.
(5) Contact, solid-point bullet. (6) Knife stab.   [Scale marks in mm.]

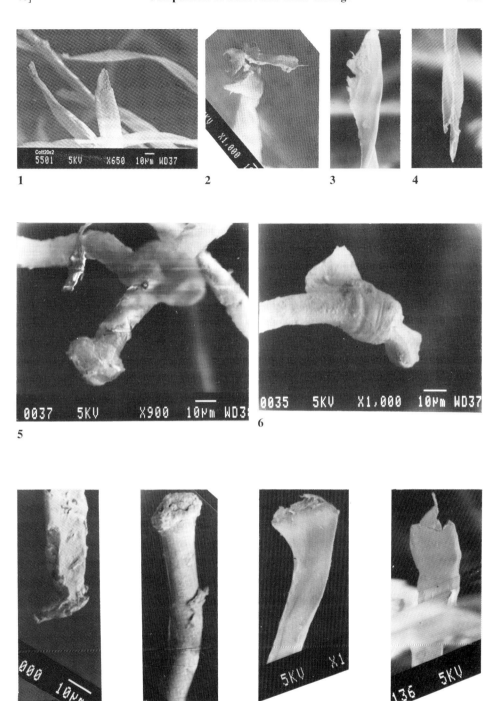

46E — SEM observations.
(1),(2) Cotton weft-knit, 20 cm, solid-point bullet. (3),(4) Cotton weft-knit, knife stab. (5),(6) Polyester weft-knit, 20 cm, solid-point bullet. (7),(8) Polyester weft-knit, contact, solid-point bullet. (9),(10) Polyester weft-knit, knife stab.

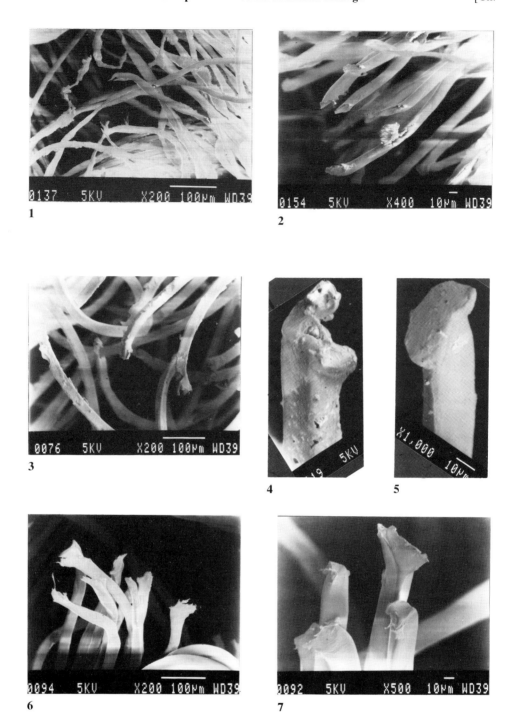

46F — SEM observations (continued).
(1) Polyester/cotton woven, contact, solid-point bullet. (2) Polyester/cotton woven, knife stab. (3) Nylon warp-knit, contact, solid-point bullet. (4),(5) Nylon warp-knit, 20 cm, solid-point bullet. (6),(7) Nylon warp-knit, knife stab.

1

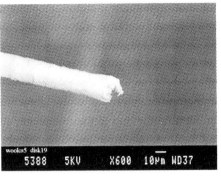

2

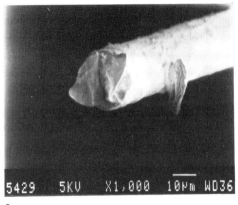

3

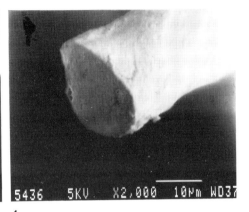

4

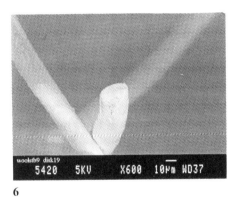

5

6

46G — SEM observations (continued).
(1),(2) Wool weft-knit, contact, solid-point bullet. (3),(4) Wool weft-knit, 20 cm, solid-point bullet. (5),(6)
Wool weft-knit, knife stab.

Part X
Medical applications

47

INTRODUCTION

Textile materials have always found medical uses, though usually in rather simple forms, such as bandages and dressings. More recently, some speciality textiles have come to be used in more demanding ways. According to Anand (1997), the medical uses of textiles can be categorised as follows:

- *non-implantable materials — wound dressings, bandages, plasters etc*
- *extracorporeal devices — artificial kidney, liver, lung etc*
- *implantable materials — sutures, vascular prostheses, artificial joints etc*
- *healthcare/hygiene products — bedding, clothing, operating room garments, wipes etc*

For disposable products, durability is of no concern and strength only has to reach minimum levels. Consequently the study of failure is not often required, though it may be needed when weak fibres, with some special properties such as wound healing, are used.

However devices and implants often need to operate for long periods subject to cyclic loading, whether due to the pumping of blood or the movements of joints. Here the study of damage and the lifetime before the product wears out are of great interest. Paradoxically, there are other situations where degradation after a certain time is desirable, because the body has grown new tissue to take over the function again.

There are a few relevant examples spread through the book. The polyester fibres in **8E(3)–(6)** were loaded with barium sulphate to make them opaque to X-rays for medical reasons. Damage to hair, as described in Chapter 19, may have medical as well as cosmetic implications. The overall fabric shown in **34G** was used in a virology laboratory. Toothbrush bristle was shown in **40H(3),(4)**. In addition, there is a link to biology through natural commercial fibres, cotton, wool and flax, which appear in many places in the book, and, more unusually, to thistledown and bacterial fibres in Chapter 21.

Because of the importance of the subject, a complete new Part has now been added. This starts with a detailed SEM examination of explanted textile ligaments from the knee and continues with some miscellaneous studies at UMIST and elsewhere.

48

FAILURE IN ANTERIOR CRUCIATE LIGAMENTS

Alan McLeod and William Cooke

Following considerable initial clinical success, the Surgicraft ABC ligament developed a pattern of performance where approximately 9% of the implanted prostheses developed signs of premature failure, i.e. a return to knee laxity or increased instability of the knee. A decision was then made to use the expertise available in UMIST to carry out SEM based fibre fracture analyses of the explanted remnants of failed ligaments, in the expectation that an understanding of the nature and cause of the break would lead to the development of a solution.

The anterior cruciate ligament (ACL) is one of the primary stabilising elements in the knee joint. The traumatic rupture of the ACL is a frequent sporting event resulting in knee laxity and episodes of the knee 'giving way'. The ABC prosthetic ligament has been in clinical use for the treatment of symptomatic deficiency of the ACL since 1985. The ABC ligament has twenty-four strands of a unit material comprising a partial polyester braid around a core that can be either polyester or carbon fibre. In order to be able to position the prosthesis in an anatomical position in the knee, it is implanted through a tibial drill hole, across the joint and up over the top of the lateral femoral condyle, **48A(1)**. Thirty ABC ligaments, which had ruptured in clinical use, were returned for mode of failure analysis. Of these eighteen had failed within the first year and eight had failed between two and four years post-operatively.

An explanted prosthesis, as received for examination, is shown in **48A(2)**. The analysis would start with a macroscopic examination with the object of identifying the point of rupture and relating that to the notes taken in the operating theatre when the ligament was explanted and the patient's history. Explanted ABC ligaments are covered by tissue ingrowth, **48A(3)**, which makes it impossible to examine the fracture morphology of the underlying polyester with an SEM. Therefore the tissue ingrown ABC ligaments were treated with a crude collagenase enzyme in an incubator; this breaks the bonds between the collagen cells allowing the tissue to be rinsed off exposing the artificial material, **48A(4)**. The entire tibial remnant and strands of unit material from the femoral remnant would then be mounted for SEM examination.

The ABC polyester ligament shown in **48A(2)** is a typical early failure, having ruptured spontaneously after only three months. This preliminary work established that the rupture was almost invariably located at the point where the prosthesis emerged from the tibial tunnel. The area of damage to the ligaments was extremely focussed. Within a few millimetres of the point of rupture, the artificial material was undamaged, **48A(5)**.

Another ABC polyester ligament, shown in **48A(6)**, failed six months after implantation. The macroscopic examination confirmed that rupture had occurred at the point where the ligament emerged from the tibial drill hole. After enzymatic cleaning, two distinct areas of damage were identified on the tibial remnant, **48A(7)**. Working from one side of the ligament, the point of rupture was at gradually increasing distances from the tibial end of the ligament up until the last three strands of ligament material which failed an additional 13 mm above the main rupture point.

In order to conduct a forensic mode of failure analysis, and so deduce the nature of the forces that cause rupture of prostheses, the investigation concentrated on the diverse modes of deformation, structural damage and breakage of the polyester filaments. Failures in the carbon fibres were uninformative, because, as shown later in **48E(3)**, they were always brittle tensile breaks. The pictures of polyester fibre damage in **48B–D**, which are categorised in the following sections, are a selection from the 30 explanted prostheses, and details of those which are illustrated are given in Table 48.1. In most cases, there was some peeling and fibrillation in addition to the main forms of damage.

CRUSHING, FLATTENING, SPLITS, FIBRILLATION

The SEM examination of the explanted material is of interest given the harshness of the situation from which the material has been taken. The prosthetic ligament, in an essentially aqueous environment at 37°C, is subjected to cyclic tensile loading varying from a negative load when the knee is flexed, to high peak loading in a sports engagement. Flexural and torsional cyclic loading is also present. High lateral loads can be present at the stress raiser created where the ligament emerges from the tibial drill hole.

The primary fracture morphology for any failure occurring within a year of implantation was flattening and crushing. A typical example is shown in **48B(1)**, where a flattened filament is looping over an undamaged filament.

In another, more unusual, example of a crushed filament, **48B(2)**, corrugations had been formed in the filament, which had been pressed against an intact bundle of polyester filaments.

Long axial splits, **48B(3)**, were frequently observed for early failures. Closer examination of the same filament revealed splitting and fibrillation within the main body of the filament, **48B(4)**. This internal splitting of a filament, **48B(5)**, creates points of weakness resulting in a fibrillated rupture **48B(6)**.

TENSILE FAILURE, SURFACE PEELING, TENSILE FATIGUE

Tensile breaks, **48C(1)**, were identified from material from six ligaments. However, tensile breaks were always a secondary fracture morphology. For example, with the ligament shown in **48A(7)**, the three projecting strands of material at one edge were found to have failed with tensile breaks as opposed to the remainder of the ligament, which failed with crushing and flattening. This pattern of fracture morphologies unambiguously demonstrated that rupture of the ligament was not the result of insufficient tensile strength for, if this was the case, the tensile breaks would have been present right across the ligament. Instead, failure was the result of focused mechanical damage working its way across the ligament until the last three strands were overloaded and failed with tensile breaks.

Some surface peeling, **48C(2)**, was identified to a greater or lesser degree in all the explanted prostheses. In general this is due to surface shear forces resulting from a lateral pressure on the prosthesis, although there is an indication that some surface peeling could result from adhesion between tissue ingrowth and the surface skin of the polyester filament, **48C(3)**. In an extreme form, the surface peeling can act layer by layer deeper into the filament, **48C(4)**. This layered surface peeling was noted at the level of the exit to the tibial drill hole for a prosthesis which had not ruptured but which had been explanted because of a reaction against wear debris released into the knee joint.

For two ligaments, a combination of a long surface peel ending in a tensile break, **48C(5),(6)**, was identified. This type of fracture morphology is associated with tensile fatigue (see **11C**). However, the surface peel could alternatively be the result of surface shear stresses and, if the ABC prosthetic ligament was subject to tensile fatigue, it is surprising that it could only be identified for two out of the thirty ligaments that were examined.

MULTIPLE SPLITTING FATIGUE, ROUNDED ENDS, KINK BANDS

For the group of ABC prosthetic ligaments which failed between two and four years after implantation, the fracture morphology identified to a greater or lesser degree on a majority of the implants was multiple splitting fatigue, **48D(1)**. The fatigue damage was not present throughout the ligament, but was isolated at the point of rupture. It has however, not been possible to establish whether or not the fatigue damage was the cause of the rupture or a result of the significant time lapse between the rupture and the explantation of the prosthesis. Both of the polyester filaments shown in **48D(1)** have fatigue damage with the more horizontal of the filaments having transverse splits indicating biaxial rotation compared to the more axially orientated splits of the other filament which indicates flexural fatigue.

The multiple splitting fatigue damage would often be present at multiple sites along individual filaments, **48D(2)**, culminating in either the characteristic brush ends seen in **48D(2)**, where the filaments have failed at fatigue damage sites, or in rounded ends, **48D(3)**, where further abrasive wear has occurred. In this example it is still possible to identify the cracks on the rounded end corresponding to the multiple splits of the original fatigue damage.

Multiple splitting fatigue was also identified in association with kink bands such as shown in **48D(4)**. The cracks of the multiple splitting fatigue can clearly be seen to start from a kink band split, **48D(5)**, and then follow a line of weakness from one line of kink bands to the next, changing direction slightly at each band. Kink bands were also identified on their own as developing bands and as bands opening into cracks, **48D(6)**.

Table 48.1 — Details of prostheses

Code **Plates**	Months *In vivo*	Location of failure of strands; main failure modes; other notes
EARLY FAILURES		
RI **A(2); E(3)**	3	most at ETDH; two pulled free of femoral remnant; C&F; spontaneous failure
TE **A(3),(4); B(3),(4)**	5	most at ETDH; rest about 10 mm back; C&F; ingrowth in patches
BO **A(6),(7); C(1)**	6	most at ETDH; 3 were tensile breaks, 13 mm above main rupture; C&F
HM **E(7)**	7	70 mm away from tibia, over lateral femoral condyle, where shielded from load, so no clinical reason for failure; sharp cut made during explant; no other damage; little tissue ingrowth; no information on why removed (may be clinical synovitis)
BB **C(4)**	8	removed due to adverse synovial reaction; mostly undamaged; some damage at ETDH; SP
DF **B(2)**	10	ragged over length of 12 mm; various; damage filaments away from point of break; return to instability after 2 months due to failure of prosthesis; removed 8 months later
TA (c/p) **E(6)**	10	C&F with fibrillation; heavy tissue ingrowth
BE **D(6)**	12	most at ETDH; 4 were 25 mm away at exit of femoral drill hole, used in operation in an attempt at isometric positioning; some strands failed at both locations; MS
HA **C(5),(6)**	20	ETDH; apparently tensile fatigue; no crushing or multiple splitting
RO(c/p) **B(1)**	20	only femoral remnant held together by tissue available; C&F; some MS&SP
LATE FAILURES		
SI (c/p) **C(2),(3); E(5)**	32	ETBD; heavily fragmented; MS and rounding; may be some high-speed breaks; rugby player's prosthesis, lasted 2 seasons, then broken after 27 months in heavy tackle; between failure and explant, tissue had formed a mat at femoral end
JO **D(1)**	33	prosthesis fragmented when received; much tissue growth; probably ETDH; tangled filaments with MS; crushing at femoral end probably occurred at removal of bollard
LE **D(3)**	35	ETDH; MS and rounding plus tensile breaks at femoral end
DI **D(4),(5); E(1,2)**	39	most at ETDH; MS&SP; no crushing; spontaneous failure
LA **D(2)**	39	most at ETDH, with MS; some projecting, 5 mm (1), 9 mm (2), 21 mm (1), tensile breaks

(ETDH is exit of tibial drill hole; C&F is crushing and flattening; MS is multiple splitting; SP is surface
peeling; c/p is carbon/polyester)

INTACT FIBRES, CARBON FIBRE, TISSUE INGROWTH, CUT FILAMENTS

Away from the point of rupture, there is occasional evidence of surface peeling from interfilament abrasion but the vast majority of the artificial material is undamaged. The undamaged polyester, **48E(1)**, and carbon fibre, **48E(2)**, are from the intra-articular portion of a
prosthesis that was implanted for thirty-nine months.

The carbon fibres used as the core element in some of the examined prostheses always
failed with a brittle tensile break, **48E(3)**, at the point of rupture of the ligament. This
repetitive brittle fracture, irrespective of the nature of the force that caused the rupture of the
prosthesis, effectively excluded the carbon fibres from the mode of failure analysis. However,
the carbon fibre was examined for any evidence of the fragmentation that has frequently been
associated with implanted carbon fibre. No evidence was identified of fragmentation away

from the point of rupture of the ligament, but there was considerable evidence of tissue ingrowth adhering to the carbon fibres, **48E(4)**.

The tissue ingrowth is more than just a surface covering; it penetrates first between the strands of unit material and then between the filaments of the polyester and carbon yarns. The ingrowth is illustrated in **48E(5)**, which is a cross-sectional view through the middle section of an ABC ligament which was ruptured in a fall 10 months after implantation. The core bundles of carbon fibre can clearly be seen; surrounding each core is the lighter polyester forming the partial braid. Filling the spaces between individual strands and individual filaments is the new tissue ingrowth.

An unexpected bonus of the SEM examination of the explanted polyester was the evidence of oriented tissue ingrowth, such as shown in **48E(6)**, in which a network of collagen fibres runs along the polyester filament. The ligament had been implanted for thirty-two months. The braided structure of the ABC prosthesis results in the polyester following the longitudinal axis of the ligament. The orientation of the collagen along the polyester indicates that the collagen will also be oriented along the longitudinal axis of the ligament.

The fracture morphology analysis was also able to identify damage that was unrelated to the rupture of the prosthesis. In particular an instance of the polyester having been cut with a scalpel, **48E(7)**, was identified. It is assumed that the cut was made during the explantation operation. The breaks are flat but unlike standard tensile breaks, they are at varying angles and are smeared into each other by the passage of the scalpel blade.

PATTERNS OF FAILURE

It was concluded from the SEM studies that early failures, within one year, resulted from crushing at the point where the prosthesis emerges from the exit to the tibial drill hole. Late failures, after two to four years, occurred at the same location, but the primary fracture morphology was multiple splitting fatigue, which is due to bending and twisting, though crushing was also found in some late failures.

There were some variations from this pattern of division into early and late failures, with failures, which occurred after more than two years having morphological features found in early failures. For example, in one carbon/polyester ABC prosthesis, an exact duplicate of the twisting football injury took place 28 months post-operatively. The rupture of the reconstructed ACL was clinically indistinguishable from the rupture of a natural ACL, with a full haematoma and pain. Examination of the ligament showed extensive tissue ingrowth, but also evidence that there had been widespread damage and failure of strands a considerable time before the final rupture of the reconstructed ACL. In particular, there was an increase in bulk of the carbon core, which would not have been possible if the polyester braid had been intact.

The mode of failure analysis of explanted prosthetic ligaments identified the point of rupture and the mechanical nature of the forces causing the ruptures. Working with clinical Research Fellows, a link was established between the mechanical damage and the impingement of the ligament resulting from a misplacement of the tibial drill hole. This led directly to the development of new instrumentation to objectively position and prepare the tibial drill hole. This has virtually eliminated the early failure of ABC ligaments.

STUDIES OF OTHER PROSTHESES

In addition to studies of ABC ligaments, a few other prostheses have been examined. The Leeds-Keio ligament is an open weave polyester tube, which is intended to act as a scaffold for new tissue ingrowth. Two ligaments were examined, one of which had been explanted within a year post-operatively and one after about two years. In the early failure, which had little tissue ingrowth, a substantial amount of the polyester had become detached, and the fibres showed crushing damage and some splitting and attenuation. The late failure was heavily ingrown by new tissue, which may have interfered with load sharing between the polyester strands, and the damage consisted of multiple splitting fatigue, including long axial splits and attenuation of fibres.

The Kennedy Ligament Augmentation Device (LAD) consists of nine strands of braided polypropylene and is used to augment autologous grafts. In one procedure, the surgeon routinely removes the LAD after 12 to 18 months, in order to avoid potential long-term problems due to artificial material in the knee. We examined a Kennedy LAD which had been used to augment a bone-patellar tendon-bone autologous graft. The LAD had been passed through a tibial drill hole and over the top of the lateral femoral condyle. After explantation, it was found to have ruptured at the exit to the tibial drill hole. There was no tissue ingrowth. SEM examination showed that the primary fracture morphologies of the polypropylene fibres were long axial splitting, fibrillation and multiple splitting fatigue, **48E(8)**. There was no sign of crushing. Although conclusions cannot be firmly drawn from one study, this does suggest that polypropylene is less fatigue resistant than polyester, which would explain the clinical reports of no statistically significant differences between augmented and non-augmented grafts.

The Graf Spinal Ligament System is an alternative to traditional metalwork prostheses. Two titanium pedical screws are linked by a band consisting of a flat braided polyester tube sewn together to make a ring. The primary advantage of the flexible system is that the surgery is less destructive, leads to earlier recovery and can be offered at an earlier stage of a degenerative condition. In bands removed from a patient after several months, there was little tissue attachment, due to the tightness of the braid. There were areas of damage on the insides of the bands and some strands had failed completely. The fracture morphology of filaments was that of crushing and flattening, with no evidence of splitting or fibrillation. The damage could be avoided if the band were positioned to avoid contact between the over-lapped section of braid and the bone screw.

Investigations of this sort are both time-consuming, in the painstaking dissection and examination of explanted prostheses, and require good collaboration with the medical team in order to maximise the information available. When this is done, the mechanisms of failure can be identified, and this, in turn, leads to ways of improving the choice and construction of material and the details of the surgical procedures.

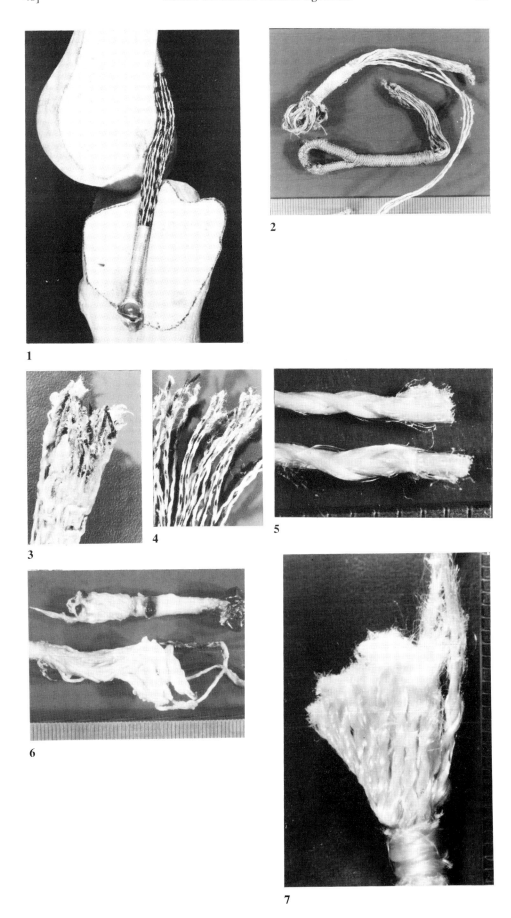

Plate 48A — ABC anterior cruciate ligament (ACL).
(1) The route taken by the ABC prosthetic ACL. (2) Prosthesis as received for examination: RI. (3) Prosthesis before cleaning: TE. (4) Prosthesis after cleaning: TE. (5) Localisation of damage: RI. (6) Tibial remnant of another prosthesis as received: BO. (7) Cleaned portion of prosthesis: BO.

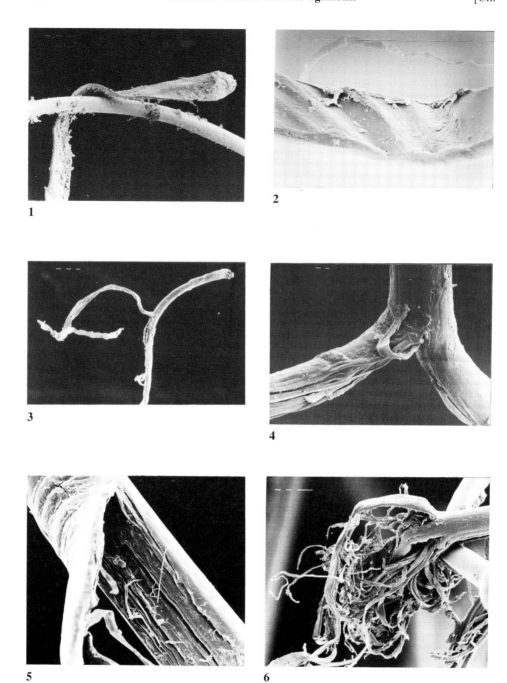

Plate 48B — Filaments from explanted ABC ligaments.
(1) Flattened filaments: RO. (2) Crushed filament: DF. (3),(4) Filament with an axial split at low and high
    magnification: TE. (5),(6) Internal splitting and fibrillation in a filament from another ligament.

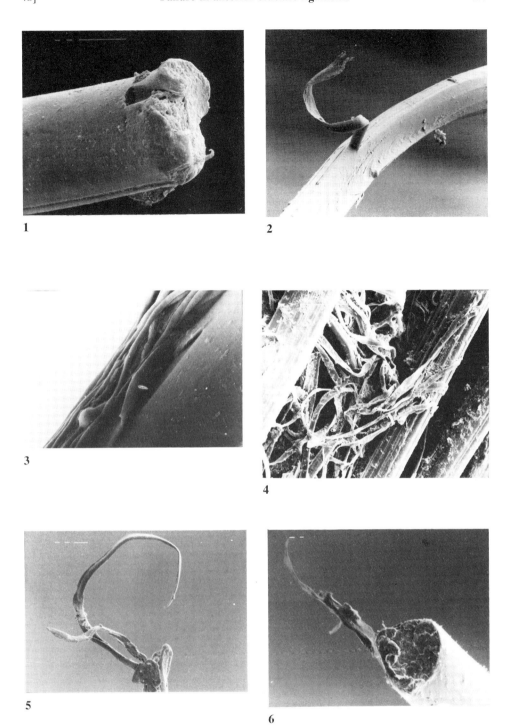

Plate 48C — Filaments from explanted ABC ligaments (continued).
(1) Tensile break: BO. (2) Surface peeling: SI. (3) Tissue adhesion with possible peeling: SI. (4) Layered surface peeling: BB. (5),(6) Possible tensile fatigue: HA.

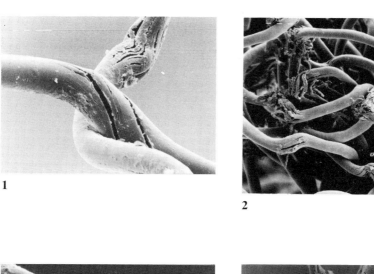

1

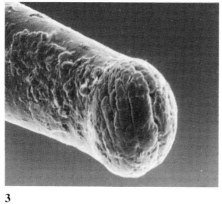

2

3

4

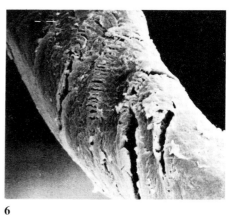

5

6

Plate 48D — Filaments from explanted ABC ligaments (continued).
(1) Multiple splitting: JO. (2) Multiple splitting and resulting breaks: LA. (3) Rounded end: LE. (4),(5)
Kink bands combined with multiple splitting: DI. (6) Kink bands opening into cracks: BE.

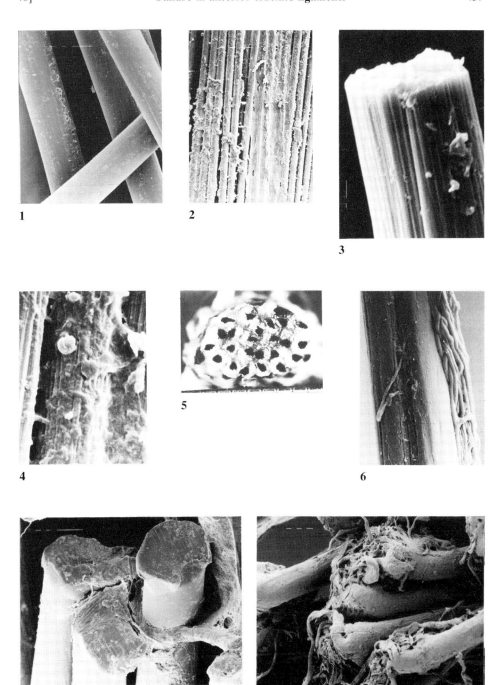

Plate 48E — Filaments from explanted ABC ligaments (continued).
(1) Hardly damaged polyester filaments from the intra-articular region away from the point of break: DI.
(2) Undamaged carbon fibres from the same region: DI. (3) Carbon fibre break.
Tissue ingrowth in ABC ligament.
(4) Tissue adhering to carbon fibres. (5) Tissue ingrowth within the prosthesis: SI. (6) Tissue ingrowth along the surface of polyester fibre: TA.
Other damage in ABC ligament.
(7) Polyester filaments cut by scalpel: HM.
Failed Kennedy LAD prosthesis
(8) Fibrillation of polypropylene fibres.

49

DRESSINGS AND IMPLANTS USING SPECIAL FIBRES

In addition to the use of regular commercial textile fibres in a variety of medical applications, special fibres with particular properties have also been developed. These may promote healing or be subject to biological degradation when only temporary inclusion within the body is needed.

A cut through a Kaltostat haemostatic wound dressing, composed of calcium alginate fibres, is shown in **49A(1)**, with the appearance of a cut fibre in **49A(2)**. A hole in such a dressing is seen in **49A(3)**. The fibre ends have rather irregular breaks, **49A(4)–(6)**, which indicate a mixture of shear splitting and tensile rupture. One example, **49A(7),(8)**, appears to be a tensile fatigue break, similar to those in Chapter 11.

Elizabeth Norton of the University of Cambridge has provided information and SEM pictures on studies of poly(glycolic acid) fibres. This material and its copolymers are semi-crystalline polymers which degrade by a hydrolytic mechanism. They are biodegradable and biocompatible allowing them to be used in biomedical applications ranging from resorbable sutures to controlled drug release devices. The exact morphology of these materials and the fundamental process of degradation is currently uncertain, Cohn *et al* (1987), but by gaining a better understanding of the mechanism of biodegradation it is hoped to develop new improved medical materials. Biodegradation of the surgical suture material, Maxon, which is a copolymer of poly(glycolic acid) and poly(trimethylene carbonate) described by Metz *et al* (1990), can be modelled using an *in vitro* system consisting of a buffered solution to maintain the physiological pH of 7.4, Reed and Golding (1981).

Undegraded Maxon has a very smooth, undamaged surface, **49B(1)**, but between 28 and 35 days of exposure to the solution long cracks begin to form in the direction of the fibre axis, **49B(2)**. On further hydrolytic attack circumferential cracking occurs, dramatically increasing with degradation time, **49B(3),(4)**. The material has then become weak and brittle fractures, as seen in **49B(5)**, can occur. The surface layer, which is rich in poly(trimethylene carbonate) can also flake off by a brittle fracture, exposing the interior, **49B(6)**.

In order to determine the effect of drying on the surface morphology of Maxon, environmental scanning electron microscopy (ESEM) was used. This technique allows wet samples to be studied, thus avoiding harsh sample preparation such as coating or drying. A degraded sample of Maxon was placed in the ESEM and slowly dried by increasing both the temperature and the vacuum. Initially, the sample appeared almost smooth with only a few longitudinal cracks. More severe cracking was seen to happen very rapidly on drying. The result depended on the rate of water removal, and **49C(1),(2)** show two pieces of the same material dried under different conditions.

This study shows that there is a major change to the surface morphology of biodegradable polymers on degradation, but the appearance is greatly affected by the techniques, such as harsh drying, used in SEM studies. The ESEM experiment indicates that there will be large differences between *in vivo* and *in vitro* conditions.

Although conventional metal plates and screws are beneficial in the early stages of fracture healing, they can cause long-term problems. This has led to research on biodegradable polymers for fracture fixation. Choueka *et al* (1995) have carried out a study of the degradation of glass fibres designed to be used as reinforcement in a composite with tyrosine-based polymers.

The fibres had a composition of 54% PO_4, 27% Ca, 12% ZnO, 4.5% Fe_2O_3, 2.5% $NaPO_3$, and were drawn from the melt by standard techniques of glass fibre formation. In order to study the effect of annealing, the fibres were tested in three forms: reheated to 420°C;

reheated to 250°C; and as first made. Chemical degradation was performed in tris-buffered HCl solution and chemically monitored. Tests were also performed in calf serum and in simulated body fluid. The fibres as made have a smooth, clean surface and showed almost no change in appearance for the first 60 days exposure to the physiologic solutions. Then, between 60 and 90 days, the fibres developed cracks and peeling away of a thin outer shell, **49C(3)**, but maintaining the overall structure of the fibre. All three forms behaved in a similar way.

In the tris-buffered HCl, the degradation was much more rapid and there was a marked difference in the modes of degradation. Those that were not annealed or were treated at 250°C delaminated in a way that destroyed the overall integrity of the fibre, **49C(4)**. This is particularly clearly seen in the cross-section in the middle of **49C(5)**. Fibres treated at 420°C developed craters on the surface, **49C(6)**, but the fibre remained intact.

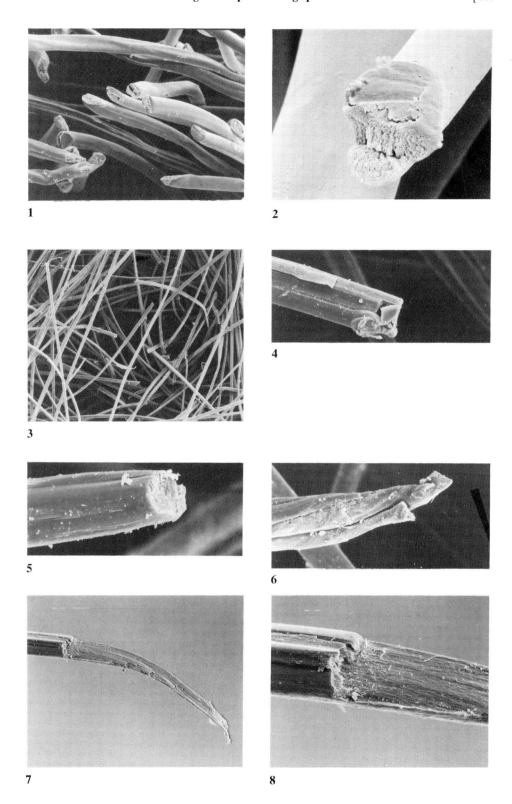

Plate 49A — Calcium alginate fibre wound dressing.
(1) Cut portion. (2) Cut fibre. (3) Hole in dressing. (4)–(6) Broken fibres from hole. (7),(8) Apparent tensile fatigue failure.

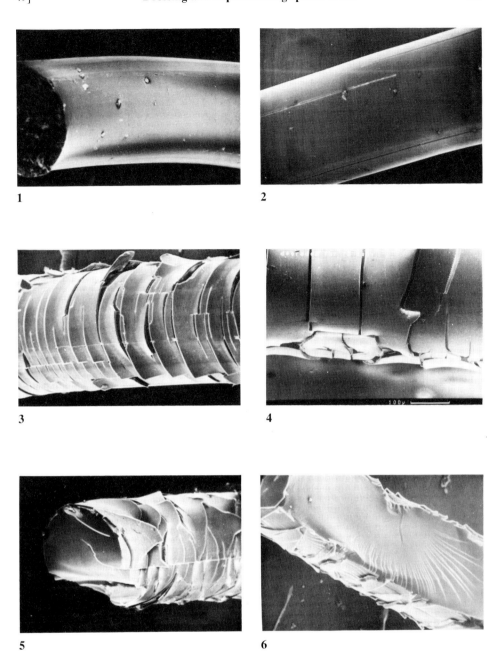

1

2

3

4

5

6

Plate 49B — Studies of Maxon suture subject to hydrolytic degradation, courtesy of Elizabeth Norton, University of Cambridge.
(1) Fibre as received; on the left is the end of the suture. (2) After about 30 days. (3)–(6) After 73 days.

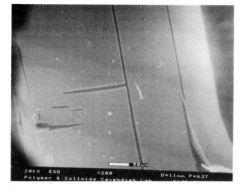

1

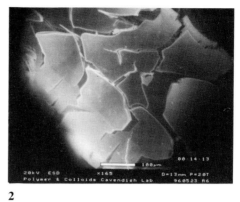

2

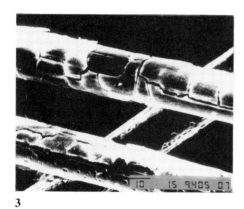

3

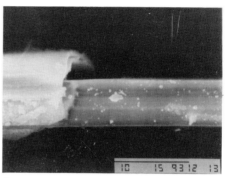

4

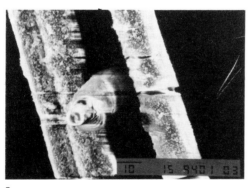

5

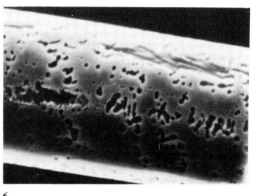

6

Plate 49C — Studies of Maxon suture subject to hydrolytic degradation, courtesy of Elizabeth Norton, University of Cambridge (continued).
(1),(2) Two samples dried under different conditions in an environmental SEM.
Degradation of calcium phosphate fibres used in absorbable implants, courtesy of Jose Charvet, Hospital for Joint Diseases Orthopaedic Institute, New York.
(3) Heat-treated fibres after 90 days in calf serum.
After 2 days in tris-buffered HCl.
(4) Non-heat-treated fibre. (5) Axial and cross-sectional views of non-heat-treated fibres. (6) Fibre annealed at 420°C.

Part XI
Conclusion

50

THE DIVERSITY OF FIBRE FAILURE

The pictures in this book really speak for themselves, and this has been augmented by explanations in the text. Only brief summary remarks are appropriate in this last chapter. The overwhelming impression of thirty years of detailed research on fibre breakage is of great diversity of form, but we can now see a pattern emerging.

Although the classification is still a bit ragged at the edges and will be improved by further research, we have identified 18 types of fibre breaks, or fibre ends resulting from other causes, as described in Chapter 1 and shown in Fig. 1.5. The classification resulted mainly from the tests on single fibres, which are reported in Chapters 3–21, but was also influenced by some supplementary observations. In laboratory tests the factors causing break are well defined, but the same types of break can also be found in the case studies of wear in use in the later chapters. In Table 50.1 we have listed the occurrence of these classified forms of broken and other fibre ends, as they appear in plates throughout the book; and some other effects in fibres are listed in Table 50.2. It must be noted that the information in the tables is limited, and reference should be made to the detailed account in the referenced chapter for full details: for example, some of the examples listed are variant forms of the classified breaks, and others occur in fibres in unusual circumstances or after degradation.

Table 50.1 — Classified breaks and other fibre ends (see Chapter 1 and Fig. 1.5)

Note:
  (i) Where four or more pictures in a plate are referenced, the individual numbers are not included, e.g. **4B** and not **4B(2),(3),(5),(6)**; but otherwise individual pictures are identified, e.g. **4C(1),(3)**.
 (ii) * — see other forms of splitting *, and comment in text.
(iii) References up to Chapter 21 indicate failure in scientific experiments on single fibres, but references after Chapter 22 show failures in textiles in testing, processing or use.

(A) TENSILE FAILURES

Type 1: brittle (not always easily distinguished from granular)

carbon	**4C(1),(3)**
carbonized acrylic	**35G(2),(4)**
carbonized rayon	**35F(5),(6)**
ceramic	**4B**
cotton	**29D(5),(6) 32D(4),(5),(6) 35C(4)**
glass	**4A 17A(1) 26C(1),(2) 26H(7)**
Monvelle	**21C(3)**
nylon	**16B(4) 16D(4)**
polyester	**16B(5) 16F(2)**
rayon	**34E(2),(3),(4)**
silicon carbide	**21C(4),(5),(6)**
silk	**43A(5),(6)**
spandex (Lycra)	**4C(4),(5),(6)**
wool	**43B(4),(5)**

Type 2: ductile

Monvelle	**21C(2)**
nylon	**5A 5B 5C 5D 5E(4) 5F(1),(2) 6A(3),(4) 12E(2) 16A 16C(1) 16D(3) 17A(2) 24A 24B(2)**
polyester	**5B(5),(6) 5E(1),(2),(3) 5F(3),(4) 21B(1),(2)**
polypropylene	**5E(5),(6)**

Type 3: mushroom (high-speed)
nylon   6A 20A(2) 24A(5),(6) 24B(3) 31C(2),(3) 37A(1),(2) 40F
                     40H(5) 45A 45B 45C 46F
Qiana   31F(3)
polyester   6A(6) 24G(6) 24H(3a) 26B(3) 37A(3) 37C(1) 39G(5) 46E

Type 4: axial split*
HMPE   7C
Kevlar   7A 7B 11E 14A(2),(3) 14C(6) 21D(1) 40H(6)
nylon   7C(5),(6)

Type 5: granular (not always easily distinguished from brittle)
acetate   8A(4),(5),(6) 17C(3),(4)
acrylic   8B 8C 8F(1),(2) 12F(3),(4) 17B 23A(4)
alginate   21F(1),(2),(3)
alumina   8D(4) 8H
bacterial fibres   21G
carbon   8D(5),(6) 8G 26H(8) 26I 40D(5)
cotton   18B(1),(2),(3) 31G(6)
flax   21A(3)
hair   19B 19C 19H(5)
jute   21A(1),(2)
Nomex   21B(3),(4)
nylon   16B(1),(2),(3) 16C(1) 16D(3),(5),(6) 26A(5) 39H
PBI   8D(3)
polyester   8F 16B(6)
PVA   8D(1),(2)
PVC   21B(5)
seed-fibre   21A(4)
silk   21A(5),(6) 31C(6)
Tencel   8E
wool   19A 23K(5) 43B(6)
rayon   8A(1),(2),(3) 12G 17C(1) 25E(3) 28E(6) 30C(6) 33F(3),(6)

Type 6: independent fibrillar*
cotton   9A
PTFE   21B(6)

Type 7: stake-and-socket
hair   19D(2),(3),(4) 19J(1),(2) 19J(1),(2)
nylon   39J(1),(2)
polyester   16D(1),(2) 16E 26B(5),(6) 34G(4),(5),(6)
polypropylene   40K(2)

FATIGUE FAILURES

Type 8: tensile fatigue*
nylon   11A 11B 24I(4),(5) 31A(6) 40G(4)
Nomex   11D(1),(2),(3)
polyester   1A(4),(5),(6) 11C 24H(3b)
polypropylene   11D(4),(5)

Type 9: bending — kinkband failure
acrylic   12F(3)
aramid   39M 39N
flax   39K(5)
HMPE   39O
nylon   12B 40G(3)
polyester   12A 39L(3)
wool   33H(7) 33I 33K 33L(1),(2)

Type 10: bending — single or multiple split*
acrylic   12F 31B(7),(8),(9) 33D(6)
aramid   39M 39N
hair   19F(6)
HMPE   12H(5),(6) 12I 39O
Kevlar   21D(5)
nylon   12B(1),(3),(5) 12C(1) 12D(1),(2) 12E(2) 25G(1),(2)
                     31A(2),(4),(5) 40B(2),(6)
polyester   12A(4),(5) 12C(2) 12D(3),(4),(5) 12E(3) 25E(5),(6) 25G(6)
polypropylene   12F(5),(6)
silk   31F(5),(6)
wool   19F(1),(3),(4) 28A(6) 33J 33L(4),(5),(6) 33M

Type 11: biaxial rotation — multiple split*
acrylic	13D(4) 33D(5)
cotton	18C 24C(2),(3),(5)
hair	19D(5),(6)
Kevlar	21D(6)
modacrylic	13D(5)
nylon	13A 13B(1),(2),(3) 13D(2),(3) 16C(4) 29E(4),(5),(6)
polyester	13A(4) 13B(5),(6) 13C 13D(1) 16C(3) 25D(5) 28E(3)
	29E(3) 31D(5),(6) 34F(3)
polypropylene	13B(4)
PVA	13D(6)
wool	19E 33B(2),(3)

Type 11a: multiple split — torsional fatigue*
nylon	17A(5),(6)
polyester	17D

Type 11b: multiple split: uncertain combinations of bending and twisting fatigue*
acrylic	23C(4) 31B(2),(5),(6) 33D(2),(3),(4)
cotton	23A(7) 25A 25C(1),(4) 27L(8) 29C(3) 30B(5),(6) 30D(4)
	31D(2),(4) 31G 32A(2),(4) 32B(5) 34B 34C(2),(3),(6)
	34D(3),(4) 35A(2),(4) 35B 35C(2),(3),(5a) 40I(6)
flax	20I(b),(e),(h) 42A(5)
hair	19H(1)
Kevlar	21D(3),(4)
Nomex	35D(5),(6)
nylon	24B(6) 24D(2) 24E(1) 24F(3) 25I(3),(5) 29F(2),(6)
	33G(5),(6) 38A(2),(3),(6) 38B(6) 40F(6) 40J(2),(3),(4)
polyester	24F(6) 24H(5a) 25F(3),(4),(5) 26B(2) 28B(5) 28E(2),(4)
	29A(5),(6) 29B(2),(3) 30B 30C(3),(4),(5) 34D(3),(4)
	34E(2),(3),(6) 34F(2) 34G(2) 34H(5),(6)
polypropylene	39I(5) 40K(2)
silk	42B(5)
unknown	28A 30B
wool	25B(4) 25D(1),(4) 28A(4) 28B 28C(4),(5),(6) 28D(3),(4),(5)
	28F(5),(6) 30A(2),(4),(5) 33A(2) 33B 33C(6) 33G 42C

Type 12: surface wear
acrylic	14E
cotton	14A(5)
hair	19D(6) 19F(5)
Kevlar	14A(1),(2),(3) 14C(5),(6)
nylon	12E(5) 14B 14D
polyester	12E(6) 14C
rayon	12G(4) 14A(6)
wool	14A(4) 19E(5) 19F(1),(2) 25B(6) 25D(1),(3) 28D(1),(2),(6)
	28G 30D(1),(2)

Type 13: surface peel and split*
cotton	23K(1) 25A(5) 32B(6) 35A(6)
flax	42A(6)
Nomex	35D(1),(2) 35E(1),(2),(4)
nylon	14B 24F(1),(2) 24I(2),(3) 25H 25I(4) 26A(4) 29F(4),(6)
	31A(6) 31E(4),(5),(6) 38A(2) 38B(3),(4) 38C 39A
	39B(4),(5),(6) 39C 39D 39E 40B(2),(3) 40E(3) 40G(1),(2)
	40H(4)
polyester	14C 23K(1),(4),(5) 24F(4),(5) 25D(6) 25E(1),(2) 25F(4),(6)
	29B(4) 34H
polypropylene	33E(5),(6)
rayon	32A(5),(6) 32D(2)
silk	31C(5) 42B
wool	25D(3)

Type 14: rounded
acrylic	33D(6)
cotton	25A(6) 25C(4) 31D(2) 34C(2) 35C(5b)
nylon	29E(4),(5) 29F(3) 38A(6) 40B(6) 40H(4)
polyester	25F(5) 29E(2) 31D(6) 34D(3),(4)
rayon	28E(5)
silk	42B(5),(6)
wool	23G(2) 28A(5) 28B(4),(6) 28D(3),(4) 28F(6) 30A(6) 30D(3)
	33M(6)

Type 15: transverse pressure — crushing, scraping, blunt cut, impact, etc.

acetate	20C(2)
acrylic	20C(4) 20F(5),(6) 23A(2),(3) 44A(5)
cotton	20B(3),(4) 23A(7) 25C(5) 28F(3) 34B(3),(4) 34C(3),(4)
flax	20I(c),(f),(i)
Nomex	35E(6a)
nylon	20A(2) 20G 23B(5) 24B(4),(5),(6) 24E(2),(3),(6) 25C(6)
	25H(5) 25I(5) 33C(4) 33E(1) 33G(3) 37B(1),(5),(6) 37C 37D
	38A(4),(6) 38D 38E(3),(6) 38F(2),(3),(6) 39D(6) 39I(4),(6)
	40F(3),(4) 40H(3) 45A 45B 45C
polyester	20A(6) 20F(4) 23A(1) 23B(5) 24D(4) 24G 37B(3) 39L(5)
polypropylene	40G(6)
rayon	20B(6) 25C(3) 33F(1),(2)
unknown	23B(3a) 35C(7)
wool	19G(1) 20C(6) 23B(6) 23I 23J

Type 16: sharp cut

acetate	20C(1)
acrylic	20C(3) 20F(3),(7)
aramid	20G(1),(2)
cotton	20B(1),(2)
glass	26C(3),(4),(5)
nylon	20A(1),(3),(4) 37B(4) 45A 45B 45C
polyester	20A(5) 20F(1),(2) 23B(4)
rayon	20B(5)
wool	20C(5) 23G(1)

Type 17: melt — bulbous ends and other melting effects

acrylic	20E(4)
cotton	20F 25J(3)
nylon	20D(1),(2) 21E 37A(5) 38F(2) 39F 39I(6) 40J(1)
polyester	20E(3) 23B(1),(2) 24E(3) 24G(3),(4) 24H(4),(5),(6)
	25J(5b),(6) 25K(3),(4),(6) 29A(1) 34D 34E(1) 34F(4),(5),(6)
	37A(4),(6) 39G
rayon	20F(1),(2)
unknown	23B(3b)
wool	20E(5),(6) 25K

Type 18: natural fibre ends

cotton	23A(5),(6)

Table 50.2 — Other effects in fibres

Note: *fibre splitting — see note in Table 50.1

1.   Adhesion between fibres

nylon	39E(3) 39H(3)

2.   Characteristic cotton break: split and tear*

cotton	18A 18B(4) 26A(3) 29D(4)

3.   Fibrillation*

cotton	18C(3) 24J 29C(2) 29D(2) 32B(2),(6) 34C(6)
flax	38G(3),(4) 42A(2),(4)
hair	19C(6) 19I(6)
Kevlar	21D(3),(4) 23D 26G(1) 39I(1),(2),(3) 39Q(2),(3)
	39R(1),(3),(4)
Nomex	35D(1),(2) 35E(1),(2)
nylon	12D(6) 39A(3),(6) 39B(4),(5) 39C(1),(5),(6) 39D(2)
silk	30E 42B
Technora	26G(1),(2),(4)
Tencel	21F(7),(8)
Vectran	39R(5)
wood-pulp	23D(1),(2)

4.   Granular break in flex fatigue

acrylic	12F(4)
rayon	12G

5.   Kinkband (without break) — see also Table 50.1 (type 9) and 50.4 (3)

aramid	39M 39N
HMPE	39O 39P

Kevlar	7B(3)
nylon	33C(5)
polyester	12A(1),(2)
rayon	12G(2)

6. Melt-tail

nylon	39F(2),(3) 40F(5),(6)
polyester	37A(3),(4) 39G(4)
spandex	4C(6)

7. Partial break — not including many examples of multiple splitting

acetate, tensile	8A(5)
acrylic, tensile	8B(4)
Kevlar, buckling	21D(2)
nylon, flex fatigue	12A(3),(4),(5)
nylon, tensile	5A(2)
nylon, tensile fatigue	11A(3),(4),(5)
polyester, flex fatigue	12A(3),(4),(5)
polyester, tensile	5B(5),(6) 5F(3),(4)

8. Sheet peel

cotton	24C(6) 29C(2) 30D(5),(6) 30E(1),(2) 31E(3) 32A(1),(2),(3) 32C 32D(1),(2),(3) 34A 35B(3)
flax	42A(2) 42E(1),(2)
HMPE	12H(4) 12I

9. Snap-back, coiling and other recovery after break

Kevlar	7B
Monvelle	21C(1)
nylon	12E(1) 25H(3),(5) 38B(4) 38C(2),(4) 39C(1)
polyester	6A(6) 14C(2),(4) 24D(4) 24G(3),(5),(6) 24H(2)

10. Step and axial split in tensile failure*

cotton	18B(2) 23K(6)
flax	21A(3)
hair	19B 19C 19I(1)
jute	21A(1),(2)
polyester	21B(2)
wool	19A(3),(4),(5)

11. Surface and transverse cracks (various causes)

acetate	8A(4),(5),(6) 17C(3),(4)
acrylic	17B(1),(2)
cotton	29D(5)
flax	42D(6)
glass	26D(3),(4)
Kevlar	21D(2)
nylon	12E(4) 16A(6) 33E(3)
polyester	16C(2) 16E(1) 16F(1) 16G(5) 24H(5),(6) 33E(4) 34G(1b),(2) 40C
polypropylene	40H(1),(2) 40K(4),(5),(6)
rayon	8A(1),(2),(3) 17C(2) 33F(4),(5)
wool	43A(4)

12. Tensile fatigue split*

acrylic	11D(6),(7),(8)
Kevlar	11E
wool	19D(1)

13. Transverse stress: axial split*

carbon	26I(1),(4)
Technora	26G(1),(2)

14. Twist anomaly

cotton	18C(4)

15. Twist split*

acrylic	17B(4),(5),(6)
cotton	18C(1),(2)
Kevlar	17C(5),(6)
nylon	5D(6) 17A(3),(4) 24B(5),(6)

16. Void formation

nylon	5A(3),(4) 21E(4),(5)

Fibre failure by splitting, which is a common mode of breakage in the wear of materials in use, is a difficult form to identify, because similar effects can arise from different causes and are classified as different types. The single-fibre test, which most clearly leads to multiple splitting breaks, is the biaxial rotation fatigue test, discussed in Chapter 13 and listed as type 11. The fibres are subject to cyclic bending plus an imposed torque. However, either single or multiple splitting can occur as a result of flex fatigue and the associated shear, without any twisting, as discussed in Chapter 12 and listed as type 10. There are differences in appearance: biaxial rotation fatigue splits are usually short, typically about five fibre diameters in length, and are twisted in opposite directions on either side of the centre point; whereas flex splits are usually much longer and have no twist. Where we can assign a break with reasonable confidence to one of these types, either from the appearance or because the test conditions are known, they are so listed in Table 50.1. Multiple splitting can also occur through pure torsional fatigue, although this has not been much studied: one example is listed as 11a in Table 50.1. Where we cannot be sure of the cause of multiple splitting, but it is regarded as due to some combination of bending and twisting fatigue in fibre asemblies, the examples are listed in 11b in Table 50.1.

But splitting can also occur in other circumstances, and all the forms of splitting are marked * in Table 50.1 and 50.2. Some types of fibre split very easily, and some types of stress promote splitting. Where the lateral intermolecular forces are weak in oriented fibres, a simple tensile failure leads to axial splitting: type 4 in Table 50.1. An extreme variant of this effect is the independent fibrillar break: type 6 in Table 50.1. Sometimes, shear stresses cause separate tensile breaks to join by an axial split: type 10 in Table 50.2. The shear stresses in twisted fibres intensify the tendency to split, and examples are as listed in Table 50.2 (12); a special case, involving untwisting at a reversal point, is the characteristic break of cotton, Table 50.2 (2). Transverse stresses can also cause axial splitting: type 13 in Table 50.2. The characteristic tensile fatigue break in melt-spun fibres, type 8 in Table 50.1, gives single, or occasionally multiple, splits; in other fibres splitting is the dominant response to tensile load cycling, type 12 in Table 50.2; and surface shear forces can also give rise to forms of splitting, listed among type 13 in Table 50.1. Finally fibrillation, Table 50.2 (3), is really an extreme form of multiple splitting.

In the interpretation of the reasons for splitting in fibres, it is therefore necessary to take account of the nature of the fibres, the circumstances of failure and the appearance of the break.

The other tensile breaks in Table 50.1 are easily identified, and are discussed in the relevant chapters. Kinkband breakage in bending fatigue is also a well-defined form, discussed in Chapter 12. Wear, type 12 in Table 50.1, is really a macroscopic manifestation of the detailed effects of surface peeling, listed as type 13.

Among the miscellaneous group, rounding is another macroscopic consequence of continued wear. Type 15 in Table 50.1 comprises a varied and poorly defined collection of failures resulting from high lateral pressures, either relatively localized, as from a blunt knife, or more distributed. In contrast to this, highly localized pressure by a sharp blade cuts through the fibre and causes type 16.

Melting, listed as Table 50.1 (17), includes both the bulbous ends of single fibres, shown in Fig. 1.5, and less localized effects in fibre assemblies.

The effects listed in Table 50.2 are rather varied. Some, namely (2,4,10,12,13 and 15), are specialized forms of fibre break, which should perhaps have been included in the main classification. Others, namely kinkbands (5), partial breaks (7), cracks (11), the twist anomaly in cotton (14) and void formation (16), show intermediate stages of deformation or damage, prior to break. Fibrillation (3) and sheet peeling in cellulosic fibres (8) are ways in which the material of a fibre can disintegrate. Snap-back (9) and the pulling out of melt-tails (6) are events which happen to fibres after break has occurred. Adhesion (1) is an effect between fibres in contact under heat or pressure.

Examples of the influence of external factors, such as various forms of degradation or attack, are listed in Table 50.3.

Table 50.3 — Special external factors

1. Biological degradation

flax	43C(2)
wool	19G 43C(1),(3)

2. Bites: insect or animal

nylon	37D 38D
wool	19G(1) 43C(4),(5)

3. Burning or carbonization

cotton	20F 25J(2a),(3),(4) 43B(1),(2),(3)
rayon	20F(2)

4.   Chemical attack: during or prior to test
glass   26D(3),(4) 49C
nylon   5C(2) 5D(2),(3) 16C(1),(4),(6) 23C(5),(6)
polyester   16C(2),(3),(5) 16D(1),(2) 16E 34G(1b),(2)
polyglycolic acid   49B 49C
polypropylene   40K

5.   Photodegradation
cotton   32D(4),(5),(6)
nylon/elastomer   40A 40B
nylon   16A 37B(2)
polyester   16F 16G 40C
polypropylene   40H(1),(2)
wool   16H

6.   Radiation damage
nylon   16D

7.   Soiling or deposits
cotton   34C(4)
Kevlar   39I(2),(3)
nylon   38C(6) 39B(3),(6) 39E(6)
polyester   24E(4)
wool   43C(6)

8.   Thermal damage
nylon   16B 38E(6) 39F(4),(5),(6) 39H
polyester   16B(5),(6) 16D(1),(2) 24H(5),(6)

9.   Smearing on surface of fabric or yarn
cotton   24J(4) 25A(5) 25C(2),(5) 34B(3),(4) 34C(3),(4)
flax   42A(3)
nylon   25H(2),(4) 38B(1),(2) 38C(1),(5) 39C(3) 40J(4),(5)
paper   25B(2)
polyester   25G(4),(5)

10.   Yarn breaks
cotton   24C(1),(4)
nylon   24B(1) 24D(1)
polyester   24D(3)
polypropylene   40K(1)

From an academic viewpoint, a lesson to be learnt from the diversity of forms of break is that it is wrong to try to explain fibre fracture or fibre fatigue in general terms. The particular type of fibre, its history, and the precise test method must be specified. We have been able to give a number of qualitative explanations of different forms of fibre failure. However, a full theoretical development of the subject will need detailed stress or deformation analysis, and this must be related to the polymer physics of fibre structure, which is itself not yet well documented for many types of fibre. Except in a few special cases, such as the brittle fracture of glass fibres, the problems are difficult for at least three reasons: firstly, rupture frequently results from the occurrence of a complex combination of stresses; secondly, large strains commonly develop over distances which spread out widely from cracks; thirdly, most fibres are composed of materials which are anisotropic and non-linear viscoelastic. An understanding of the essential morphology of the failure, which has been opened up by the studies in this book, is necessary if the right approximations are to guide the analysis.

An understanding of single fibre failure is only part of the story. Fibres are used in assemblies containing millions of fibres or fibre elements. The mechanics of stress distribution within the assembly also needs to be understood, although no more than some qualitative and descriptive comment has been appropriate in this book. But once again the observation of the forms of failure, reported in Chapters 22–43, will provide guidance to future theoretical work, as well as giving immediate practical information on the nature of wear in textile materials and their durability. Some effects in fibre assemblies, involving many fibres, are listed in Table 50.4. They range from the appearance of yarn breaks, through indications of the progressive breakdown in fabrics, to delamination in composites.

From a practical viewpoint the most important conclusion from the studies of wear in use is that the commonest form of failure, in most types of fibre in most applications, is multiple splitting leading to a bushy end, with a subsequent rounding in further wear. Such failures are caused by flexural and torsional fatigue of fibres, although, as discussed above, it is not always possible to identify the particular combination of bending and twisting which leads to failure; and, sometimes, there can be confusion with splitting due to other causes. Other common causes of failure in use are surface peeling and the crushing or mangling effects of transverse pressure.

Table 50.4 — Effects in structures

1. Adhesion loss and delamination in composites	
carbon/epoxy	**40D(4),(6)**
carbon/nylon	**26E(1)**
carbon/PEEK	**26F 26H(4),(5),(6) 26I(6)**
glass/nylon	**26E(3),(4)**
glass/PET	**26C(4),(5) 26E(4),(6) 26H(1),(2),(3)**
nylon/elastomer	**26A(1),(2) 40A(2),(3),(4)**
Technora/epoxy	**26G(5),(7)**

2. Coating failure	
nylon/elastomer	**26B(1)**

3. Fabric damage	
ballistic impact	**40L 46B 46C 46D**
braid break	**24H(1) 24I(1)**
break, hole or severe wear	**28A(1) 29B(1) 29C(1),(4) 29D(1) 31A(1) 31B(1),(4) 31G(1) 32D(4) 33A(1) 34C(1) 34F(1),(4) 35A(1) 35F(2),(3),(4) 35G(5)**
broken fibres in interstices	**25M(1),(2) 35C(1)**
knife cuts	**46B 46C 46D**
tear	**25L(5),(6)**
wear on yarn crowns	**25A(1) 25G(3),(4),(5) 28E(5) 29C(5) 29F(5) 34D(2) 34E(5)**
yarn breaks	**25L**

4. Kinkbands	
composite	**26G(6)**
yarn	**39E(1)**

5. Pilling	
acrylic	**31B(3)**
cotton	**25L(7) 31D(1b) 31E(2a)**
cotton/PET blend	**29E(1) 30B(1) 30C(1),(2)**
nylon	**31C(1)**
Qiana	**31F(2),(3)**
wool/PET blend	**28C(3),(4)**
wool	**30A(1)**

6. Plastic flow in composite matrix	
carbon/epoxy	**26H(1),(2),(3)**
carbon/PEEK	**26F(3),(5) 26H(4),(5),(6)**
glass/PET	**26D**
Technora/epoxy	**26G(5)**

Almost all the examples of tensile failure in Table 50.1 are from the chapters in Parts II and IV of the book, and occur in laboratory tests. Textiles rarely fail in use as a result of the direct application of too large a load, and so the infrequency of tensile breaks is not surprising.

As with single fibre breakage itself, the deformation and sequence of damage within fibre assemblies is diverse and complicated. In order to try and develop understanding, it is necessary to be specific about the product and how it is used. Careful study, starting with little-worn regions and going on to locations of failure, can then elucidate in qualitative terms the mechanisms of wear and ultimately lead to predictions of durability.

APPENDIX 1
SOURCES

UMIST: DEPARTMENT OF TEXTILES

Most of the work described in this book comes from research in the Department of Textiles, UMIST, under the direction of Professor John Hearle. It started with the purchase of a scanning electron microscope with a grant from the Science Research Council in 1967, together with five-year funding for an experimental officer and a technician. Since 1972, the staff have been supported by general UMIST funds; a second grant from SERC enabled a replacement SEM to be bought in 1979; industrial sponsors, listed below, have contributed through membership of the Fibre Fracture Research Group; special research grants have been made by the Ministry of Defence (SCRDE, Colchester, and RAE, Farnborough) and jointly by the Wool Research Organization of New Zealand (WRONZ) and the Wool Foundation (IWS); and other research programmes and contract services have contributed indirectly to our knowledge.

Pat Cross was the first SEM experimental officer and she was followed in 1969 by Brenda Lomas, who retired in 1990. Trevor Jones then took on responsibility for microscopy in the Department of Textiles in addition to photography. Over the years, many staff and students have contributed to the research. Their names are given below. Some have worked wholly on fibre fracture problems. Others have used fracture studies as an incidental element in their work.

PERSONNEL

The following people at UMIST have contributed to the research.

Academic staff

J.D. Berry	Aspects of fibre breakage
C.P. Buckley	Mechanics of tensile fracture, general direction
C. Carr	Fabric studies
W.D. Cooke	Pilling in knitwear, conservation studies
G.E. Cusick	Abrasion testing, some wear studies
K. Greenwood	Ballistic impact
W.Y. Hamad	Wood-pulp fibres, confocal microscopy
J.W.S. Hearle	General direction
I. Karacan	Studies of Kevlar
M.A. Wilding	Tensile fracture, general direction

Professional staff

D.J. Clarke	Instrument development
C.R. Cork	Archaeological textiles and other studies
P.M. Cross	Microscopy techniques, single fibre fractures
B. Lomas	Microscopy and other techniques, wear studies

Technicians

D. Ball	Instrument developer
S. Butt	Instrument developer
L. Crosby	Scanning electron microscopy
C. Green	Scanning electron microscopy
J.T. Jones	Photography
R.E. Litchfield	Scanning electron microscopy

A. McClellan	Instrument development
J.T. Sparrow	Scanning electron microscopy
H. Splitt	Instrument developer
A. Williams	Scanning electron microscopy

Academic visitors

N.E. Dweltz	Abrasion study
Bhuvenesh C. Goswami	Comparative study of nylon fracture
A. Rast-Eicher	Archaeological textiles
M.A.I. Sultan	Ballistic impact
C.Y. Zhou	Twist and tension in Kevlar

Research students

R. Ahmed	Cotton yarn abrasion
I.A.O. Bahari	Biaxial rotation fatigue
K. Banks	Archaeological textiles
J.A. Batchelor	Archaeological textiles
B.H. Bhutta	Heat-set fibres
A.R. Bunsell	Tensile fatigue
S.F. Calil	Biaxial rotation fatigue
I.E. Clark	Biaxial rotation fatigue
C.R. Cork	Ballistic impact
F. Dadashian	Breaks of Tencel
M. Goksoy	Biaxial rotation fatigue of cotton; surface abrasion
Binode C. Goswami	Flex fatigue
R. Greer	Heat setting
M. Habibullah	Thermomechanical properties
N. Hasnain	Biaxial rotation fatigue of cotton
B.C. Jariwala	Flex fatigue
D.C. Jones	Photodegradation of wool, atomic force microscopy
F.S.H. Kassam	Rupture of tear webbing
L. Konopasek	Tensile fatigue
T. Laksophee	Wool/mohair abrasion
K. Liolios	Biaxial rotation fatigue
R. Mandal	Heat-treated fibres
M. Miraftab	Flex fatigue
B. Parvizi	Flex fatigue
A. Sengonul	Fatigue of HMPE; carpet wear
J.T. Sparrow	Tensile failure of cotton
R. Stanton	Kinkbands
E.A. Vaughn	Tensile fatigue
B.S. Wong	Rotation over a pin
Y. Zhu	Oxidised PAN fibres

Undergraduates and summer student workers

A. Chan
A. Clarke
J. Griffiths
A.D. Hearle
A. Kaynak
S.R. Moore
P. Noone
S.J. Watson

FIBRE FRACTURE RESEARCH GROUP

The following organizations have joined in sponsorship for part or all of the period 1972–1987:

Albany International Research (formerly FRL), USA
Allied Fibers, USA
AKZO, Netherlands
Carrington-Viyella, UK
Celanese, USA
Chemiefaser Lenzing, Austria
Courtaulds, UK
Du Pont, USA
H and T Marlow (formerly Hawkins and Tipson), UK
Hoechst, Germany
Imperial Chemical Industries (ICI), UK
International Institute for Cotton (IIC), UK

International Wool Secretariat (IWS), UK
Marks and Spencer, UK
Monsanto, USA
Phone-Poulenc, France
Strattwell Developments, UK
TEFO, Sweden
Teijin, Japan
Toray, Japan
US Army European Office, UK

In the provision of materials, information, valuable advice, and discussion, the following have made particular contributions:

G.A. Carnaby, WRONZ (carpets)
M. El-Masri, Strattwell (rental textiles)
J. Flory, Exxon (ropes)
M.R. Parsey. H and T Marlow (ropes)
W. Puchegger, Chemiefaser Lenzing (rayon)
G. Stevens, RAE (tensile fatigue)
R.A. Stocks, SCRDE (Service clothing)
J. Swallow, RAE (service items)
R.D. Van Veld, Du Pont (fibres)
M. Webb, RAE (service items)

MATERIALS FOR EXAMINATION

Specimens for examination have come from a variety of sources. Some items — worn clothing, hair, etc. — have been given casually by people in contact with the laboratory. Others have come in for special investigations.

Particularly well-documented samples have been supplied by:

Du Pont (light-degraded nylon)
H and T Marlow (ropes)
Hampton Court Palace Conservation Studio, London
IWS (worn trousers)
OCIMF, through John Flory, Exxon (ropes)
P and S Filtration (filter fabrics)
RAE (service items)
SCRDE (service clothing)
Scapa-Porritt (felts)
Shirley Institute (sheets, shirts and samples of machine abraded fabrics)
Strattwell Developments (rental textiles and clothing)
Textile Conservation Centre (Hampton Court), London
Victoria and Albert Museum, London
Vindolanda Trust, Cumbria
Whitworth Gallery, Manchester
G.M. Vogelsang-Eastwood, Leiden
WRONZ and IWS (carpets)

PROVISION OF PICTURES AND INFORMATION

Stephen Banfield, TTI Ltd, UK
Edward Boyes and Gerald Lavin, DU PONT, USA
T.R. Burrow, Courtaulds, UK
Anthony Bunsell, Ecole des Mines de Paris, France.
Robert Burling-Claridge, WRONZ, New Zealand.
Jose Charvet, Stevens Institute, USA
Hawthorne Davis, North Carolina State University, USA.
Karlheinz Foos, Bayeischen Kriminalamtes, Germany.
Bhuvenesh Goswami, Clemson University, USA
Sarah Holmes, University of Texas, USA
Michael Huson, CSIRO, Geelong, Australia.
Elizabeth Norton, University of Cambridge, UK
David Salem and Sigrid Ruetsch, TRI, Princeton, USA.
H.S.S. Sharma, Queen's University, Belfast, UK
Alan Swift, Consultant, UK
Hiroko Yokura, Shiga University, Japan

APPENDIX 2
BIBLIOGRAPHY

GENERAL REFERENCES: BOOKS

Atkins, A. G. and Mai, Y.-W. (1985) *Elastic and Plastic Fracture*, Ellis Horwood, Chichester.

Booth, J. E. (1961) *Principles of Textile Testing*, National Trade Press, London.

Farnfield, C. A. and Perry, D. R. (eds) (1975) *Identification of Textile Materials*, 7th edn, Textile Institute, Manchester.

Goldstein, J. I. and Yakowitz, H. (eds) (1975) *Practical Scanning Elecron Microscopy*, Plenum Press, New York.

Hearle, J. W. S., Cross, P. M., and Sparrow, J. T. (eds) (1972) *The Use of the Scanning Electron Microscope*, Pergamon, Oxford.

Hearle, J. W. S., Thwaites, J. J., and Amirbayat, J. (eds) (1980) *Mechanics of Flexible Assemblies*, Sijthoff and Noordhof, Alphen aan den Rijn.

Morton, W. E. and Hearle, J. W. S. (1993) *Physical Properties of Textile Fibres*, 3rd edn, Textile Institute, Manchester.

Wells, O. C. (1974) *Scanning Electron Microscopy*, McGraw-Hill, New York.

Williams, J. G. (1983) *Stress Analysis of Polymers*, 2nd edn, Ellis Horwood, Chichester.

Williams, J. G. (1984) *Fracture Mechanics of Polymers*, Ellis Horwood, Chichester.

THESES PRESENTED TO UNIVERSITY OF MANCHESTER OR TO UMIST

MPhil

Mannering, U. (1994) Analyses of textiles from a 6th-8th century cemetery ay Nørre Sandegaad Vest-Bornholm.

MSc

Bahari, I. A. (1978) A comparative study of the fatigue of some synthetic fibres in various aqueous and non-aqueous environments.

Bhutta, B. H. (1973) The effect of heat-setting parameters on linear density and mechanical properties of nylon 6.

Clark, I. E. (1978) The fatigue testing of fibres by biaxial rotation over a pin.

Cork, C. R. (1980) The response of textile materials to transverse ballistic impact.

Cross, P. M. (1968) Load deformation effects in bonded fibre fabrics.

Goksoy, M. (1981) Fatigue of resin treated cotton fibres.

Goswami, B. C. (1971) Bending behaviour of man-made fibres (a microscopic study).

Jariwala, B. C. (1971) Kink-bands in nylon and polyester.

Konopasek, L. (1975) Tensile fatigue properties of textile fibres subjected to cyclic loading.

Lomas, B. (1973) Scanning electron microscope study of the breakdown of textile fibres in use.

Sparrow, J. T. (1970) A study of the morphology of cotton fracture and nylon fatigue by scanning electron microscopy.

PhD

Booth, A. J. (1963) The fatigue properties of twisted continuous filament yarns.

Bunsell, A. R. (1972) The fatigue of nylon 66 and other synthetic fibres.

Calil, S. F. (1977) A study on the fatigue and abrasion properties of fibres.

Clark, I. E. (1980) Fibre fatigue in various environments.

Cooke, W. D. (1982) The attrition mechanisms of knitted fabrics.

Cork, C. R. (1983) Aspects of ballistic impact onto woven textile fabrics.

Goksoy, M. (1986) A study of yarn-on-yarn abrasion.

Greer, R. E. (1970) Torsional and heat-setting characteristics of visco-elastic materials.
Habibullah, M. (1979) Physical properties of heat-set synthetic fibres.
Hasnain, N. (1978) Fatigue of cotton fibres and yarns.
Jariwala, B. C. (1974) The study of kinkbands and flex failure in nylon 6.6 and polyester fibres.
Jones, D. C. (1995) Photodegradation of wool and wool blend fabrics in relation to their use in automotive upholstery.
Madhura Nath, A. D. (1967) Fatigue testing of textile fibres.
Mandal, R. (1976) Thermo-mechanical properties of visco-elastic fibres and their relation to structure.
Miraftab, M. (1986) The influence of temperature and humidity on the flex fatigue life of nylon 6, nylon 6.6 and polyester fibres.
Ozsanlav, V. (1971) Mechanical properties of bonded fibre fabrics.
Peacock, E. (1994) Freeze drying and the conservation of wet preserved textiles.
Sengonul, A. (1993) Studies of the time-dependence and fatigue behaviour of gel-spun polyethylene fibres.
Sparrow, J. T. (1973) The fracture of cotton.
Vaughn, E. A. (1969) On the fatigue properties of textile fibres.
Wong, B. S. (1975) A comparative study of the fatigue of some synthetic fibres in various environments.
Zhu, Y. (1994) A study of the structure-property relationships of PAN precursor fibres during thermo-oxidative stabilization.

PAPERS FROM UMIST GROUP

Booth, A. J. and Hearle, J. W. S. (1965) The fatigue behaviour of textile yarns, *Proc. 4th Int. Congress on Rheology*, Interscience, New York, 203–225.
Buckley, C. P. and Habibullah, M. (1980) Traverse fracture versus axial splitting in the tensile failure of twist-set fibres, *J. Textile Inst.* **71** 317–320.
Bunsell, A. R. and Hearle, J. W. S. (1971a) Mechanics of failure in nylon parachute materials. *Royal Aeronautical Society Symp. on Parachutes and Related Technologies.*
Bunsell, A. R. and Hearle, J. W. S. (1971b) A mechanism of fatigue failure in nylon fibres, *J. Materials Sci.* **6** 1303–1311.
Bunsell, A. R. and Hearle, J. W. S. (1974) Fracture and fatigue of fibres, *Rheologica Acta* **13** 711–716.
Bunsell, A. R., Hearle, J. W. S., and Hunter, R. D. (1971) An apparatus for fatigue testing of fibres, *J. Phys. E: Sci. Instrum.* **4** 868–872.
Bunsell, A. R., Hearle, J. W. S., Konopasek, L., and Lomas, B. (1974) A preliminary study of the fracture morphology of acrylic fibres, *J. Appl. Polymer Sci.* **18** 2229–2242.
Calil, S. F., Clark, I. E., and Hearle, J. W. S. (1989) The sequence of damage in biaxial rotation fatigue of fibres, *J. Materials Sci.* **24** 736–748.
Calil, S. F., Goswami, B. C., and Hearle, J. W. S. (1980) The development of torque in biaxial rotation fatigue testing of fibres, *J. Phys. D.* **13** 725–732.
Calil, S. F. and Hearle, J. W. S. (1977) Fatigue of fibres by biaxial rotation over a pin, *Fracture 1977, ICF4 Conference, Waterloo, Canada* **2** 1267–1271.
Cannon, S. L., Statton, W. O., and Hearle, J. W. S. (1975) The mechanical behaviour of springy polypropylene, *Polymer Eng. Sci.* **15** 633–645
Clark, I. E. and Hearle, J. W. S. (1979) The development of the biaxial rotation test for fatiguing fibres, *J. Phys. E: Sci. Instrum.* **12** 11–14.
Clark, I. E. and Hearle, J. W. S. (1980) Anomalous breaks in the biaxial-rotation fatigue-testing of fibres, *J. Textile Inst.* **71** 87–95.
Clark, I. E. and Hearle, J. W. S. (1982) The influence of temperature and humidity on the biaxial rotation fatigue of fibres, *J. Textile Inst.* **73** 273–280.
Clark, I. E. and Hearle, J. W. S. (1983) A statistical study of biaxial-rotation fatigue of fibres, *J. Textile Inst.* **74** 168–170.
Clark, I. E., Hearle, J. W. S., and Taylor, A. R. (1980) A multi-station apparatus for fatiguing fibres in various environments, *J. Phys. E: Sci. Instrum.* **13** 516–519.
Cooke, W. D. (1981) Torsional fatigue and the initiation mechanism of pilling, *Textile Res. J.* **51** 364–369.
Cooke, W. D. (1982) The influence of fibre fatigue on the pilling cycle. Part I: Fuzz fatigue, *J. Textile Inst.* **73** 13–19.
Cooke, W. D. (1983) The influence of fibre fatigue on the pilling cycle. Part II: Pill growth, *J. Textile Inst.* **74** 101–108.
Cooke, W. D. (1984) The influence of fibre fatigue on the pilling cycle. Part II: Pill growth, *J. Textile Inst.* **74** 101–108.
Cooke, W. D. (1985) Pilling attrition and fatigue, *Textile Res. J.* **55** 409–414.
Cooke, W. D. (1988a) Creasing in ancient textiles, *Conservation News* No. 35, 27–30.
Cooke, W. D. (1988b) Fibre preservation in two carbonised textiles from Soba, *Archaeological Textiles Newsletter* No. 6, 8.

Cooke, W. D. and Howell, D. (1988) Diagnosis of deterioration in a tapestry using scanning electron microscopy, *The Conservator* No. 12, 47–51.

Cooke, W. D. (1990) The mechanics of tapestry wall-bangings, *The Conservator*, NO. 14 (to be published).

Cooke, W. D. and Arthur, D. (1981) A simulation model of the pilling process, *J. Textile Inst.* **72** 111–120.

Cooke, W. D. and Lomas, B. (1987) Ancient textiles — modern technology, *Archaeology Today* No. 3, 21–25.

Cork, C. R., Cooke, W. D. and Wild, J. P. (1996) The use of image analysis to determine yarn twist levels in archaeological textiles, *Archaeometry* **38** 337–345.

Cork, C. R., Wild, J. P., Cooke, W. D. and Fang-Lu, L. (1997) Analysis and evaluation of a group of early Roman textiles from Vindolanda, Northumberland, *J Archaeological Sci* **24** 19–32.

Cross, P. M., Hearle, J. W. S., Lomas, B., and Sparrow, J. T. (1970) Study of fibres in the scanning electron microscope, *Proc. 3rd Annual SEM Symp., IITRI Chicago*, 81–88.

Cross, P. M., Hearle, J. W. S., Cross, J. D., and Stands, A. (1971) A method for studying fabric extension in a scanning electron microscope, *Textile Res. J.* **41** 629–631.

Dweltz, N. E., Hearle, J. W. S., Cusick, G. E., and Lomas, B. (1978a) The surface abrasion of cotton fabrics as seen in the scanning electron microscope, *J. Textile Inst.* **69** 250–261.

Dweltz, N. E., Hearle, J. W. S., Cusick, G. E., and Lomas, B. (1978b) A study of the attrition of some abradants with the scanning electron microscope, *J. Textile Inst.* **69** 294–298.

Elmasri, M. T., Hearle, J. W. S., and Lomas, B. (1983) Product quality and durability in workwear, *Papers of 67th Annual Conference, Quality, design, and the purchaser, Textile Institute, Manchester.*

El-Gaiar, M. N. and Cusick, G. E. (1975) A study of various mechanisms of attrition of fibres as a result of abrasion, *J. Textile Inst.* **66** 426–430.

Flory, J. F., Goksoy, M., and Hearle, J. W. S. (1987) Abrasion resistance of polymeric fibres in marine conditions, *Inst. Marine Engineers Conf., Polymers in a Marine Environment.*

Flory, J. F., Goksoy, M., and Hearle, J. W. S. (1988) Yarn-on-yarn abrasion testing of rope yarns. Part I: The test method, *J. Textile Inst.* **79** 417–431.

Goksoy, M. and Hearle, J. W. S. (1988a) Yarn-on-yarn abrasion testing of rope yarns. Part II: The influence of machine variables, *J. Textile Inst.* **79** 432–442.

Goksoy, M. and Hearle, J. W. S. (1988b) Yarn-on-yarn abrasion testing of rope yarns. Part III: The influence of aqueous environments, *J. Textile Inst.* **79** 443–450.

Goswami, B. C. and Hearle, J. W. S. (1976) A comparative study of nylon fibre fracture, *Textile Res. J.* **46** 55–70.

Greenwood, K. and Cork, C. R. (1984) Ballistic penetration of textile fabrics, *UMIST Report to SCRDE.*

Greer, R. and Hearle, J. W. S. (1970) Snarl splitting in polythylene, *Polymer* **11** 441–442.

Hearle, J. W. S. (1967) Fatigue in fibres and plastics: a review, *J. Materials Sci.* **2** 474–488.

Hearle, J. W. S. (1973) Bonding, friction and failure in nonwoven fabrics, *Proc. Symp. Manmade Fibres in Paper-making, IUPAC* 303–310.

Hearle, J. W. S. (1975) Fibre fracture and its relation to fibre structure and fibre end-uses, *Textile Inst. and Inst. Textile de France Conf., Contributions of Science to the Development of the Textile Industry, Textile Institute, Manchester* 60–75.

Hearle, J. W. S. and Bunsell, A. R. (1974) The fatigue of synthetic polymer fibres, *J. Applied Polymer Sci.* **18** 267–291.

Hearle, J. W. S., Bunsell, A. R., and Lomas, B. (1973) The classification of fibre fracture morphology, *Proc. Int. Fracture Conf., Munich* vi, **321** 1–5.

Hearle, J. W. S. and Clark, I. E. (1981) The fatigue of polymeric textile fibres, *Proc. Int. Fracture Conf., Cannes*, D. Francois, *et al.*, eds, Pergammon, Oxford.

Hearle, J. W. S. and Cross, P. M. (1970) The fractography of thermoplastic textile fibres, *J. Materials Sci.* **5** 507–516.

Hearle, J. W. S. and Goksoy, M. (1986) Einflusseiner Kunstharzausrustung auf das Ermudungs und Zugfestigkeit von Baumwollfasern, *Melliand Textilberichte* **2** 129.

Hearle, J. W. S. and Hasnain, N. (1979) Textile Inst. Annual Conf., *Cotton in a Compelitive World*, P. W. Harrison, ed., pp. 163–183.

Hearle, J. W. S., Lomas, B., Goodman, P. L., and Carnaby, G. A. (1983) Microscopic examination of worn carpets: I. Comparison of the modes of fibre failure in floor trials, *WRONZ (Wool Research Organization of New Zealand) Communication No. C79.*

Hearle, J. W. S. and Lomas, B. (1974) Natural weathering of polyester filaments, *J. Materials Sci.* **9** 1388–1389.

Hearle, J. W. S. and Lomas, B. (1977a) The fracture of nylon 66 yarns which have been exposed to light, *J. Appl. Polymer Sci.* **21** 1103–1128.

Hearle, J. W. S. and Lomas, B. (1977b) Breakdown of shirts during use, *Clothing Res. J.* **5** 47–60.

Hearle, J. W. S. and Lomas, B. (1985a) Wear of wool and wool blend fibres in carpets, *Proc. 7th Int. Wool Textile Res. Conf., Tokyo* **3** 317–326.

Hearle, J. W. S. and Lomas, B. (1985b) What fibre properties should be studied in relation to wear in use? *Objective Measurement: Applications to Product Design and Process*

Control, S. Kawabata, R. Postle and M. Niwa, ed, *The Textile Machinery Society of Japan, Osaka* 93–99.

Hearle, J. W. S., Lomas, B., and Bunsell, A. R. (1974) The study of fibre fracture, *Appl. Polymer Symp. Dallas, Texas* **23** 147–156.

Hearle, J. W. S., Jariwala, B. C., Konopasek, L., and Lomas, B. (1976) Aspects of fracture of wool and hair fibres, *Proc. Int. Wool Textile Research Conf., Deutsches Wollforschungsinstitut, Aachen* **11** 370–380.

Hearle, J. W. S., Lomas, B., and Wilding, A. M. (1985) Forms of fracture and fatigue of polymeric fibres, *PRI Conf. on Deformation, Yield and Fracture of Polymers* 82.1–82.4.

Hearle, J. W. S. and Parsey, M. R. (1983) Fatigue failures in marine ropes and their relation to fibre fatigue, *PRI Int. Conf. on Fatigue in Polymers, London.*

Hearle, J. W. S. and Simmens, S. C. (1973) Electron microscope studies of textile fibres and materials, *Polymer* **14** 273–285.

Hearle, J. W. S. and Sparrow, J. T. (1971) The fractography of cotton fibres, *Textile Res. J.* **41** 736–749.

Hearle, J. W. S. and Sparrow, J. T. (1979a) Tensile fatigue of cotton fibres, *Textile Res. J.* **49** 242–243.

Hearle, J. W. S. and Sparrow, J. T. (1979b) Further studies of the fractography of cotton fibres, *Textile Res. J.* **49** 268–282.

Hearle, J. W. S. and Vaughn, E. A. (1970) Fatigue studies of drawn and undrawn fibre materials. *Rheologica Acta* **9** 76–91.

Hearle, J. W. S. and Wong, B. S. (1977a) A comparative study of the fatigue failure of nylon 6.6, PET polyester and polypropylene fibres, *J. Textile Inst.* **68** 89–94.

Hearle, J. W. S. and Wong, B. S. (1977b) The effects of air, water, hydrochloric acid and other environments on the fatigue of nylon 6.6 fibres, *J. Textile Inst.* **68** 127–132.

Hearle, J. W. S. and Wong, B. S. (1977c) Statistics of fatigue failure, *J. Textile Inst.* **68** 155–157.

Hearle, J. W. S. and Wong, B. S. (1977d) A fibre fatigue tester, based on rotation over a wire, for use in different environments, *J. Phys. E. Sci. Instrum.* **10** 448–450.

Hearle, J. W. S. and Wong, B. S. (1977e) Flexural fatigue and surface abrasion of Kevlar-29 and other high-modulus fibres, *J. Materials Sci.* **12** 2447–2455.

Hearle, J. W. S. and Zhou, C. Y. (1987) Tensile properties of twisted para-aramid fibers, *Textile Res. J.* **57** 7–13.

Howell, D., Bilson, T. and Cooke, W. D. (1997) Mechanical aspects of lining 'loose-hung' textiles, *Textile Symposium '97*, Canadian Conservation Institute, Ottawa, to be published.

Jones, D. C., Carr, C. M., Cooke, W. D., Mitchell, R., and Vickerman, J. C.(1995) Photodegradation of wool and wool blend fabrics in relation to their use in automotive upholstery, *Proc. 9th Int.Wool Textile Research Conf., Biella.*

Konopasek, L. and Hearle, J. W. S. (1977) The tensile fatigue behaviour of para-oriented aramid fibres and their fracture morphology, *J. Appl. Polymer Sci.* **21** 2791–2815.

Mukhopadhyay, S. K. and Zhu, Y. (1995) Structure-property relationships of PAN precursor fibers during thermo-oxidative stabilization, *Textile Res. J.* **65** 25–31.

Parsey, M. R., Hearle, J. W. S., and Banfield, S. J. (1987) Life of polymeric ropes in the marine environment, *Inst. Marine Engineers Conf., Polymers in a Marine Environment.*

Rast-Eicher, A., Windler, R. and Mannering, U. (!995) Nessel und flachs — textilefunde aus einem frühmittelalterlichen mädchengarb in Flurlingen (canton Zurich), *Archaeology Suisse* **18** 155–161.

Sengonul, A. and Wilding, M. A. (1994) Flex fatigue in gel-spun high-performance polyethylene fibres, *J. Textile Inst.* **85** 1–11.

Sengonul, A. and Wilding, M. A. (1996) Flex fatigue in gel-spun high-performance polyethylene fibres at elevated tenperatures, *J. Textile Inst.* **87** 13–22.

Zhu, Y., Wilding, M. A. and Mukhopadhyay (1996) Fibre-on-fibre abrasion in oxidised polyacrylonitrile, *J Textile Inst* **87** 417–431.

OTHER REFERENCES

Anand, S. (1997) Editor, Medical Textiles 1996, Woodhead Publishing Ltd, Cambridge, England.

Anderson, C. A. and Hoskinson, R. M. (1970) Scanning electron microscopy of insect-damaged wool fibres, *J. Textile Inst.* **61** 355–358.

Anderson, C. A., Leeder, J. D., and Robinson, V. N. (1971) Morphological changes in chemically treated wool fibres during abrasion, *J. Textile Inst.* **62** 450–453.

Anderson, C. A. and Robinson, V. N. (1971) Morphological changes in wool fibres during fabric wear and abrasion testing, *J. Textile Inst.* **62** 281–286.

Ansell, M. P. (1983) The degradation of polyester fibres in a PVC-coated fabric exposed to boiling water, *J. Textile Inst.* **74** 263–271.

Backer, S., Mandell, J. F. and Williams, J. H., principal investigators (1983) Deterioration of synthetic fibre rope during marine usage, *Summary Progress Report Project R/T-11, January 1982-June 1983*, Sea Grant Office, MIT, Cambridge, Massachusetts, USA.

Backer, S. and Seo, M. (1985) Mechanics of degradation in marine rope, *Objective measurement: applications to product design and process control*, 653–664, The Textile Machinery Society of Japan.

Barish, L. (1987) The structure of sunlight-degraded polypropylene textile fibres. *Proc. 45th Annual Meeting of EMSA*. 470–471.

Barish, L. (1989) Sunlight-degradation of polypropylene textile fibres: A microscopical study. *J. Textile Inst.* (in press).

Billica, H. R., Van Veld, R. D. and Davis, H. A. (1970) Electron microscopy staining technique for oriented polymers, *Septième Congrès International de Microscopie Electronique*, Grenoble, France.

Boyes, E. (1994) High resolution imaging and microanalyis in the SEM at low voltage, *ICEM 13*, Paris, France 51–54.

Brecht, G. (1988) Mode III interlaminar fracture of composite materials, MS thesis, University of Delaware.

Brown, A. C. and Swift, J. A. (1975) Hair breakage: the scanning electron microscope as a diagnostic tool., *J Soc Cosmetic Chem* **26** 289–297.

Burling-Claridge, R. (1997), private communication, WRONZ, New Zealand.

Carnaby, G. A. (1981) The mechanics of carpet wear, *Textile Res J* **51** 514–519.

Carnaby, G. A. (1984) Theoretical prediction of the durability of wool carpets under laboratory and use conditions, *WRONZ Communication* No. C91.

Carnaby, G. A. (1985) The mechanics of carpet wear processes, *Proc 7th Int Wool Textile Res Conf, Tokyo*, Vol. III, 327–336.

Carroll, G. R. (1992) Forensic fibre microscopy, in *Forensic examination of fibres*, editor: Robertson, J., Ellis Horwood, Chichester.

Choudry, M. (1987) The use of scanning electron microscopy for identification of cuts and tears in fabrics: observations based on criminal cases, *Scanning Microscopy*, **1** 119–125.

Choueka, J., Charvet, J. L., Alexander, H., Oh, Y.H., Joseph, G., Blumenthal, N.C. and LaCourse, W. C. (1995) Effect of annealing temperature on the degradation of reinforcing fibers for absorbable implants, *J Biomedical Materials Research* **29** 1309–1315.

Cohn, D., Younes, H. and Maron, G. (1987) Amorphous and crystalline morphologies in glycolic acid and lactic acid polymers, *Polymer* **28** 2018–2022.

Crick, R. A., Leech, D. C., Meakin, P. J., and Moore, D. R. (1987) Interlaminar fracture morphology of carbon-fibre/PEEK composites, *J. Materials Sci.* **22** 2094–2104.

Crispin, K. (1987) The Crown versus Chamberlain, Alabatross Books, Sydney, Australia.

Davis, H. A. (1989) Electron microscopy investigation of structure and damage in polyethylene terephthalate fibers, *Ars Textrina* **12** 23–49, 335–336.

Duckett, K. E. and Goswami, B. C. (1984) A multi-stage apparatus for characterizing the cyclic torsional fatigue behavior of single fibers, *Textile Res J* **54** 43–46.

Duerden, I. J. and Dance, D. M. (1984) Recognition of restraint failure modes, *Proc. Canadian Multi-disciplinary Road Safety Conf. III, London, Ontario* 174–198.

Duerden, I. J., German, A., Lidzbarski, E. S., and Nowak, E. S. (1983) Identification of failure modes in occupant-restraint webbing. Vol. I: *Methodology and analysis*. Vol. II: *Atlas of failure types*. Road and Motor Vehicle Traffic Safety, Transport Canada (Surface), Ottawa.

Echlin, P. (1975) Sputter coating techniques for SEM, *Proc. Annual SEM Conf., IITRI Chicago* 217–224.

Egerton, G. S. (1948) Some aspects of the photochemical degradation of nylon, silk and viscose rayon, *Textile Res. J.* **18** 659–669.

Ellison, M. S., Zeronian, S. H., and Fujiwara, Y. (1984) High energy radiation effects on ultimate mechanical properties and fractography of nylon 6 fibres, *J. Materials Sci.* **19** 82–98.

FIBRE TETHERS 2000 (1994), Joint Industry Project: Phase IIA — Axial Compression Studies for United States Navy, Noble Denton Europe Ltd, London.

FIBRE TETHERS 2000 (1995) Joint Industry Study: High Technology Fibres for Deepwater Tethers, Noble Denton Europe Ltd, London.

Foos, K. (1993) Trennstellen an textilfasern im rasterelektronenmicroskop, *Arch. Krim.* **181** 26–32.

Friedrich, K. (1983) Fracture of polymer composites, *Report CCM-83-18, University of Delaware*.

Fu-Min, L., Goswami, B. C., Spruiell, J. E. and Duckett, K. E. (1985) Influence of fine structure on the torsional fatigue behavior of poly(ethylene terephthalate) fibers, *J Appl Polymer Sci* **30** 1859–1874.

Gharehaghaji, A. A. (1994) Wool fibre microdamage due to contact stresses in opening processes, *PhD thesis*, University of New South Wales, Australia.

Goswami, B. C., Duckett, K. E., and Vigo, T. L. (1980) Torsional fatigue and the initiation mechanism of pilling, *Textile Res. J.* **50** 481–485.

Goynes, W. R. and Trask, B. J. (1985) Effects of heat on cotton, polyester and wool fibres in blended fabrics — a scanning electron microscope study, *Textile Res. J.* **55** 402–408.

Goynes, W. R. and Trask, B. J. (1987) Effects of heat on structures of cotton, polyester and wool fibres in a tri-blended fabric with and without flame retardant, *Textile Res. J.* **57** 549–554.

Granger-Taylor, H. (1988) The significance of creases in archaeological textiles. *UKIC Archaeology Section Conf., York, April 1988* (to be published).

Guenon, V. A. F. (1987) Interlaminar fracture toughness of a three-dimensional composite, MS thesis, University of Delaware.

Hamad, W. Y. (1995) Microstructural cumulative material degradation and fatigue-failure micromechanisms in wood-pulp fibres, *Cellulose* **2** 159–177.

Hammers, I., Arns, W., and Zahn, H. (1987) Scanning electron microscopic investigations of the breakdown of wool by insect larvae, *Textile Res. J.* **57** 401–406.

Hepworth, A., Buckley, T. and Sikorski, J. (1969) Dynamic studies of fibres and polymers in the specimen chamber of the scanning electronn microscope, *J Sci Instruments (J Physics E) Series 2* **2** 789–796.

Heuse, O. (1982) Beschädigungen der kleidung durch stichwerkzeuge, *Arch. Krim.* **170** 130–145.

Hobbs, R. E., Overington, M. S., Hearle, J. W. S. and Banfield, S. J. (1997) Element buckling within ropes and cables, *J Textile Inst,* in press.

Ishizu, H., Doi, Y., Hayakawa, S., Kaneko, H., Funatsu, Y., Seno, M., Yoshino, T., Nakanishi, K., and Mikami, Y. (1974) Relationship between impingement on textile fibres and causative instruments: a scanning electron microscopic observation, *Japanese J Legal Medicine* **28** 104–108.

Janaway, R. (1983) Textile fibre characteristics preserved by metal corrosion: the potential of S.E.M. studies, *The Conservator No.* 7, 48–52.

Johnson, N. and Stacey, A. (1991) Forensic textiles: the morphology of stabbed fabrics, *Australian Textiles* (3) 32–33.

Kawabata, S. and Niwa, M. (1980) Formulas KN-101 and KN-201 for the translation of basic mechanical properties of fabric into hand values and KN301 from the hand value into total hand value, *J Textile Machinery Soc Japan* **33** 164–169.

Kemmenoe, B. H. and Bullock, G. R. (1983) Structure analysis of sputter-coated and ion-beam sputter-coated films a comparative study, *J. Microscopy* **132** 153–163.

Knudsen, J. P. (1963) The influence of coagulation variables on the structure and physical properties of an acrylic fiber, *Textile Res. J.* **33** 13–20.

Lavaste, V., Berger, M. H., Bunsell, A. R., and Besson, J. (1995) Microstructure and mechanical characteristics of alpha-alumina-based fibres, *J. Materials Sci.* **30** 4215–4225.

Lewis, J. (1975) 'The biodeterioration of wool by microorganisms'. The microbal aspects of the deterioration of materials. Eds D.W. Lovelock and R.J. Gilbert. The society for applied bacteriology. Technical series No 9. Academic Press, London.

Matsuda, T. (1987) Mechanical properties of Technora T200/Epikote 828 composites, *CCM Report,* University of Delaware.

Mendelson, N. H. (1992) Production and initial characterization of bionites: materials formed on a bacterial backbone, *Science* **258** 1633–1636.

Mendelson, N. H. (1994) Bacterial macrofibers and bionites: materials of natural and synthetic design, *Mat. Res. Soc. Symp. Proc.* **330** 95–111.

Metz, S. A., Chegini, N. and Masterson, B. J. (1990) In vivo and in vitro degradation of monofilament absorbable sutures, *Biomaterials* **11** 41–45.

Monahan, D. L. and Harding, H. W. J. (1990) Damage to clothing — cuts and tears, *J. Forensic Sciences, JFSCA* **35** 901–912.

Morling, T. (1987) Royal Commission of Inquiry into the Chamberlain Convictions — Report of the Commissioner, Government Printer of the Northern Territory, Darwin, Australia. *Also*: Commonwealth of Australia/Northern Territory Government (1986–87), Transcript of the Proceedings before the Commissioner, Mr Justice T. Morling, at Darwin, NT, and Sydney, NSW, Australia.

Nee, P. B. (1975) Specimen coating techniques for SEM — a comparative study, *Proc. Annual SEM Conf., IITRI Chicago* 225–232.

Niwa, M. and Kawabata, S. (1988) The three mechanical components of fabric relating to suit appearance, *Proc Advanced Workshop on the application of mathematics and physics in the wool industry,* Canterbury, New Zealand, 404–416.

Ogata, N. Dougasaki, S., and Yoshida, K. (1979) Fractography of nylon 6 yarn, *J. Appl. Polymer Sci.* **24** 837–852.

Oudet, Ch., Bunsell, A. R., Hagege, R. and Sotton, M. (1984) Structural changes in polyester fibres during fatigue, *J Appl Polymer Sci* **29** 4363–4376.

Paplaukas, L. (1973) The scanning electron microscope: a new way to examine holes in fabrics, *J Police Sci and Admin,* **1** 362–365.

Peacock, E. (1988a) Freeze-drying archaeological textiles: the need for basic research, *UKIC Archaeological Section Conf., York, April 1988* (to be published).

Peacock, E. (1988b) The svalbard textile conservation project. J. P. Wild and P. Walton, eds, *Textiles in Northern Archaeology,* NeSAT (in press).

Pelton, W. (1995) Distinguishing the cause of textile fiber damage using the scanning electron microscope, *J Forensic Sci,* **40** 874–882.

Pelton, W. and Ukpabi, P. (1995) Using the scanning electron microscope to identify the cause of fibre damage, Part II: an explanatory study, *Canadian Soc Forensic Sci J* **28** 189–200.

Petrina, P., Phoenix, S. L. and Head, E. (1994) Effect of braided jacket on the lifetime of 3/8 inch Kevlar strands under a loaded sheave and high tension, *MTS Proc, Washington DC, September 1994.*

Petrina, P., Phoenix, S. L., Leban, F. and Pappas, V. (1995) Lifetime studies of synthetic cables subjected to lateral loads from sheaves, *OCEANS 95, MTS/IEEE Proc, San Diego, October 1995.*

Phoenix, S. L., Petrina, P. and Deyhim, A. (1993) Lifetime studies of specially jacketed 3/8 inch Kevlar strands under a loaded sheave and high tension, *Special Report, Department of Theoretical and Applied Mechanics, Cornell University, Ithaca, NY.*

Poole, F. (1996) Fibre fracture morphology: ballistic versus severance, *Project Report,* Diploma of Applied Science in Forensic Investigation (NSW Police), Canberra Institute of Technology, Australia.

R v Chamberlain and Chamberlain (1982) SCC No. A/S 19–20, Supreme Court of the Northern Territory, Transcript of Proceedings before Acting Chief Justice Muirhead at Darwin, NT, Australia.

Reed, A. M. and Golding, D. K. (1981) Biodegradable polymers for use in surgery — poly(glycolic)/poly(lactic acid) homo and copolymers: 2. *In vitro* degradation, *Polymer* **22** 494–518.

Salem, D. R. and Ruetsch (1997) Influence of microstructure on the ultra-violet degradation of poly(ethylene terephthalate) fibers, *Proc Fiber Society Conf, Mulhouse, April 1997* 35–38.

Scelzo, W. A., Backer, S. and Boyce, M. (1994) Mechanistic role of yarn and fabric structure in determining tear resistance of woven cloth. Part I: Understanding tongue tear, *Textile Res J* **64** 291–304.

Seo, M. (1988) Mechanical deterioration of synthetic fibre rope in marine environment, *PhD thesis*, Massachusetts Institute of Technology, USA.

Seo, M., Realff, M. L., Pan, N., Boyce, M., Schwartz, P. and Backer, S. (1993) Mechanical properties of fabric woven from yarns produced by different spinning technologies: yarn failure in woven fabric, *Textile Res J* **63** 123–134.

Sharma, H. S. S, Faughey, G. and McCall, D. (1996) Effect of sample preparation and heating rate on the differential thermogravimetric analysis of flax fibres, *J Textile Inst* **87** 249–257.

Stowell, L. J. and Card, K. A. (1990), Use of scanning electron microscopy (SEM) to identify cuts and tears in nylon fabric, *J. Forensic Sciences* **35** 947–950.

Tandon, S. K., Carnaby, G. A. and Wood, E. J. (1990) Evaluation of a formula for predicting carpet wear, *Proc 8th Int wool Textile Res Conf, Christchurch*, Vol V, 429–438.

Thwaites, J. J. and Mendelson, N. H. (1985) Biomechanics of bacterial walls: studies of bacterial thread made from *Bacillus subtilis, Proc Natl Acad Sci USA* **82** 2163–2167.

Thwaites, J. J. and Surana, U. C. (1991) Mechanical properties of *Bacillus subtilis* cell walls: effects of removing residual culture medium, *J Bacteriol* **173** 179–203.

Thwaites, J. J., Surana, U. C. and Jones, A. M. (1991) Mechanical properties of *Bacillus subtilis* cell walls: effects of ions and lysozyme, *J Bacteriol* **173** 204–210.

Trethewey, B. R. (1986) Mode II interlaminar fracture of composite materials, MS thesis, University of Delaware.

Ukpabi, P. and Pelton, W. (1995) Using the scanning electron microscope to identify the cause of fibre damage, Part I: a review of related literature, *Canadian Soc Forensic Sci J* **28** 181–187.

Valentin, D., Paray, F., and Guetta, B. (1987) The hygrothermal behaviour of glass fibre reinforced PA66 composites: a study of the effect of water absorption on their properties, *J. Materials Sci.* **22** 46–56.

Van der Vegt, A. K. (1962) Torsional fatigue of fibres, *Rheologica Acta* **2** 17–22.

Walker, N., Haward, R. N., and Hay, J. N. (1979) Plastic fracture in poly(vinyl chloride), *J. Materials Sci.* **14** 1085–1094.

Webster, J. (1996) Effects of ageing on selected mechanical properties of ISO-301 stitched seams, *PhD thesis*, University of Otago, New Zealand.

Wong, Y. (1982) Forensic applications of SEM/EDX analyzer in Hong Kong, *Scanning electron microscopy* **11** 591–597.

Yan, Y. (1991) Wool fibre damage caused by clothed surfaces, *PhD thesis*, University of New South Wales, Australia.

Yang, X., Sun, W. and Yan, H. (1988) Deformation and fracture of fibres. Part1: The dynamic study of the deformation and fracture of keratin fibres in tension, *Journal of China Textile University (Chinese Edition)* **14** 9–17.

Yokura, H. and Niwa, M. (1990) Durability of fabric handle and shape retention during wear of men's summer suits, *Textile Res J* **60** 194–202.

Yokura, H. and Niwa, M. (1990) Analysis of mechanical fatigue phenomens in wool and wool blend suiting fabrics, *Textile Res J* **61** 1–10.

Young, M. D. and Johnson, N. A. G. (1988) Wool fibre damage in feeding devices for open-end spinning, *Proc Textile Inst Annual World Conf, Sydney, Australia,* 389–396.

Young, M. D. and Johnson, N. A. G. (1990) Abrasion of wool fibres by pinned surfaces, *Proc. 8th International Wool Text. Res. Conf., Christchurch*, Vol III 389–398.
Young, N. (1989) Innocence regained, Federation Press, Sydney, Australia.

INDEX

Note: Readers are also referred to Chapter 50, "The Diversity of Fibre Failure", for a classification of fibre breaks according to type, cause and structure.